Sustainability Applied to Unconventional Oil and Gas Field Exploration and Development

Jesus Samuel Armacanqui
National University of Engineering of Peru, Peru

Susan Smith Nash
American Association of Petroleum Geologists, USA & University of Oklahoma, USA

Luz Eyzaguirre Gormendia
National University of Engineering of Peru, Peru

Rouzbeh Moghanloo
University of Oklahoma, USA

A volume in the Practice, Progress, and Proficiency in Sustainability (PPPS) Book Series

Published in the United States of America by
    IGI Global
    Engineering Science Reference (an imprint of IGI Global)
    701 E. Chocolate Avenue
    Hershey PA, USA 17033
    Tel: 717-533-8845
    Fax: 717-533-8661
    E-mail: cust@igi-global.com
    Web site: http://www.igi-global.com

Copyright © 2024 by IGI Global. All rights reserved. No part of this publication may be reproduced, stored or distributed in any form or by any means, electronic or mechanical, including photocopying, without written permission from the publisher.
Product or company names used in this set are for identification purposes only. Inclusion of the names of the products or companies does not indicate a claim of ownership by IGI Global of the trademark or registered trademark.

                Library of Congress Cataloging-in-Publication Data

Names: Armacanqui Tipacti, Jesus Samuel, 1961- editor.
Title: Sustainability applied to unconventional oil and gas field
   exploration and development / edited by Jesus Samuel Armacanqui Tipacti,
   Susan Nash, Luz de Fátima Eyzaguirre Gorvenia, Rouzbeh Moghanloo.
Description: Hershey, PA : Engineering Science Reference, [2024] | Includes
   bibliographical references and index. | Summary: "The book presents in
   specific examples the application of the concept of the sustainable
   development to the exploration and development of unconventional shale
   oil and gas"-- Provided by publisher.
Identifiers: LCCN 2024002185 (print) | LCCN 2024002186 (ebook) | ISBN
   9798369307403 (h/c) | ISBN 9798369307410 (eISBN)
Subjects: LCSH: Petroleum reserves. | Sustainable engineering. |
   Petroleum--Prospecting--Environmental aspects.
Classification: LCC TN864 .S87 2024 (print) | LCC TN864 (ebook) | DDC
   622/.1828--dc23/eng/20240205
LC record available at https://lccn.loc.gov/2024002185
LC ebook record available at https://lccn.loc.gov/2024002186

British Cataloguing in Publication Data
A Cataloguing in Publication record for this book is available from the British Library.

All work contributed to this book is new, previously-unpublished material.
The views expressed in this book are those of the authors, but not necessarily of the publisher.

For electronic access to this publication, please contact: eresources@igi-global.com.

Practice, Progress, and Proficiency in Sustainability (PPPS) Book Series

Ayman Batisha
International Sustainability Institute, Egypt

ISSN:2330-3271
EISSN:2330-328X

MISSION

In a world where traditional business practices are reconsidered and economic activity is performed in a global context, new areas of economic developments are recognized as the key enablers of wealth and income production. This knowledge of information technologies provides infrastructures, systems, and services towards sustainable development.

The **Practices, Progress, and Proficiency in Sustainability (PPPS) Book Series** focuses on the local and global challenges, business opportunities, and societal needs surrounding international collaboration and sustainable development of technology. This series brings together academics, researchers, entrepreneurs,

Coverage

- Environmental informatics
- ICT and knowledge for development
- Knowledge clusters
- Socio-Economic
- Technological learning

IGI Global is currently accepting manuscripts for publication within this series. To submit a proposal for a volume in this series, please contact our Acquisition Editors at Acquisitions@igi-global.com or visit: http://www.igi-global.com/publish/.

The (ISSN) is published by IGI Global, 701 E. Chocolate Avenue, Hershey, PA 17033-1240, USA, www.igi-global.com. This series is composed of titles available for purchase individually; each title is edited to be contextually exclusive from any other title within the series. For pricing and ordering information please visit http://www.igi-global.com/book-series/practice-progress-proficiency-sustainability/73810. Postmaster: Send all address changes to above address. Copyright © IGI Global. All rights, including translation in other languages reserved by the publisher. No part of this series may be reproduced or used in any form or by any means – graphics, electronic, or mechanical, including photocopying, recording, taping, or information and retrieval systems – without written permission from the publisher, except for non commercial, educational use, including classroom teaching purposes. The views expressed in this series are those of the authors, but not necessarily of IGI Global.

Titles in this Series

For a list of additional titles in this series, please visit:
http://www.igi-global.com/book-series

Impact of Societal Development and Infrastructure on Biodiversity Decline
Ashok Kumar Rathoure (Saraca Research Inc., India)
Engineering Science Reference • copyright 2024 • 374pp • H/C (ISBN: 9798369369500)
• US $185.00 (our price)

Exploring Waste Management in Sustainable Development Contexts
Chandra Mohan (K.R. Mangalalm University, Gurgaon, India) Shobhna Jeet (K.R. Mangalam University, Gurugram, India) Saurav Dixit (Division of Research and Innovation, Uttaranchal University, Dehradun, India & Peter the Great St. Petersburg Polytechnic University, Saint Petersburg, Russia) and Sónia A.C. Carabineiro (NOVA University of Lisbon, Portugal)
Engineering Science Reference • copyright 2024 • 291pp • H/C (ISBN: 9798369342640)
• US $295.00 (our price)

Promoting Multi-Sector Sustainability With Policy and Innovation
Aftab Ara (University of Hail, Saudi Arabia) and Renuka Thakore (University of Central Lancashire, UK)
Engineering Science Reference • copyright 2024 • 286pp • H/C (ISBN: 9798369321133)
• US $290.00 (our price)

The Role of Female Leaders in Achieving the Sustainable Development Goals
Mercia Selva Malar Justin (Xavier Institute of Management and Entrepreneurship, India) and Joycia Thorat (Churches Auxiliary for Social Action, India)
Business Science Reference • copyright 2024 • 341pp • H/C (ISBN: 9798369318348) • US $285.00 (our price)

701 East Chocolate Avenue, Hershey, PA 17033, USA
Tel: 717-533-8845 x100 • Fax: 717-533-8661
E-Mail: cust@igi-global.com • www.igi-global.com

Table of Contents

Preface ....
xxiv

Chapter 1
The Concept of Sustainable Exploration, Development, and Operation of Unconventional Oil and Gas Fields ....
1

    Jesus Samuel Armacanqui-Tipacti, National University of Engineering of Peru, Peru

Chapter 2
The Muerto Shale: Stratigraphic Distribution of Carbonate Content and Organic Geochemistry in the Corcobado Outcrop ....
40

    Jose Alfonso Rodriquez-Cruzado, National University of Engineering of Peru, Peru
    Jorge Luis Ore-Rodriguez, National University of Engineering of Peru, Peru

Chapter 3
Volumetric Estimation of the Gas in the Generator Rock Through an Evaluation of the Adsorption in the Northeast of Peru ....
61

    Heraud Taipe- Acuña, National University of Engineering of Peru, Peru
    Victor Huerta Quiñones, National University of Engineering of Peru, Peru
    Ali Tinni, University of Oklahoma, USA
    Hugo Valdivia Ampuero, National University of Engineering of Peru, Peru
    Isabel Moromi Nakata, National University of Engineering of Peru, Peru

Chapter 4
Characterization of the Structure of Mesopores in the Generating Rock of Northeastern Peru: Muerto Formation... 84

Kevin Chipana-Suasnabar, National University of Engineering of Peru, Peru
Heraud Taipe Acuna, National University of Engineering of Peru, Peru
Joseph Sinchitullo Gomez, National University of Engineering of Peru, Peru
Ali Tinni, University of Oklahoma, USA
Israel Chavez Sumarriva, National University of Engineering of Peru, Peru
Marco Tejada Silva, National University of Engineering of Peru, Peru

Chapter 5
The Shale of the Cabanillas Formation: Integration Methodology and the Presence of the Gas Resource (Shale Gas) in the Marañón-Peru Basin............ 100

Walter Jacob Morales-Paetan, National University of Engineering of Peru, Peru
Pedro Zegarra Sanchez, National University of Engineering of Peru, Peru

Chapter 6
First Insights in the Estimation of the Petroleum Generation Potential of the Muerto Source Rock From el Cortado Outcrop, Northwestern Peru 123

Jose Alfonso Rodriguez-Cruzado, National University of Engineering of Peru, Peru
Jorge Oré-Rodriguez, National University of Engineering of Peru, Peru

Chapter 7
Lithological Facies Classification, Surface Gamma Ray, and Toc Analysis Using an Outcrop of Unconventional Rock Field: Case in the Muerto Formation, Peru, for the Identification of New Areas With Hydrocarbon Potential 146

    Brayan Nolasco Villacampa, National University of Engineering of Peru, Peru
    Jorge Luis Oré- Rodriguez, National University of Engineering of Peru, Peru
    José Alfonso Rodriguez-Cruzado, National University of Engineering of Peru, Peru
    Israel J. Chavez-Sumarriva, National University of Engineering of Peru, Peru
    Manuel Lopez Reale, LCV Group, USA
    Humberto Chiriff-Rivera, National University of Engineering of Peru, Peru
    Jesus Samuel Armacanqui-Tipacti, National University of Engineering of Peru, Peru
    Luz Eyzaguirre-Gorvenia, National University of Engineering of Peru, Peru
    Alfredo Vazquez-Barrios, National University of Engineering of Peru, Peru

Chapter 8
Evaluation of Geological Boundary Conditions of a Water Well for Aquifer Integrity Conformance by Pressure Transient Analysis: Presence of Hydraulic Fracturing, the Operation of Wastewater, CO2 Storage, and Geothermal Wells 173

    Gustavo Enrique Rodriguez-Robles, National University of Engineering of Peru, Peru
    Jesus Samuel Armacanqui-Tipacti, National University of Engineering of Peru, Peru

Chapter 9
Cost-Effective Advanced Remote Diagnostics of Sucker Rod Pumping Wells From Dynamometric Charts: A Deep Learning Approach.............................. 188

> *Joel Hancco Paccori, National University of Engineering of Peru, Peru*
> *Manuel Castillo Cara, Polytechnic University of Madrid, Spain*
> *Jesus Samuel Armacanqui-Tipacti, National University of Engineering of Peru, Peru*

Chapter 10
Use of Limonene as a Biodegradable Surfactant for the Inhibition and Removal of Paraffins in Oil Production Operations 220

> *Maria Rosario Viera-Palacios, National University of Engineering of Peru, Peru*
> *Miguel Ángel Guzmán, Peruvian University of Applied Sciences, Peru*
> *Jesus Samuel Armacanqui-Tipacti, National University of Engineering of Peru, Peru*
> *Cesar Lujan Ruiz, National University of Engineering of Peru, Peru*
> *Guillermo Prudencio, National University of Engineering of Peru, Peru*
> *Luz Eyzaguirre-Gorvenia, National University of Engineering of Peru, Peru*

Chapter 11
Using H2 to Increase the Energy Efficiency and Reduce the Containment Emissions in the Operation of Generators 236

> *Franco A. Cassinelli-Cisneros, National University of Engineering of Peru, Peru*
> *Jesus Samuel Armacanqui-Tipacti, National University of Engineering of Peru, Peru*
> *Ciro Ormeño-Aquino, National University of Engineering of Peru, Peru*
> *Jose Carlos Rodriguez, San Luis Gonzaga University of Ica, Peru*
> *José Rosendo Campos Barrientos, San Luis Gonzaga University of Ica, Peru*
> *Mohamed Yehia, Suez Oil Company, Cairo, Egypt*
> *Nelson Michael Villegas-Juro, National University of Engineering of Peru, Peru*
> *Alfredo Vazquez-Barrios, National University of Engineering of Peru, Peru*
> *Ricardo Hector Rodriguez-Robles, National University of Engineering of Peru, Peru*

Chapter 12
Water Footprint Reduction in Engineering, Laboratory, and Administrative Buildings: A Field Case of the Scientific University of the South 289
 Tiffany Krisel Billinghurst, Scientific University of the South, Peru
 Alfredo David Lescano Lozada, Alwa, Peru

Compilation of References 313

About the Contributors 341

Index 345

Detailed Table of Contents

Preface...
xxiv

Chapter 1
The Concept of Sustainable Exploration, Development, and Operation of
Unconventional Oil and Gas Fields ....
1

    Jesus Samuel Armacanqui-Tipacti, National University of Engineering
       of Peru, Peru

The current daily life of society is founded upon an abundance of affordable energy that covers the residential, commercial, and industrial energy needs. More recently, the pursuit to reduce the CO2 emissions has been directly linked to the use of fossil fuels, triggering a reactive response that calls for the complete cease of it. While the objective conditions on the ground for such undertaken are not available in the short term, a few decades ahead, the oil and gas industry will significantly improve best practices. This includes the activities related to the production of oil and gas from unconventional fields, also called hydraulic fracturing; however, still there is a room for improvement. Nearly 80% of the current US hydrocarbon production comes from these fields, which has made it the top producer with a total production of 16 million barrels of oil equivalent. In the present work, a proposal is presented towards a more sustainable exploration, development, and operation of unconventional fields.

Chapter 2
The Muerto Shale: Stratigraphic Distribution of Carbonate Content and Organic Geochemistry in the Corcobado Outcrop.. 40

>*Jose Alfonso Rodriquez-Cruzado, National University of Engineering of Peru, Peru*
>
>*Jorge Luis Ore-Rodriguez, National University of Engineering of Peru, Peru*

The stratigraphic distribution of carbonate content (%CaO), total organic carbon (TOC), and pyrolysis parameters of a 277-meter section of the Muerto Formation, which was measured in the Corcobado outcrop, located in the Lancones Basin, Northwest Peru is characterized in the chapter. This is an organic-rich, calcareous sequence, with a thermal maturity in the oil window (Tmax = 455 °C and 1%Ro), which is also supported by the presence of bitumen in fractures. The authors identified an inverse relationship between %CaO and TOC, where it was observed that the highest organic enrichment (on average 2.42 wt.%) is associated with a carbonate content in the 30-40% range.

Chapter 3
Volumetric Estimation of the Gas in the Generator Rock Through an Evaluation of the Adsorption in the Northeast of Peru 61
> *Heraud Taipe- Acuña, National University of Engineering of Peru, Peru*
> *Victor Huerta Quiñones, National University of Engineering of Peru, Peru*
> *Ali Tinni, University of Oklahoma, USA*
> *Hugo Valdivia Ampuero, National University of Engineering of Peru, Peru*
> *Isabel Moromi Nakata, National University of Engineering of Peru, Peru*

In Peru, energy security is based on gas. The decrease in natural gas reserves between 2016 and 2019 was from 16.1 to 10.1 TCF, which reflects a negative replacement rate in this period and a decrease in exploration. However, the evaluation of shale gas reservoirs could influence the increase of gas resources in Peru. Technological advances in the exploitation of these resources made it possible to transform gas resources into economic reserves. The study of unconventional shale gas reservoirs in South America has been developing with greater interest, especially in the business of exploration, characterization, and pilot tests in these reservoirs. This research is the first to evaluate the volume of gas adsorbed in the source rock (Murder Formation) in the Lancones basin in Northeastern Peru, through isotherms obtained in the laboratory, and aims to volumetrically quantify the total gas content (scf/ton) in this formation.

Chapter 4
Characterization of the Structure of Mesopores in the Generating Rock of Northeastern Peru: Muerto Formation.. 84

Kevin Chipana-Suasnabar, National University of Engineering of Peru, Peru
Heraud Taipe Acuna, National University of Engineering of Peru, Peru
Joseph Sinchitullo Gomez, National University of Engineering of Peru, Peru
Ali Tinni, University of Oklahoma, USA
Israel Chavez Sumarriva, National University of Engineering of Peru, Peru
Marco Tejada Silva, National University of Engineering of Peru, Peru

The contribution of organic matter to the total porosity in shale formations is mainly adsorbed in micropores and mesopores, which have a strong influence on the productive life of the reservoir, prolonging the productive life of the reservoir. An incorrect understanding of the total porosity and factors that influence it could lead to the rejection of potential productive zones during the initial stage of exploration. This research aims to characterize pores from 2 to 50 nm by low pressure N2 adsorption in conjunction with geochemical and mineralogical analysis of 5 samples from the Muerto Formation, acquired from the Lancones basin. The results indicated that the specific surface area is in the range of 0.58 to 3.41 m2/g values according to the oil and condensate generation window. In the pore size distribution, the presence of micropores was not observed; for the mesopores their distribution consists of two main ranges from 4 to 10 nm and from 20 to 50 nm, the porosities found were from 1.8 to 8.2%, being the average contribution of the mesopores to the total porosity is 35%.

Chapter 5

The Shale of the Cabanillas Formation: Integration Methodology and the Presence of the Gas Resource (Shale Gas) in the Marañón-Peru Basin 100

> Walter Jacob Morales-Paetan, National University of Engineering of Peru, Peru
>
> Pedro Zegarra Sanchez, National University of Engineering of Peru, Peru

Peru is economically dependent on hydrocarbons, so exploring gas shale in unconventional systems can be a new source of energy which has stimulated the interest of major oil companies and countries with greater energy dependence. The present study is based on the evaluation of the shales of the Cabanillas formation as a hydrocarbon generating rock and to be considered in the unconventional system and that encompasses the theoretical and conceptual framework, for which a collection and bibliographic review (digital data of exploratory well logs, seismic information, reports, and data analysis but not the state of the art) in the southern part of the Marañon Basin that exploits hydrocarbons with the objective of considering the shales as an unconventional reservoir was carried out.

Chapter 6

First Insights in the Estimation of the Petroleum Generation Potential of the Muerto Source Rock From el Cortado Outcrop, Northwestern Peru 123

> Jose Alfonso Rodriguez-Cruzado, National University of Engineering of Peru, Peru
>
> Jorge Oré-Rodriguez, National University of Engineering of Peru, Peru

The Muerto Formation is generally defined as an oil prone source rock in the Peruvian Northwest. As other worldwide source rocks, it could also act as a self-contained source-reservoir system; however, there are not quantitative estimations about the generated petroleum volume and its oil/gas composition. The data for this study includes TOC and Rock-Eval pyrolysis analyzes, which were performed by Petroperu in 1986 and Perupetro in 1999, on rock samples from the El Cortado outcrop, located in the Lancones Basin. Although both datasets contain lab analyzes performed more than 20 years ago, modern calculation techniques were applied to assess the petroleum generation potential. It was calculated that this sequence generates predominantly oil (45-73%) and cogenerates gas (27-55%), in the oil window; however, the Muerto Formation is most probably a shale gas resource, due to its advanced thermal maturity in the Northern Lancones Basin.

Chapter 7
Lithological Facies Classification, Surface Gamma Ray, and Toc Analysis Using an Outcrop of Unconventional Rock Field: Case in the Muerto Formation, Peru, for the Identification of New Areas With Hydrocarbon Potential 146

    Brayan Nolasco Villacampa, National University of Engineering of Peru, Peru
    Jorge Luis Oré- Rodriguez, National University of Engineering of Peru, Peru
    José Alfonso Rodriguez-Cruzado, National University of Engineering of Peru, Peru
    Israel J. Chavez-Sumarriva, National University of Engineering of Peru, Peru
    Manuel Lopez Reale, LCV Group, USA
    Humberto Chiriff-Rivera, National University of Engineering of Peru, Peru
    Jesus Samuel Armacanqui-Tipacti, National University of Engineering of Peru, Peru
    Luz Eyzaguirre-Gorvenia, National University of Engineering of Peru, Peru
    Alfredo Vazquez-Barrios, National University of Engineering of Peru, Peru

In the present work, the stratigraphic distribution of the Muerto Formation was evaluated, with the objective of identifying the lithological facies, using petrography studies through thin sections and gamma ray analysis for an elevation. The section under study crops out on the eastern margin of the Amotape Mountains. Stratigraphic, facies, and sedimentological analyzes are methods of vital importance to evaluate the potential of an unconventional reservoir. These studies make it possible to predict the spatial distribution of facies, as well as to identify the areas with the most favorable petrophysical properties for the prospecting and exploitation of hydrocarbons, in addition to being able to estimate with greater certainty the volume of hydrocarbons in situ and the optimal completion of a well.

Chapter 8
Evaluation of Geological Boundary Conditions of a Water Well for Aquifer Integrity Conformance by Pressure Transient Analysis: Presence of Hydraulic Fracturing, the Operation of Wastewater, CO2 Storage, and Geothermal Wells ....
173

    Gustavo Enrique Rodriguez-Robles, National University of Engineering
        of Peru, Peru
    Jesus Samuel Armacanqui-Tipacti, National University of Engineering
        of Peru, Peru

It is very important to determine the characteristics of the nature of geological faults in the stage of production and exploration of hydrocarbons. However, there is no cost-effective methodology to determine the sealing or non-sealing character of a geological fault. There are methods to determine these characteristics, which have some drawbacks, e.g., it is required to maintain the production conditions of an oil well at a constant flow rate or to shut down the production of the well, which entails operating expenses for the period of time that the well is not in production. The objective of this study is to develop a cost-effective and environmentally sustainable methodology in order to determine the effect of geological faults (i.e., boundary effect). Therefore, a water well was located near the geological fault and the pressures that were measured during transient pressure tests were recorded, the benefit obtained by performing a water well was the speed of the pressure response, which It is transmitted more quickly, due to the characteristic of the fluid.

Chapter 9
Cost-Effective Advanced Remote Diagnostics of Sucker Rod Pumping Wells From Dynamometric Charts: A Deep Learning Approach................................ 188

Joel Hancco Paccori, National University of Engineering of Peru, Peru
Manuel Castillo Cara, Polytechnic University of Madrid, Spain
Jesus Samuel Armacanqui-Tipacti, National University of Engineering of Peru, Peru

There is a growing number of oil production wells in the world that use rod pump units as an extraction system. In fact, this lifting method is become the preferred one for unconventional wells that are producing at the late-stage period, yet with still attractive rates of a few hundred bopd. Dynamometry basically consists of the visual interpretation of the shape of the load graph based on the position of the piston of the subsoil pump. This task is carried out by an operator and his experience is used for the correct interpretation that can be contrasted. With additional tests, diagnosis becomes very important because it allows optimizing production by adjusting rest and production times, reducing operation and maintenance costs, avoiding failures and unscheduled stops, but many times there are not enough trained personnel for interpretation. In this context, dynamometry has been refined both in the acquisition and in the interpretation of dynamometric records.

Chapter 10
Use of Limonene as a Biodegradable Surfactant for the Inhibition and Removal of Paraffins in Oil Production Operations 220

    Maria Rosario Viera-Palacios, National University of Engineering of Peru, Peru
    Miguel Ángel Guzmán, Peruvian University of Applied Sciences, Peru
    Jesus Samuel Armacanqui-Tipacti, National University of Engineering of Peru, Peru
    Cesar Lujan Ruiz, National University of Engineering of Peru, Peru
    Guillermo Prudencio, National University of Engineering of Peru, Peru
    Luz Eyzaguirre-Gorvenia, National University of Engineering of Peru, Peru

Paraffin buildup is one of the most common problems in oil production operations. The agglomeration of these alkanes in the production tubing and the surface flow lines not only lowers the production rate but also results in an additional damage repair cost. It also represents a threat to both the operation of the equipment (rods and subsoil pumps) and to the environment, and it is a risk for the personnel who carry out the paraffin removal work, since the conventional chemicals used, such as xylene and toluene, are highly toxic and polluting. The presence of paraffins in crude oil is frequent in unconventional fields or share oil fields, which have API gravity values greater than 35. In conventional fields, cases of paraffinic crude oil can be found; especially in the final part of the field's life cycle. In the present work, crude oil samples were used from a conventional field located in the Talara Basin, in Northern Peru, which had an API gravity (ASTM D1298) of 36.

Chapter 11

Using H2 to Increase the Energy Efficiency and Reduce the Containment Emissions in the Operation of Generators 236

Franco A. Cassinelli-Cisneros, National University of Engineering of Peru, Peru
Jesus Samuel Armacanqui-Tipacti, National University of Engineering of Peru, Peru
Ciro Ormeño-Aquino, National University of Engineering of Peru, Peru
Jose Carlos Rodriguez, San Luis Gonzaga University of Ica, Peru
José Rosendo Campos Barrientos, San Luis Gonzaga University of Ica, Peru
Mohamed Yehia, Suez Oil Company, Cairo, Egypt
Nelson Michael Villegas-Juro, National University of Engineering of Peru, Peru
Alfredo Vazquez-Barrios, National University of Engineering of Peru, Peru
Ricardo Hector Rodriguez-Robles, National University of Engineering of Peru, Peru

The application of gaseous hydrogen in combustion internal (CI) diesel dual-fuel (DDF) engine is not new, and it does not yet have broad commercial applications that ensure the increase in performance and consequent reduction in costs to which it is normally associated. The effect of hydrogen on diesel internal combustion engines can be analyzed through the different parameters that an CI engine has grouped into four main aspects: performance, combustion characteristics, polluting emissions, and future challenges. The effect of hydrogen application is consequently a multivariate analysis that this study addresses individually on each parameter, explaining its probable causes and a necessary theoretical framework for it. The effect on each parameter is finally quantified in the best approximate framework for application in the petroleum industry, which is defined with a hydrogen contribution of between 10-20% and load conditions of at least 50%. The importance of this study lies in the empirical analysis and verification of the supply of hydrogen to DDF engines.

Chapter 12
Water Footprint Reduction in Engineering, Laboratory, and Administrative Buildings: A Field Case of the Scientific University of the South 289

    Tiffany Krisel Billinghurst, Scientific University of the South, Peru
    Alfredo David Lescano Lozada, Alwa, Peru

The water footprint is a tool that indicates the direct and indirect consumption of fresh water, whether in a production process, product, service, building, institution, geopolitical area, economic sector, or a person. In the headquarter of the oil and gas (O&G) corporations, also called exploration and production (E & P) companies, typically there are administration, engineering, and laboratory activities related to the business bottom line a close analogy to these types of buildings are educational buildings, such as universities. Therefore, the outcome of the presented work could be applied to the O&G buildings as well. The objective of this research was to calculate the water footprint of the Villa Campus of the Scientific University of the South for its activities in 2019, using the methodology of the Water Footprint Network, which is considered a world standard for the evaluation of the water footprint and based mostly on the work of Professor Dr. Arjen Hoekstra, who introduced the concept of the water footprint in 2002.

Compilation of References 313

About the Contributors 341

Index 345

Foreword

Shale gas and liquid production have created a radical change in land-based oil and gas operations that can only be solved by technical understanding and scientific application to achieve a balance between profitability and the responsibilities of sustainable recovery of a limited mineral reserve. The chapters in this book explain the ways in which new technologies, tools and techniques have extended our abilities to reduce environmental impact such as water usage, additive improvements and reducing methane emission and flaring.

The knowledge of conventional oil and gas production stretched over one hundred and seventy years, limping painfully from cable tool rigs and gushers to modern secondary and tertiary recovery methods. This trial-and-error compilation of development learnings has been somewhat blindly applied to shale gas and liquids production, resulting in a slow improvement in reserve recovery over a thirty-year span. These incremental improvements are the result of focusing work on individual problems. Mechanical changes were initially favored above increasing the geologic, geophysical and geo-chemical understanding that actually controlled production from the wide variety mud-stone composition in rocks that are labeled we know as "shale". This book pushes that technical know-how further in many areas, and should increase recovery while reducing environmental impact.

The initial high flush production from shale wells advanced the vision of fast pay-back of well cost and following profits that quickly attracted investment, often by investors unfamiliar with oil and gas operations and environmental issues. Stated simply, the financial pressure to quickly recover costs lead to consequences in some shale developments that impacted emissions and resulted in leaving 80 to 90% of the hydrocarbon in the rock after the initial flush production. In short, the operational methods of conventional hydrocarbon production were, in some cases, reducing the long-term profitability of a shale well. "Net present value of money" was a common driver in well development.

The large hydrocarbon reserves in these source rocks is undeniable, but the results of early and even some current shale production and development methods fall short of achieving responsible recovery. Challenges involve hold-over issues from conventional hydrocarbon recovery and new problems based on the changes in shale formations resulting from production operation methods as well as the chemical, physical and stress changes to the often-ductile shale formations.

As a bit background, not every shale contains recoverable hydrocarbon and understanding the conditions that lead up to the presence of hydrocarbons involves geological depositional history of potentially producible shales and both field and laboratory studies. First, most shales with hydrocarbons are deep marine deposits where organic material was a part of the sediment and the water was deep enough to create a very low oxygen content that would favor organic maturity of specific organics with pressure and temperature into gas or oil. Some shales have little or no hydrocarbon content. Second, following deposition, the occurrence of natural fractures over geologic time enables movement of hydrocarbon. In protecting the stability of these flow channels through the shale is critically important to achieving better recovery and drives responsible development toward sustainable recovery.

Shales are commonly very low permeability, having one or two orders of magnitude lower permeability than conventional reservoir rocks that trap the hydrocarbons generated by the shales. Shales have only moderate porosity due to the small size of the sediments and higher clay content. Shales with no natural fractures or fissures are usually not productive. Avoiding these shales or using new technologies to effectively develop stranded reserves is key to future development.

Among the needs to moving shale operations nearer to sustainable in a mineral production operation, start with changes applied in the past twenty years and several from this book highlighted innovations, as its chapters provide and explain new insights into responsible shale operations and techniques vital to improving water reuse, frac fluid efficiency, and improving flow through the formation by reducing formation damage. In addition, chapters address advances in operational efficiency by reusing produced water, developing fit for purpose efficient frac fluids, avoiding formation damage, and monitoring operations for methane emissions, all of which are vital in complying with regulations. The level of innovation, potential creativity, and new technology development are truly impressive.

June 27, 2024

George E. King
Amoco Research Center, USA

George E. King is a registered professional engineer with more than 40 years of experience since joining Amoco Research Center in 1971. His technical work has provided advances in foam fracturing, production from unstable chalk, underbalanced perforating, sand control reliability, and shale gas completions and complex fracturing. Currently, King is working with new technologies for the oil and gas industry. He holds a BS degree in chemistry from Oklahoma State University, and a BS degree in chemical engineering and an MS degree in petroleum engineering, both from the University of Tulsa, where he also taught completions and workovers for 11 years as an adjunct professor. King has written 65 technical papers, was awarded the 2004 SPE Production Operations Award, and was recognized as the 2012 Engineer of the Year by the Greater Houston Chapter of the Texas Society of Professional Engineers. After 37 years, he retired in 2008 from BP as a distinguished advisor and is currently Apache's global technical consultant. King lives in Katy, Texas. One of his hobbies is rebuilding vintage Ford Mustangs.

Preface

In the landscape of energy extraction, the familiar narrative of prioritizing profits over environmental stewardship is rapidly losing ground. The dichotomy between economic gains and ecological responsibility, once considered immutable, now stands as a relic of outdated thinking. Today, as the imperatives of sustainability loom large and Environmental, Social, and Governance (ESG) criteria reign supreme, the oil and gas industry find itself at a crossroads. More and more the statement that if something is beneficial for the nature/environment, then it is good for mankind, is prevailing in in both the public eyes and in the Board Rooms.

Sustainability Applied to Unconventional Oil and Gas Field Exploration and Development is a pioneering volume conceived to drive a seismic shift in how we approach energy extraction with environmental and social responsibility in the light of the latest technological advances and industry best practices. Authored by Jesus S. Armacanqui Tipacti, Susan Smith Nash, Luz Eyzaguirre Gorvenia, and Rouzbeh Moghanloo, this book stands as a beacon for those seeking to reconcile profitability with ecological and social imperatives.

At its core lies a holistic framework, one that shatters the confines of antiquated thinking to forge a path where economic prosperity and environmental stewardship are not adversaries but allies. This volume serves as a guiding light for Oil and Gas Operators of unconventional shale fields, Regulatory Bodies and Academia, and interested Researchers, offering a roadmap to navigate the intricate web of compliance and sustainability mandates that define the industry today.

Drawing upon a wealth of expertise, the authors delve into specific technologies and strategies designed to foster the swift adoption of economically viable, technologically sound, and ecologically harmonious practices. Through fostering dialogue among stakeholders – from R&D teams to regulatory bodies – this book envisions a collaborative approach that benefits not only corporate interests but also the communities and environments impacted by energy extraction.

Preface

Tailored for a diverse audience, including oil and gas operators and of unconventional fields, (field and office staff as well as Managers and C-Level Executives), product and service companies, research teams, governmental entities, and students in the fields of petroleum engineering and earth sciences and oil and gas fields economics, this volume beckons all who are vested in both the energy transition and in the future of energy. From governmental regulators to product and service providers, all are invited to embark on this transformative journey.

Within these pages lie a treasure trove of insights covering every facet of unconventional oil and gas field exploration, development, and operation. From technological innovations to circular economy applications, from the challenges of climate change to the intricacies of Environmental Impact Assessment (EIA) and Health Safety and Environment (HSE) protocols. This book leaves no stone unturned towards building a fit for purpose Process Centered and Life Cycle Approach that is presented in a very comprehensive manner as the basis for very specific and meaningful Key Performance Indicators - KPIs.

As editors, it is our fervent hope that this volume serves as a catalyst for change – a rallying cry for an industry poised on the brink of transformation. We invite you to join us on this journey towards a future where profitability and environmental responsibility are not mere aspirations but fundamental tenets of our collective ethos.

CHAPTER 1: THE CONCEPT OF SUSTAINABLE EXPLORATION, DEVELOPMENT AND OPERATION OF UNCONVENTIONAL OIL AND GAS FIELDS

In the intricate tapestry of modern society, energy stands as its lifeblood, fueling residential, commercial, and industrial domains alike. Yet, this vitality comes at a cost, with the pervasive use of fossil fuels casting a long shadow over environmental sustainability. As calls for reducing CO_2 emissions reverberate, the spotlight falls on unconventional energy sources, notably Hydraulic Fracturing (HF), driving both innovation and scrutiny within the Oil and Gas Industry. Amidst this landscape, a crucial question emerges: How can we navigate the energy transition towards sustainability?

In this chapter, Armacanqui-Tipacti addresses the evolving landscape of energy consumption and the imperative of sustainability in unconventional oil and gas field operations. He highlights the significance of improving industry practices to align with environmental concerns while maintaining energy production. By examining the current state of unconventional fields, Armacanqui- Tipacti proposes a path towards more sustainable exploration, development, and operation methods of unconventional fields. Environmental concerns loom large, with contamination risks, seismic activities, and water resource consumption in focus. Against this backdrop, the integration of Environmental, Social, and Governance (ESG) criteria emerges as a viable alternative, steering the industry towards a more conscientious path. However, in to get there practical guidelines that needed to the bottom line business needs derived from the exploration, development and operation activities.

Drawing upon real-world examples, this chapter advocates for a holistic, lifecycle-based approach, championing executive commitment, technological innovation, and community engagement as cornerstones of sustainable practice. Through nuanced analysis and practical insights, it seeks to illuminate a path forward, where energy abundance harmonizes with environmental stewardship.

This chapter introduces the basis for a new way of working in the sense that it sets a well-defined and functional relationship of the Sustainable Development Concept applied to activities in Unconventional Shale Oil/Gas Fields, linking the specific working tools/procedures to a Process Centered Workflow and to a Measuring of Results that translates to specific meaningful KPIs.

CHAPTER 2: THE MUERTO SHALE: STRATIGRAPHIC DISTRIBUTION OF CARBONATE CONTENT AND ORGANIC GEOCHEMISTRY IN THE CORCOBADO OUTCROP, NOTHWESTERN PERU

Rodriguez-Cruzado and Ore-Rodriguez delve into the geological characteristics of the Muerto Formation, focusing on carbonate content, organic carbon distribution, and geochemical parameters. Through their study of the Corcobado outcrop, situated in the Lancones Basin of Northwest Peru. The focus lies on analyzing the carbonate content (%CaO), total organic carbon (TOC), and pyrolysis parameters across a 277-meter section. This organic-rich, calcareous sequence exhibits thermal maturity within the oil window, evidenced by a Tmax of 455°C and 1%Ro, corroborated by the presence of bitumen in fractures. Notably, an inverse relationship between %CaO and TOC is observed, with the highest organic enrichment (averaging 2.42 wt.%) associated with 30-40% carbonate content. Three distinct intervals emerge: the Lower Interval (0 – 55 meters), Middle Interval (55 – 140 meters), and Upper Interval (140 – 278 meters), each revealing nuanced variations in %CaO, TOC, and hydrocarbon potential. This analysis sheds light on the complex interplay between lithology, organic richness, and hydrocarbon generation potential within the Muerto Formation, renaming it as a potential unconventional resource.

Preface

CHAPTER 3: VOLUMETRIC ESTIMATION OF GAS IN THE GENERATOR ROCK THROUGH AN EVALUATION OF ADSORPTION IN NORTHEAST OF PERU

In the landscape of Peru's energy security, gas reserves have undergone a notable shift, experiencing a decline from 16.1 to 10.1 TCF between 2016 and 2019. This decline has raised concerns, accentuated by a negative replacement rate and a slump in exploration activities. However, a ray of hope emerges with the exploration of shale gas reservoirs, offering the potential for a resurgence in gas resources. Technological advancements have rendered the extraction of unconventional resources economically viable, sparking interest in their exploration and characterization. This chapter the authors delves into groundbreaking research evaluating gas content in the Muerto Formation of Peru's Lancones basin, shedding light on promising reserves and their potential implications. As resulted from the key parameters to determination to quantify the gas in place

CHAPTER 4: CHARACTERIZATION OF THE STRUCTURE OF MESOPORES IN THE GENERATING ROCK OF NORTHEASTERN PERU: MUERTO FORMATION

Understanding the intricate interplay between organic matter and porosity in shale formations is paramount for optimizing reservoir productivity. Micropores and mesopores, repositories for organic matter, wield considerable influence over reservoir lifespan. Misinterpreting total porosity and its influencing factors risks dismissing potentially lucrative exploration zones. This chapter Chipana-Suasnabar and colleagues delves into a comprehensive analysis of pore characterization, employing low-pressure N_2 adsorption alongside geochemical and mineralogical assessments of Muerto Formation samples from the Lancones basin. Results unveil a nuanced landscape, showcasing mesopores as primary contributors to porosity, with organic matter content at its core. Such insights are crucial for informed decision-making in reservoir management and exploration endeavors.

CHAPTER 5: THE SHALE OF THE CABANILLAS FORMATION: INTEGRATION METHODOLOGY AND THE PRESENCE OF THE GAS RESOURCE (SHALE GAS) IN THE MARAÑÓN-PERU BASIN

In the quest for energy diversification, Peru's reliance on hydrocarbons prompts exploration into unconventional sources like gas shale. This chapter Morales-Paetan and Zegarra-Sanchez delves into the geological assessment of the Cabanillas formation within the Marañón Basin's southern region. This was achieved using old log. eg seismic data. utilizing new interpretation methods in the Peruvian jungle. As a significant hydrocarbon source rock, it holds promise as an unconventional reservoir. Drawing from exploratory well logs, seismic data, and comprehensive analysis, this study unveils insights crucial for Peru's energy future.

CHAPTER 6: FIRST INSIGHTS IN THE ESTIMATION OF THE PETROLEUM GENERATION POTENTIAL OF THE MUERTO SOURCE ROCK FROM EL CORTADO OUTCROP, NORTHWESTERN PERU

Rodriguez-Cruzado and Oré-Rodriguez examine the petroleum generation potential of the Muerto Formation in northwestern Peru. The Muerto Formation, a recognized oil-prone source rock in the Peruvian Northwest, presents intriguing possibilities as a self-contained source-reservoir system, akin to similar formations worldwide such as Eagle forth and Vaca Muerta Formations However, despite its potential, precise quantitative estimations regarding the volume and composition of generated petroleum remain elusive. Utilizing data from TOC and Rock-Eval pyrolysis analyses conducted by PetroPeru in 1986 and PeruPetro in 1999, this study focuses on rock samples from the El Cortado outcrop within the Lancones Basin. Applying modern calculation techniques to these decades-old datasets, it's revealed that the Muerto Formation primarily yields oil (45-73%) alongside gas (27-55%) within the oil window. Notably, its advanced thermal maturity suggests a shale gas resource potential, particularly in the Northern Lancones Basin, with an estimated gas generation potential of 903-1278 mcf/ac-ft. Considering typical reservoir conditions, the anticipated storage capacity ranges from 270-540 mcf/ac-ft, significantly lower than the generation potential. Consequently, saturation by gas at depth is expected within the Muerto Formation. This chapter presents a show case about the use of modern calculation techniques to decode -old data sets.

Preface

CHAPTER 7: LITHOLOGICAL FACIES CLASSIFICATION, SURFACE GAMMA RAY AND TOC ANALYSIS USING AN OUTCROP OF UNCONVENTIONAL ROCK – FIELD CASE IN THE MUERTO FORMATION, PERU, FOR THE IDENTIFICATION OF NEW AREAS WITH HYDROCARBON POTENTIAL

Within this chapter, we delve into a meticulous examination of the Muerto Formation's stratigraphic distribution. Nolasco-Villacampa and colleagues examine lithological facies and gamma ray analysis, this chapter represents a show case for using surface gamma ray measurement to relate to more expensive down hole gamma ray recording, by means of utilizing the surface outcrop. Situated along the eastern margin of the Amotape Mountains – La Brea, Piura, Peru, the section under scrutiny offers invaluable insights.

CHAPTER 8: EVALUATION OF GEOLOGICAL BOUNDARY CONDITIONS OF A WATER WELL FOR AQUIFER INTEGRITY CONFORMANCE BY PRESSURE TRANSIENT ANALYSIS IN THE PRESENCE OF HYDRAULIC FRACTURING, AND THE OPERATION OF WASTEWATER, CO2 STORAGE, AND GEOTHERMAL WELLS

Rodriguez-Robles and Armacanqui-Tipacti present a methodology for evaluating geological fault characteristics to evaluate aquifer integrity in hydrocarbon production areas, specifically around the proposed wells to be drilled.

In the realm of hydrocarbon production and exploration, understanding the nature of geological faults is paramount. Yet, the absence of a cost-effective method to determine whether a fault seals or partially sealing poses a significant challenge. Existing approaches, while available, come with drawbacks such as the need for maintaining constant flow rates or temporarily shutting down production wells, incurring additional operational costs. This study aims to bridge this gap by developing a methodology that is both economical and environmentally sustainable, focusing on delineating the boundary effect of geological faults. Through transient pressure tests conducted near a water well, the study leveraged the fluid's rapid pressure response to unveil crucial insights. Results revealed the masking effect of well filling, obscure visibility within reservoir zones, and highlighted the predominant influence of active aquifers over other boundary effects. This innovative approach holds promise for enhancing fault characterization in hydrocarbon exploration endeavors. This chapter shows a case the use of a cost-effective method with a low footprint to interpret the data.

CHAPTER 9: COST EFFECTIVE ADVANCED REMOTE DIAGNOSTICS OF SUCKER ROD PUMPING WELLS FROM DYNAMOMETRIC CHARTS: A DEEP LEARNING APPROACH

In recent years, the global landscape of oil production has witnessed a surge in the utilization of rod pump units as primary extraction systems. This trend is particularly pronounced in unconventional wells operating during their late-stage periods, maintaining robust production rates of a few hundred barrels of oil per day (bopd). Dynamometry, the method of interpreting load graphs based on subsoil pump piston positions (looked vs positions), plays a pivotal role in this domain. However, despite its significance, the accurate interpretation of dynamometric records often relies heavily on the expertise of operators. This dependence underscores the pressing need for more sophisticated and reliable diagnostic techniques that can be accurate and be carried out automatically and remote.

In response to this demand, recent advancements in dynamometry have seen substantial refinement in both data acquisition and interpretation. Automatic diagnosis, boasting an impressive accuracy surpassing 99%, represents a notable breakthrough. Among the explored methodologies, classification algorithms and convolutional neural networks have emerged as frontrunners, facilitating the identification of individual operating conditions on dynamometric charts. Yet, the true essence of effective diagnosis lies not merely in pinpointing isolated conditions but in synthesizing the entirety of information encapsulated within these charts.

Against this backdrop, Hancco-Paccori, Armacanqui-Tipacti and Castillo-Cara, proposes a novel approach: the development and evaluation of deep learning architectures tailored for diagnosing one or multiple operating conditions depicted on dynamometric charts an opposed to the common use of 1 chart and 1 condition, given that real cases then present 2 or even 3 conditions occurring at the sometime. Embracing a one-to-many architecture, this innovative framework aims to offer a comprehensive diagnostic solution, harnessing the rich trove of information embedded within dynamometry charts.

CHAPTER 10: USE OF LIMONENE AS A BIODEGRADABLE SURFACTANT FOR THE INHIBITION AND REMOVAL OF PARAFFINS IN OIL PRODUCTION OPERATIONS

In the realm of oil production operations, paraffin buildup stands out as a pervasive challenge. The accumulation of these alkanes within production tubing and surface flow lines and even inside the downhole pumps not only diminishes production rates but also escalates repair costs and it can lead to premature failures.

Preface

Moreover, it poses threats to equipment functionality, environmental integrity, and the safety of personnel tasked with paraffin removal, given the toxicity and pollution associated with conventional chemical treatments like xylene flashing. This issue is particularly prevalent in Unconventional and Share Oil Fields with API gravity values surpassing 35, and in Conventional fields, especially towards the end of their operational lifespan. In this Chapter Viera-Palacios and colleagues investigate the use of limonene as a biodegradable surfactant for paraffin inhibition and removal in oil production operations. Through experimental simulation and analysis, this work reveals promising results, demonstrating the efficacy of limonene-based solutions in both paraffin inhibition and removal, offering a safer, eco-friendly, and cost-effective alternative to traditional methods.

CHAPTER 11: USING H2 TO INCREASE THE ENERGY EFFICIENCY AND REDUCE THE CONTAMINANT EMISSIONS IN THE OPERATION OF GENERATORS

The utilization of gaseous hydrogen in Combustion Internal (CI) Diesel Dual-Fuel (DDF) engines represents an area of ongoing exploration. Despite its historical presence, widespread commercial adoption remains elusive, hindering anticipated performance enhancements and cost reductions. This chapter Cassinelli-Cisneros, Armacanqui-Tipacti and colleagues delves into the multifaceted impact of hydrogen on CI engines, encompassing performance, combustion characteristics, emissions, and forthcoming challenges. Through meticulous analysis, each parameter's response to hydrogen application is scrutinized, elucidating causal relationships and requisite theoretical underpinnings. Ultimately, empirical insights culminate in an optimal framework, advocating for a hydrogen blend of 10-20% under load conditions exceeding 50%. This study's significance lies in its potential to revolutionize a major oil-consuming industry through empirical validation and exploration of hydrogen supply to DDF engines. In addition to the use of dual diesel-gas engines, the use of h2 means reducing exhaust gas emission.

CHAPTER 12: WATER FOOTPRINT REDUCTION IN ENGINEERING, LABORATORY AND ADMINISTRATIVE BUILDINGS: A FIELD CASE OF THE SCIENTIFIC UNIVERSITY OF THE SOUTH

This chapter Billinghurst-Vargas and Lescano-Lozada delves into the intricacies of water footprint assessment, a vital tool illuminating the intricate web of direct and indirect freshwater utilization. Beyond mere production processes, its scope extends to diverse realms like services, institutions, and geopolitical landscapes. Drawing parallels between the operational dynamics of oil and gas corporations and educational institutions like universities, it embarks on an enlightening journey. Focused on the Villa Campus of the Scientific University of the South in Lima Peru, the research employs the esteemed methodology of the Water Footprint Network, pioneered by Professor Dr. Arjen Hoekstra. Through meticulous analysis and sustainability scrutiny, crucial insights are unraveled for optimizing water consumption. This chapter unfolds a specific roadmap towards a more sustainable water future, underpinned by empirical data and strategic measures.

CONCLUSION

As editors of this comprehensive volume on *Sustainability Applied to Unconventional Oil and Gas Field Exploration and Development* we are gratified to present a diverse array of chapters that collectively contribute to a deeper understanding of the challenges and opportunities facing the energy industry, especially in this energy transition period. From the conceptual framework of sustainable exploration to the practical applications of advanced diagnostics and environmental mitigation strategies, each chapter offers valuable insights gleaned from rigorous research and practical experience, in the light of the latest technological advances.

Through the collective efforts of our esteemed authors, this book serves as a testament to the collaborative spirit driving innovation in the unconventional shale oil and gas sector. By fostering dialogue and sharing knowledge across disciplines, we pave the way for a more sustainable future where economic prosperity and environmental responsibility are not mutually exclusive but mutually reinforcing, more over generate new opportunities.

As we navigate the complexities of energy transition and environmental stewardship, the insights presented in these pages serve as guiding beacons for industry professionals, researchers, policymakers, and students alike. We are confident that the ideas and methodologies explored within these chapters will inspire further inquiry, dialogue, and action towards a more sustainable energy landscape.

Preface

In closing, we extend our heartfelt gratitude to all the contributors who have enriched this volume with their expertise and dedication. May the insights contained herein serve as catalysts for transformative change, ushering in an era where sustainability is not merely a goal, but a fundamental principle guiding the exploration, development and operation of unconventional shale oil and gas fields.

Jesus Samuel Armacanqui Tipacti

National University of Engineering of Peru, Peru

Susan Smith Nash

American Association of Petroleum Geologists, USA & University of Oklahoma, USA

Luz Eyzaguirre Gorvenia

National University of Engineering of Peru, Peru

Rouzbeh Moghanloo

University of Oklahoma, USA

Section 1
Development Module

Chapter 1
The Concept of Sustainable Exploration, Development, and Operation of Unconventional Oil and Gas Fields

Jesus Samuel Armacanqui-Tipacti
National University of Engineering of Peru, Peru

ABSTRACT

The current daily life of society is founded upon an abundance of affordable energy that covers the residential, commercial, and industrial energy needs. More recently, the pursuit to reduce the CO_2 emissions has been directly linked to the use of fossil fuels, triggering a reactive response that calls for the complete cease of it. While the objective conditions on the ground for such undertaken are not available in the short term, a few decades ahead, the oil and gas industry will significantly improve best practices. This includes the activities related to the production of oil and gas from unconventional fields, also called hydraulic fracturing; however, still there is a room for improvement. Nearly 80% of the current US hydrocarbon production comes from these fields, which has made it the top producer with a total production of 16 million barrels of oil equivalent. In the present work, a proposal is presented towards a more sustainable exploration, development, and operation of unconventional fields.

DOI: 10.4018/979-8-3693-0740-3.ch001

1. INTRODUCTION

1.1. The Objective of the Book

The objective of the present book is to propose a process centered Life Cycle Concept for the Exploration, Development and Operation of Unconventional Oil and Gas Fields containing technical sustainable practices and procedures to ensure a smooth Energy Transition, which is realistically expected to be spanned over a few decades.

1.2. The Importance of the Unconventional Oil and Gas in the US Industry and its Worldwide Impact

The current daily life of the society is founded upon an abundance of affordable energy that is used in factories and homes, derived from hydrocarbons. Hydrocarbons do also provide the fuel for most of the transportation on sea, land and air, and the raw material for a variety of industries such as the chemical, food processing, medical-pharmaceutical industry, and agriculture as well. In fact, even the renewable sources of energy are in need of hydrocarbons, considering the extraction, processing and transport of the key minerals used to make renewable energies viable, such as Cooper, Nickel, Steel and rare earths.

In the year 2022, oil production in the United States was 11.911 million of BDP. While dry gas production was around 5,833.33 million BDP, according to the EIA annual report, a growth in gas production is estimated due to the increase in international demand for natural gas exports (Energy Information Administration, 2023).

According to the United States Energy Information Administration (EIA), in 2022, about 2.84 billion barrels of crude oil were produced directly from Shale Oil or Unconventional fields. This is equivalent to approximately 66% of the total crude oil production in the United States in 2022. Similarly goes for the gas production. This figure increases year by year to cope with the current and expected demand, as well as due to the increase in oil production from the Unconventional fields and decrease of oil and gas from the Conventional Fields.

The following Figure 1 shows the evolution of total crude oil production in the United States between the years 1982 and 2022, in the areas of New Mexico, Texas and the rest of the states in the United States. A simple lineal extrapolation indicates that the production of the US Unconventional Fields has spared the import of nearly 3 out of 4 barrels, while the production of Texas and New Mexico would have already ceased at all before 2020, while the produced oil from there is currently at around 6 million bopd.

The Concept of Sustainable Exploration, Development, and Operation

Figure 1. Crude oil and natural gas production (Energy Information Administration, 2009)

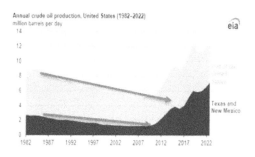

The above trend has three main impacts for the US Oil and Gas Industry, both positive:

- Cover domestic oil and gas demand.
- Achieve full Independence in oil and gas.
- Revive the activities of both states related to drilling, production, gathering and transport, that have a significant impact on the associated services.

While these above impacts are being taken for granted, they are actually not, rather they represent a technological quantum leap in the global Oil and Gas Industry.

1.2.1. Oil Peak: A Forgotten Strategical Worry

In 1956 King Hubbert coined the term "peak oil". According to his work oil production in a particular region would approximate a bell curve, followed by a terminal decline. coming up with the point in time when the maximum rate of global crude oil production is reached. His prediction for the USA Fields was that oil production would peak between 1965 and 1970. In fact, the US Oil production peaked in 1970 (Energy Information Administration, 2009). Later Colin J. Campbell predicted that global oil production would peak by 2007 (Campbell & Laherrère, 1998) and (TU Clausthal & Campbell, 2003). While above predictions were off, as they did not consider the technological advances that resulted in improved recovery techniques, these works served to increase awareness on the topic of the hydrocarbon resource depletion, as we are dealing with no-renewable resources. Below, Figure 2 shows the oil production expectations in conventional and unconventional reservoirs.

Figure 2. Expected oil production of conventional and unconventional crude oil and natural gas (Rogner et al., 2012)

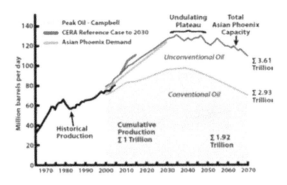

1.2.2. Untapping Vast Domestic Reserves and Supporting US Energy Security, and Oil and Gas Price Stability

Conventional hydrocarbon reservoirs are defined as permeable geological formations, that have an impermeable sealing layer above and below and where the pressure conditions have enabled the accumulation of hydrocarbons in so-called geological traps. These terms also make reference to the fact that such occurrences are uncommon due to the geological deposition process. The final proof of the existence of such a conventional reservoir is drilling a well that results in hydrocarbons flowing up to the surface, via this flow conduit. As indicated earlier, the number of hydrocarbons found in conventional reservoirs is in steady decline, since day one of its production, due to the natural pressure decline that results from the hydrocarbon extraction. Based on the downhole pressures and the extracted and injected volumes (e.g., water or gas injection for pressure maintenance purposes), this decline can be calculated and predicted, by means of the Material Balance equations.

In the case of the US, one of the most prolific conventional basins, the Permian one, would have cease oil yield before 2020, as shown in Figure 1. The reality shows an opposite trend, hydrocarbon production is actually booming in the Permian, due to the Unconventional Reservoirs. In fact, current estimations indicate that the hydrocarbons will last over 100 years (Pimentel & Burgess, 2017).

Objectively this has led to the Energy Security of the US, to the point that it has become independent of foreign hydrocarbons, with the derived geopolitical impact. In economic terms it also has translated in a more stable oil and gas prices, as the OPEC is no longer the sole price regulating entity. In fact, due to its nature unconventional oil and gas projects can respond considerably faster than conven-

The Concept of Sustainable Exploration, Development, and Operation

tional projects to price signals, resulting in more stable oil price markets (Robert Kleinberg & Sergey Paltsev, 2015).

Extracting hydrocarbons from unconventional rocks was made possible by a technical breakthrough, that basically converts almost impermeable rocks with very low porosity that contain hydrocarbons, in reservoirs. If the hydrocarbon material oil Kerogen contained in the mother or source rock has certain geochemical characteristics such as TOC (Total Organic Content) and Maturation, and the containing layer has a sizable thickness and areal spread, it has the potential to be developed as an Unconventional Play – see also Figure 3 and Figure 4.

Therefore, the unconventional projects are able to be fast tracked, without the lengthy and costly exploration phases that is very typical for conventional projects, giving that these types of projects are moving more and more to depth sea and to pristine areas and remote areas such as the artic.

In the following figures below, which compares the values of Total Organic Carbon (TOC), production thickness (f), porosity (%) and GIP (bcf/section) in the shale production formations at level Worldwide, the Muerto formation located in the Lancones basin northwest of Peru is considered the generating formation of shale oil.

Figure 3. Total Organic Carbon (TOC) values (Own elaboration)

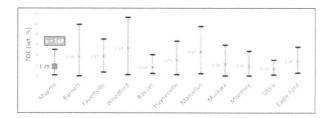

Figure 4. Production thickness (f), porosity (%) and GIP (bcf/section) values (Own elaboration)

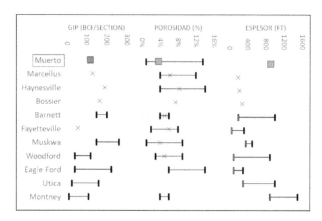

1.2.3. Potential for Covering the global needs of Hydrocarbon Resources

The below **Figure 5** shows the Basins with potential for Unconventional Oil and Gas resources.

Figure 5. Map of global shale basins (Energy Information Administration, 2015)

While the US has turned the tide in terms of covering its own consumption and even having a surplus for export, by the intensive development of its Unconventional Oil and Gas resources, Argentina is striving in the same direction in the "Vaca Muerta" Unconventional Basin. The falling production trend of both Oil and Gas production has reversed and is achieving record production levels. China is also making strides in the Ordos, Jungar and Gulong basins. On the other side many counties among them Mexico, Germany, France have blocked unconventional shale gas and oil production, mainly because of the perceived involved environmental risks and social concerns. These legal restrictions aiming at banning hydraulic fracturing, have prevented a more global development of shale gas worldwide.

According to (Pimentel & Burgess, 2017) there is sufficient oil and gas for resources to cover the global needs for the next 100 to 200 years by using both the conventional and unconventional resources.

In fact, there is a growing oil consumption of key countries like USA, China and India Is shown in below Figure 6 considering the significant consumption oil growth from 2012 to 2022 of China and India that increased by 42% and 41%, respectively (ZeroHedge, 2023).

Figure 6. Countries with the highest oil consumption in 2022 (Statista, 2023)

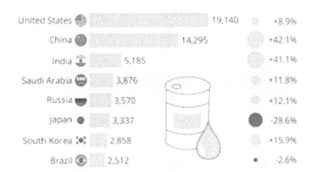

According to the above Figure 6 much praised Energy Transition – estimated to last for a few decades, appears to be realistic if it is not bridged by the of Unconventional Oil and Gas production.

The review of the historical development and projection of the hydrocarbon resources and primary sources is shown in the below Figure 7.

Figure 7. Petroleum resource triangle (Zou, 2017)

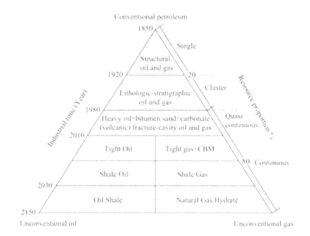

1.3 New Worries on Anthropogenic CO2 Emissions and the Global Warming

1.3.1 Anthropogenic and Native CO2 Emissions

Carbon dioxide is naturally present in the atmosphere as part of the Earth's carbon cycle. Tons of CO_2 are absorbed by oceans and living biomass and are emitted to the atmosphere annually through natural processes (EPA, 2023). While CO2 emissions come from a variety of natural sources, anthropogenic emissions are responsible for the increase that has occurred in the atmosphere since the industrial revolution (IPCC, 2013) (**Table 1**).

Table 1. Anthropogenic and natural CO2 emissions

Anthropogenic Sources	Natural Sources
Energy supply	Outgassing from the ocean or ocean-atmosphere exchange
Transport sector	Decomposing vegetation and other biomass
Industrial sector	Venting volcanoes
Residential and commercial buildings	Naturally occurring wildfires
Agriculture	Plant and animal respiration
Forestry	Soil respiration and decomposition
Waste and wastewater	

The Concept of Sustainable Exploration, Development, and Operation

1.3.2. Challenges Related to the Global Warming

The phenomenon, experienced in recent decades, where the average temperature of the Earth's near-surface air and oceans increased, has been defined as Global Warming - according to the Intergovernmental Panel on Climate Change (IPCC) of the United Nations. This has been directly attributed to the Green House Effect Gases – GHEG, among them the CO2. Where the Green House Effect- GHE is the water vapor, carbon dioxide (CO_2), methane and other atmospheric gases absorb outgoing infrared radiation leading to a rise in the global temperature.

Further, the international community has defined CO2 as the reference gas to have a sort of metric that is easy to relate, even though there are other gases that have a manyfold effect on it, such as methane, having 85 times more impact than CO2. The CO_2 is considered to be the main factor because it is the most important anthropogenic greenhouse gas (IPCC, 2005). However, the role of atmospheric CO_2 in relation to temperature increase is not a closed chapter.

In fact, the evaluation of the Earth's life through the geological eons may indicate contradictory results. A debate has arisen on the accuracy of temperature reconstructions as well as on the exact impact of CO2. A specific regression analysis to forecast the correlation between CO2 concentration and temperature was carried out by (Florides & Christodoulides, 2009). Concluded that it heavily depends on the choice of data used, "and one cannot be positive that indeed such a correlation exists (for chemistry data) or even, if existing (for ice-cores data), whether it leads to a "severe" or a "gentle" global warming". Further, it is highlighted based on data from paleoclimatology, that the CO2 content in the atmosphere is at a minimum in this geological eon. Moreover, it is concluded that "scientific knowledge is not at a level to give definite and precise answers for the causes of global warming".

There are also a variety of factors that have to be considered, such as occurring changes in the Earth's orbit, in the Sun's intensity, the ocean currents, ocean CO2 solubility, volcanic emissions and, also changes in greenhouse-gas concentrations, among them CO2. In prior geological eons, CO2 increase has favored the growth of the plants, as well it has modified its physiology.

Based on the above-cited research on the analysis of available climate data, as well as on CO_2's palaeo-history, global warming may not be as crucial as widely believed. With no doubt Earth's climate system is complex, and there is some work to do in the available climate models, to consider the change in above mentioned factors. Modeling attempts are not fully reproducing the relationship between GHEG, such as CO2 or methane, and global earth temperature rise, in terms of spatial distribution (Chen et al., 2023), (Killgoar et al., 2020) further, complementing these research a more recent work (Harde, 2023a) simulates the data of the last century concluding that the calculations indicate that the temperature raise and its variations over the last 140 years can best be explained by combined CO2 and solar radiative forcing. The used model considers the influence of both the increasing CO2 concentrations on global warming and the impact of solar variations on the climate, as shown in Figure 8.

Figure 8. Two-layer climate model for the Earth-atmosphere system with the main parameters (Harde, 2023b)

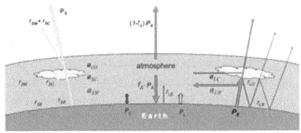

The modeling was based on an advanced energy-radiation transfer-balance model from (Harde, 2019) According to that the contribution of CO2 to global warming is below a third while solar variations over this period can well explain two thirds of the increase. The calculations indicate that the temperature increase and its variations over the last 140 years can best be explained by combined CO2 and solar radiative forcing. For the calculations the combined land-ocean-temperature composite of the Northern Hemisphere as derived by Soon & Connolly is used (Soon et al., 2015).

On the other hand, the natural sources of atmospheric carbon dioxide are considered to be the bulk (85%) volume, as compared to the anthropogenic CO2, (15%) derived from human activities (Harde & Salby, 2021). Interestingly, the IPCC assumptions assume that most of the temperature rise since the 1950s is due to changes in atmospheric greenhouse gas concentrations (IPCC, 2013), (IPCC, 2021).

1.4. Challenges of the Energy Transition to Mitigate Global Warming by Reducing CO2 Emissions From Fossil Fuels

1.4.1. Renewable Power Generation

Over the last few years there has been a strong drive for energy transition to renewable energies, particularly from sun and wind, however, according to the EIA, the projected figures indicate that by 2050 the share of renewable energies in the US Energy Basket will be in the range of 20 to 30% only. Moreover, the recent war in Ukraine shows that even Germany, being the most advanced country in terms of the use of renewable energy, cannot switch on the go to renewables, as this is a gradual process that won´t occur overnight. While renewable energy generation is moving forward, its transport and distribution presents important challenges.

The following figure shows the projection of the share of renewable energy in power generation in the United States. It shows a projected increase from 21% in 2021 to 44% by 2050 – less than half of the Energy Mix, driven primarily by wind and solar energy. The contribution of hydropower remains almost constant, while other renewable sources represent less than 3%.

Figure 9. Projected EIA energy share up to 2050 (Energy Information Administration, 2022)

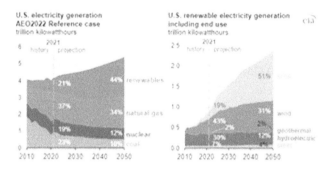

According to the above Figure 9 much praised Energy Transition – estimated to last for a few decades, appears to be realistic if it is not bridged by the of Unconventional Oil and Gas production.

The intensive introduction of renewable energies requires a significant increase in the demand of mineral resources. Specifically, the construction of both wind turbines and solar panels requires from rare earths, that are considered as scarce resources.

The below 17 rare earth elements - the 15 lanthanides plus scandium and yttrium are elements critical to the energy transition (Renee Choo, 2023).

Figure 10. Rare earth element table (Vtorov I. P., 2017)

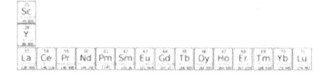

While other minerals such as copper, Nickel, Aluminum, iron, silicon, cobalt, lithium, graphite, and manganese are not rare earth elements, yet are also critical ones. Provision of the required growing quantities of the above minerals in a sustainable manner will be a real challenge to enable the replacement of fossil fuels. At the present time it remains as a pending homework – How to cover this huge demand on minerals, whose exploration, extraction, transport, and processing demands the use of fossil fuels.

Further, wind turbines need for permanent magnets, built with the rare earth elements neodymium and praseodymium to strengthen them, as well as dysprosium and terbium, to resist demagnetization. For solar panels chromium, copper, manganese and nickel are critical, while rare earth minerals are critical to enhance the light absorption capabilities. It is estimated that 400 new mines will be needed to satisfy the demand (Deslances, 2022).

1.4.2. EV: Electrical Vehicles for Transport Electrification

While the European Union estimates that a quarter of the CO2 Emission is caused by fossil fuel driven vehicles, there is an active push to promote the electrification of the transport to mitigate greenhouse gas emissions (GHG).

Taking the case of the E-Vehicles the below picture compares the mineral consumption to the conventional cars. In the case of the use of Cupper, it increases more than 100%, and for Nickel, it is more than 500%. The Figure 11 shows the minerals required for energy production in (Kg/MW). In the same graph the amount of minerals used to produce both an electric car and a conventional car in (Kg/Vehicle) is shown, where to produce an electric car a large amount of nickel, cobalt, graphite and lithium is required.

Figure 11. Minerals used in selected clean energy technologies and fossil fuel technologies (International Energy Agency, 2022)

Regarding the Batteries, at present time large storage capacities are needed, and the existing technologies are over expensive and even hazardous to operate due to the involved very high operating temperatures. Here also a critical gap is the amount of minerals needed to build them, Lithium, Nickel, Manganese and Cobalt. Specifically, Lithium and Cobalt are scarce and will need to be first explored, discovered, and developed. Cobalt ensures the safe operation of the battery as it provides its thermal resistance, however it is mainly found in the Democratic Republic of Congo – with the best quality, yet the current Cobalt extraction practices, including infant labor, hazardous working conditions, among others, are becoming a sort of "dark side" of the Energy Transition. Given that the inhabitants of the cities surrounding the Cobalt mining sites live with less than 2 USD per day, lacking hospitals, schools and unable to use the river waters that are polluted. Likewise. in Baotouin Inner Mongolia, China— considered the world's rare earth capital— the soil and water are polluted with arsenic and fluorite due to mining (Renee Choo, 2023).

Besides the fact that currently, lithium batteries are disposal objects after one-time use only, and the undeniable fact that the global Energy Transition is very mineral-intensive. Given the huge number of new mines required, mining's footprint on the planet, under the current linear economy conditions will be exacerbated, therefore the question arises: is mankind on the verge of creating the objective conditions for a new environmental disaster?

Some authors are suggesting an increased use of the Circular Principles (Toledano et al., 2023), while this is a valid proposal, it only will cover for a fraction of the actual needs. Certainly, the key stakeholders are challenged to come up with viable alternatives to these emerging associated challenges.

1.4.3. Electrical Distribution Constrains

On the other side, the existing distribution power grid will have to be fully rebuilt, to create the required capacity, that is currently not available, to replace a system that has taken nearly a century of maturation.

Assuming that the entire energy mix is based on renewables and batteries, and the fossil fuels are completely phased out, then what would be a reasonable (not exorbitantly costly) backup in case of any contingency resulted from severe storms or consecutive cloudy weather??

Therefore, from practical, technical-technological, resource based, financial, and contingency standpoint, it is expected that the need for fossil fuels will be around for a few more decades, and even then, its presence as an effective backup may still be required. In this context, it is not fully responsible to jump into ventures that present a degree of uncertainty when it is about securing the energy needs of the population and the industry.

1.5. On How the Current Unconventional Oil and Gas Industry Can Support the Energy Transition

1.5.1. Help Maturing Geothermal Energy

All the massive investment allocated to mature the drilling and completing of Unconventional Wells is contributing to pushing the technological frontiers of Geothermal Energy. The main burden of costs is benefitted from the skills gathered in the unconventional fields. This is twofold. On one side the drilling of deeper wells and the followed horizontal steering at such depths. On the other side the fracturing process itself creates the flowing pads for the cold water to flow to and get heated on its way to the producing well, making the whole process much more cost-effective.

1.5.2. Support CO2 Sequestration

Another key limitation of Energy Transition derives from the existing constraints regarding CO2 Sequestration. Among them are the technical difficulties, excessive costs, storage capacity, regulatory frame, and the concerns on safe leak free storage. Interestingly the same unconventional wells may present a unique opportunity to help solve the above limitation. As it is known the Enhanced Oil recovery - EOR of the Unconventional Wells is far less than of the Conventional Wells. One method to enhance it, that is currently being tested is the injection of CO2, given its properties. Among those is the dissolution of the gas in the oil making it easier to flow. On the other side it is the good absorption of the rock material that enables a better displacement of the oil.

1.6. Producing Hydrocarbons From Impermeable Shale Rock Strata

Not long ago the shale rocks from where the Unconventional oil and gas are produced were considered just as the rock source or motherstone of the hydrocarbons. Given its impermeable nature, tighter than cement and with very low porosity values, below 6%, nobody ever tried to produce from it – see below Figure 12 .Nobody could imagine that the shale rock would ever be able to behave as a reservoir. The classical definition of an Oil or Gas Reservoir requires certain values for the properties of the rock, such as porosity and permeability, above 12% and a few millidarcies, at least. Well, having just a fraction of the above values, the Shale rock strata was considered as an impermeable rock and were therefore discharged at all as a producing source, since the beginning of the conventional oil and gas exploration and production.

The Concept of Sustainable Exploration, Development, and Operation

Figure 12. Tight oil rock permeability fundamentals (Salama et al., 2017)

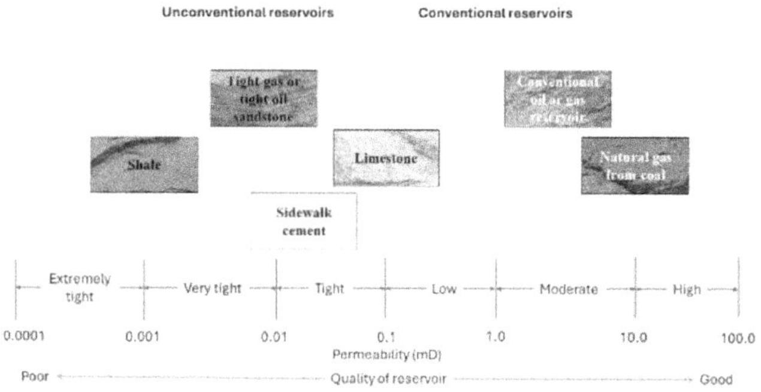

1.6.1. Shale Oil and Gas: Enabled by a True Disruptive Technology: Horizontal Drilling + Multi-Stage Hydraulic Fracturing (HF)

It all started in the 90´s with the Pioneer work of George Mitchell owner of Mitchell Energy Mitchell Energy & Development Corp, a company producing gas based in the Barnett Shale Area, North of Texas. Given the reducing gas production of his wells, he tried to find out ways on how to produce more gas and keep using the existing facilities. While Hydraulic fracturing (one stage, that progressed in the following decades to more stages) was used since the late 1940s, matured horizontal drilling can be traced back to the early 80s in the US. Mr. George Mitchell combined these two previously known technologies: horizontal drilling and multistage hydraulic fracturing, giving birth to the shale revolution in the United States.

By the 2000´s, following the Barnett Shale of Texas (Marc Airhart, 2007), it then started to wide spread to the point that a prediction for Shale Gas indicated that the reserves to that time, would last for 105 years (Potential Gas Committee, 2008). In fact, at present time nobody talks about the Oil Peak Theory anymore. While these oil peak worries have been appeased, it should not be taken for granted.

Given that HF-Technology takes on the root causes of the no-flow condition of the shale formations, and resolves it, it has a few far-reaching implications and complex interactions. Further, it is a truly disruptive technology that is evolving and improving at a fast pace, in terms of costs, oil and gas yield, environmental impact, and energy efficiency. The rapid development of the HF has led to its complete transformation, avoiding or reducing the initial shortcomings, as described further below.

It is to note that other recent technologies that have become disruptive are Artificial Intelligence - AI, and Genome DNA Editing with CRISPR. In both cases there are pros and cons, yet its advantages are undeniable. In both cases there are calls to enforce a level of regulation.

1.6.2. Challenges of Producing Oil and Gas From Unconventional Fields

Commonly, talking about Unconventional Fields is referred to the use of Hydraulic Fracking HF. This is the technology that enables impermeable hydrocarbon source rocks, become producing formations, by artificially inducing permeability. It is achieved by the combination of two technologies, horizontal drilling, and hydraulic multistage fracking. While both have been already around for several decades, it was its combined use that originated the oil and gas production boom that has taken place in the US shale basins for over a decade now, making it a world top producer in 2023.

The main problems associated with the unconventional fields, specifically the HF, are related to environmental concerns. Among them, the contamination in aquifers, induced earthquake or tremors, a massive consumption of water, and the intensive use of chemicals. Further the intensive drilling causes a large footprint, as the new wells that enter into production lose nearly 70% of their initial production in the first year, therefore new wells are continuously required, to maintain and increase the production. On the other side the actual fracturing job requires a large fleet of tracks, that carry the equipment composed of generators, motors, and pumps that are used to perform the fracturing jobs with nearly 50,000 hp on the well site. This is required to pump large fluid volumes comprised of water, chemical additives and sand, at large pumping pressures. The fracking fluid is pumped down the well at pressures that can exceed 9,000 pounds per square inch (psi), generating also high noise levels during the fracturing process.

The Figure 13 shows the routes in which an aquifer could be contaminated, such as poor operation in the cementation of the casing pipe, migration of fracking fluids through geological faults towards shallower aquifer.

Figure 13. Fracking infrastructure (Howarth et al., 2011)

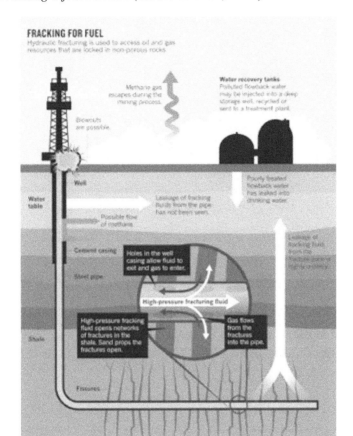

(Du et al., 2022) mentions that the use of water in hydraulic fracturing increases the dispute over local water resources and alters its supply. This implies that the treatment and reuse of wastewater becomes essential to preserve the viability of water resources in areas where oil and gas are produced by means of HF, given the growth of this industry.

The following Figure 14 shows the volume of water used in millions of gallons in an arid climate in the states where the oil and gas industry is developed.

Figure 14. Water use under arid climates (Du et al., 2022)

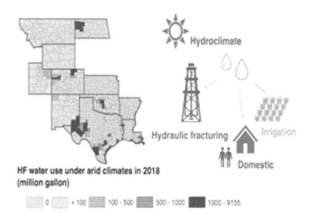

1.6.3. Achieved Improvements in the Hydraulic Fracturing (HF)

Since the beginning of the 2000s, several remarkable improvements have been implemented in HF, while others are in the development stage. This has resulted in important improvements in the areas of well cost, oil and gas production, environmental impact, as follows:

Well Costs: It has got down from the 12 mln to 7 mln range and with it the production cost that started with 60 $/barrel and it has lowered to 20 $/barrel.

Oil and Gas Yield: The first wells produced around 500 bopd, while the recent productions are in the range of 3,000 bopd. However, it declines 70% within the first year.

Health problems: Increasing cases of respiratory problems (asthma) and an increased risk in mild, moderate and severe cases were reported (Rasmussen et al., 2016). Toxic chemical substances were identified that could affect the skin, eyes, respiratory system, etc (Colborn et al., 2011).

Environmental impact: Change in air quality due to the composition of chemicals, or that would cause health problems (Colborn et al., 2011)high values of concentrations of methane, ethane, and propane in aquifers near the hydraulic fracturing zones (Jackson et al., 2013).

Tremors or earthquake events: Space-time studies were identified in the Oklahoma area, between the years 2006-2015 and 2016-2017, in which there was a correlation of the recorded earthquakes and the dates of drilling (water injection) (Hong et al., 2018)

High Water Consumption: The consumption of water for the hydraulic fracturing process, in areas and times of low water availability, turns out to be significant (Environmental Protection Agency, 2016).

1.6.4. Current Attempts to Pursue Sustainability in the Extractive Activities of the Oil and Gas Fields

More recently, the drive for the energy transition has accelerated the strive for sustainable development of resources, resulting in the introduction of new requirements for the Exploration and Production E&P Companies, involving concepts such as ESG – that stands for an Environmental, Sustainability and Governance criteria. while this is an attempt to do something about it, by incorporating a set of requirements and some KPIs to support both the E & P Companies as well as the regulatory entities to be on the same page on the subject.

Table 2 shows the criteria and elements covered by the concept of ESG.

Table 2. Criteria of ESG (Alameda García et al., 2021)

Environmental	Social	Governance
- Impact on climate/ greenhouse gas emissions - Sustainability - Climate change risks - Energy efficiency - Air and water pollution - Water scarcity and waste management - Site rehabilitation - Biodiversity and habitat protection	- Human rights - Community impact - Respect for indigenous people. - Employee relations - Working conditions - Discrimination - Gender diversity - Child and forced labor. - Health and safety - Consumer relations	- Alignment of interests between executives and shareholders - Executive compensation - Board independence and composition. - Board accountability. - Board diversity. - Shareholder rights - Transparency and disclosure - Anti-corruption measures - Financial policies - Protection of property rights

While the ESG is a well-meant tool, it does not cover critical process related aspects, nor provides it the required procedures and specific metrics that are required to implement and monitor it.

The commitment of oil companies to ESG policies is very well encompassed to Energy Transition. This in turns requires to pay more attention to the objective analysis and quantification of the dimensions of the ESG index to ensure a higher quality of Corporate Social Responsibility (CSR) engagement of the companies (Ramírez-Orellana et al., 2023)

Among the existing challenges for ESG are the disparate definitions among organizations, the different utilization, and combinations of measurement methods among rating providers, and the insufficient transparency during the process (Liu, 2023).

2. METHODS AND MATERIALS

2.1. The Technically Responsible Alternative: The Concept of the Sustainable Exploration, Development, and Operation of Unconventional Fields

A much broader, comprehensive, process centered, and overarching solution proposal is needed, to pursue the sustainable exploration, development and operation of the shale oil and gas fields. The present work is an attempt in this direction, with the idea of inspiring the stake holders to adopt a more innovative approach. Certainly, the scientific community is committed to coming up with viable alternatives, rather than to sit and wait for that process alien decisions are taken decisions on behalf of the entire Oil and Gas Industry.

The core question to answer is:

How can we supply the energy transition more sustainably?

As explained above, it is more realistic to envision that the need for fossil fuels will be around for a few more decades. The present work supports the implementation of a specific workflow that is comprised of specific techniques, procedures, and best practices to achieve a more sustainable exploration, development, and operation of unconventional fields, instead of waiting for it - while carrying out business as usual.

It is about utilizing a self-specific approach based on efficient and environmentally friendly Solutions applied to the exploration, development and operation of Unconventional Oil and Gas Resources.

A systematic pursue for a sustainable process requires a more structured workflow that incorporates actionable specific measures that are process centered and aligned to the core business, beyond the mere report writing to appease the public and the regulatory bodies.

In the light of the prior works from(Armacanqui, 2015, 2016; Armacanqui et al., 2016; Samuel Armacanqui Tipacti, 2013) the present work presents the concept of the sustainable exploration, development, and operation of unconventional hydrocarbon resources, in the context of a Life Cycle Approach.

2.2. The Life Cycle Concept

A Life Cycle Approach favors an unbiased analysis of the subject and enables a) to map out the complete process as well as the associated subprocess. Further it enables both a b) multidisciplinary process view and an c) in deep analysis, and properly applied. Another inherent ingredient of this approach is that it enables to apply d) Preventive Measures to avoid unwanted events, such as incidents or accidents, and process related failures. Further it favors the e) introduction of Innovations and New Technology, to resolve current technical challenges, as well as the incorporation of f) Industry Best Practices.

These features results in step change enhancements, in the applied areas as described in (Armacanqui, 2015) and summarized in a few examples below:

Pressure Transient Interpretation – PTA Life Cycle: In the first case a Life Cycle Approach was used in the giant Furrial field in Venezuela in the beginning of the 90´s, to significantly enhance the Interpretation results of gathered pressure transient data, that is a costly work that is carried out in oil wells. This approach removed commonly occurring errors affecting the final interpretation by means of incorporating a QC – Quality Control Method, that spanned from the time prior to the test, the test design, the field data recording, and the confirmation of the results obtained from the interpretation.

ESP life cycle: In the second case the life cycle concept was applied in the early 2000´s Middle East to the massive use of the artificial lift – ESP pumps (electrical submersible pumps), resulting in multimillion us$ savings and reverting the severe production decline, by equipment failure reductions and by utilizing advance tendering techniques.

Figure 15. ESP life cycle (Samuel Armacanqui Tipacti, 2013)

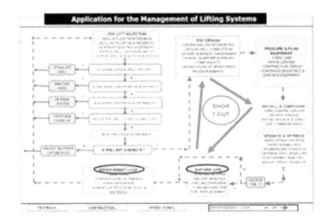

2.3. Innovations Life Cycle

2.3.1. The Life Cycle Approach for the Effective Introduction of Innovations and New Technology

In the third case the life cycle concept was applied to the introduction of innovations and new technologies. The correct application of this resulted in fast tracking of innovation projects and in the introduction of new technology for the Oil and Gas Industry in North Africa (Egypt), bringing to a new level of cooperation among all the partners of the Join Company, laying the ground for an effective collaboration to drive innovative initiatives.

That was applied to other different industries as well, as it touches a fundamental approach that starts not with the innovation itself, yet with the actual business need, which in turn requires an in-depth knowledge on the specific process.

In the Figure 16 shows below lines the Life Cycle Approach is extended to work out the subject of the present work.

The Concept of Sustainable Exploration, Development, and Operation

Figure 16. The life cycle approach (Armacanqui, 2015)

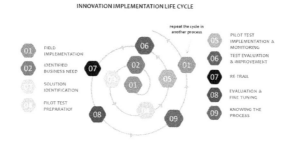

2.3.2. The Life Cycle Concept for the Exploration, Development, and Operation of Unconventional Fields

The life cycle concept for the exploration, development and operation of unconventional fields is based on the following key elements, that can be considered as enablers:
a) buying in of the executive c-level
b) Finding the bottlenecks and the solutions within the company
c) use of the best practices of HSE standards
d) the right preparation of the environmental impact assessment study - EIA
e) fit for purpose technological improvements
f) goal alignment between the E & P Companies and the Oil Field Service Companies
g) process procedures that support a preventive approach
h) Contractual and Procurement Related Aspects
i) best practices in the participation of local communities

The aim is that the said concept serves to enhance the standards on the subject by means of a full alignment to the core business, rather than to represent another set of reports to be filled out.

2.3.2.1. a) Buying in of the Executive C-Level

It appears that the energy transition efforts are unstoppable, however this is not going to occur in a few years, rather in a few decades. the active participation of the oil and gas industry to steer this transition in an orderly manner is of outmost importance, instead of assuming a denial or an "ostrich like attitude". This transition is occurring in an environment where it is more media than process driven, therefore there is a need to steer it, to prevent leaving it to outsiders to decide.

Currently the CeO´s are facing a mounting pressure to steer into the Energy Transition at fast pace, coming from different angles. They know what is in the Plan of the present year and the next few years. Moreover, they are directly responsible for their outcome. Fortunately, the Boards of Directors are getting more and more conscious of the existing challenges. E.g., nowadays they are clear about the importance of investing in Innovation, as part of the solution. Sitting with the CeO and the relevant executive Leaders opens the ground to start identifying the bottlenecks, formulating the solutions, and conducting their implementation.

2.3.2.2. b) Finding the Bottlenecks and the Solutions Within the Company

After paving the way with the blessing of the C-Level Executives the task is to get the key parties within the company involved. This can occur through an in-depth review of the key processes, both at the office and in the field. It is particularly important to listen to the staff that is at the forefront, they know a lot about the process and often already have alternatives in mind, however, often are not even asked for. Any outside Expert needs to validate his suggestions with the staff, as they are the ones that know the most about the existing problems and gaps. Therefore, bringing key people together to talk about the key problems under the guidance of a Subject Matter Expert yields the best results.

The use of tools that enable the participation of the Company Staff is of paramount importance. Among those are the Suggestion Box that rewards fostering innovative initiatives, the Self Specific Workshops where key stake holders meet to talk on key items., related to the Sustainable Approach.

2.3.2.3. c) The Best Practices of HSE Standards

The best HSE performance in a company is translated in zero fatalities and zero LTI´s. Those companies that have achieved this have created a "culture of safety" within their staff. A key learning of the best HSE practices is that it gives priority to preventive measures. the following sentence is familiar to them: "1 gram of preventive measures is worth of 1 ton of reactive measures". This learning is the first ingredient of the presented concept. another line that has been proven true is: "in a company an accident can occur for the first time, however the occurrence of a similar second accident is negligence". in this sense the learnings of the accident investigation is very crucial.

2.3.2.4. d) The Right Preparation of the Environmental Impact Assessment Study: EIA

Often the environmental impact assessment – EIA, is seen as mere paperwork that is part of the permitting process for a project in order to get started. However, digging into the basics of the EIA Elaboration is much more than this. in fact, it is to be applied to the different stages of the project, starting with the planning or design phase.

While the common practice is first to complete the project plan and ask thereafter the relevant party to go ahead with the EIA. In some instances, the EIA is done by a team that lacks the technical skills and knowledge on the fundamentals of the process. This results in a useless exercise when it comes to preventing environmental events. The learning here is to apply the EIA in the early phases of the project whereby the EIA team must be led by qualified professionals who are technically qualified to issue recommendations that will impact in the type of technologies and procedures to be used in the project construction, and later in the project operation and even project abandonment, thus yielding a solution that is environmentally friendly.

It is about applying the EIA at the earlier phase of the Project, at the Planning Stage, as shown in the below figure. This will impact throughout the entire project life, because well sounded procedures, methods, and technology – that supports a sustainable approach will be incorporated.

Figure 17. Inclusion of the EIA at planning stage (Own elaboration)

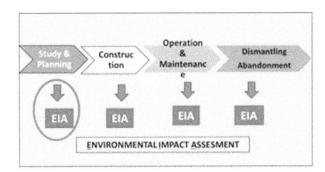

2.3.2.5. e) Fit for Purpose Technological Improvements

The latest technological developments have resulted in specific solutions to the key problems associated with the use of hydraulic fracturing, such as the local earth tremors, leaks into the aquifers, massive use of water, etc. In fact, the said solutions are being applied by the top E & P Companies with good results. The details are described further below. What is lacking is the standardized use of them, as part of the best practice sharing, which is also the aim of the present work. It is about avoiding "the business as usual" approach, and to incorporate new ways of working that use the said technologies. Here it is referred to the work (Armacanqui, 2015). Along these lines the replacement of Xylen by biodegradable environmentally friendly chemicals to remove and inhibit paraffine depositions (Viera Palacios, 2020), can be named, given that unconventional wells are affected by paraffin depositions.

Figure 18. Pipe filled with paraffin (Viera Palacios, 2020)

2.3.2.6. f) Goal Alignment Between the E&P Companies and the Oil Field Service Companies

The use of new technologies can lead to a significant reduction of the environmental load in hydraulic fracturing jobs; however, this implies associated sizable capital investments. For their development. A proper alignment between the User (the Company) and the Service Provider is the basis of the field roll out of them. A good example is E-Fracking.

Among the significant benefits for the E & P Companies are the reduction of the footprint on the well, and of the fuel consumption by replacing the conventional diesel driven motor generators with gas driven turbines. However, changing the conventional fracking fleet is a real challenge for the service providers. in this context we cite Patterson-UTI CEO William Hendricks on September 3, 2020, one of the largest oil field service companies "The E-Technology is beneficial to the E&P Companies, yet detrimental to us" (Hendricks, 2020).

2.3.2.7. g) Process Procedures That Support a Preventive Approach

A few E & P Companies have established best practices in terms of the responsible handling of the well abandonment strategy. e.g., Shell considers this a part of the Asset Integrity check, whereby as part of the daily operation there is a dedicated Team that has the required Tools and Engineering and operational staff in charge of this activity. The test and evaluation of the wells that are in shut in status are carried out and based on the results the decision is taken to abandon the well or to perform a work over job to recomplete the well. if the abandonment is considered a part of Operations, the problems with having to deal with orphan wells later, are eliminated.

The main reason for that is that operational activities do have a budget. Another example worth mentioning is the use of the "Walking Rig" by Continental Resources. This does not only reduce the associated rig mobilization costs but also accelerates the drilling start in the next well within a cluster and favors the construction of several wells in the same cluster location. These two examples show that if environmentally friendly measures are adopted as part of process procedures in a preventive manner, sustainable solutions are smoothly implemented.

Figure 19. Walking rig from continental resources (Columbia Industries, 2024)

2.3.2.8. h) Contractual and Procurement Related Aspects

Any disclosure of both OPEX and CAPEX expenditures of the E & P Companies show that nearly 80% of the total budget is expended to pay Contractors that provide Products and Services. These only carry out what is specified in the respective

The Concept of Sustainable Exploration, Development, and Operation

Contracts, not more and not less. All the contract schedules contain the scope, the technical specs, the payment – including penalties and bonus. The most advanced contracts, called Performance Oriented Contracts, do include other schedules such as a Matrix to monitor the performance in HSE and in the use of Sustainable Technologies and Procedures, or ESG. If this is not stated upfront, they will adopt "the business as usual" working mode, and the E & P Companies will miss important opportunities to transition to more sustainable ways of working.

2.3.2.9. i) Best Practices in the Participation of Local Communities

This is about the effective involvement of the surrounding communities. In order to achieve this in a credible way, the E & P Companies can resort to three measures: i) to adopt an approach that is process centered, rather than public relationship based (PR), and drive a comprehensive approach such as the one presented in this work, in order to eliminate or minimize the risk of negative impact in the environment resulted from their activities. ii) assign a single focal point to deal with the information request - that comes from the public and the media, pertaining to their operational activities.

This is especially critical when an HSE or environmental incidents occur. It is assumed that this person is both technically knowledgeable and with good communications skills. iii) support spreading of best practices on the subject via professional institutions that have reached out to both the public and the industry on the states that have unconventional oil and gas fields, such as the Petroleum Alliance, based in Oklahoma.

When it comes to Sustainability, Environment and Governance comes, in the present work four key Aspects are identified. Those are illustrated in the below Figure 20.

Figure 20. Stakeholders for the right Interaction of the process, environment, government and communities (Own elaboration)

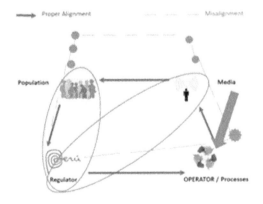

The starting point is the process knowledge that is supposed to be mastered by the Operator. If an Incident occurs the Operator should provide objective information to the inhabitants in the surrounding areas. Thereafter an objective investigation should be initiated with the goal of identifying the underlying causes and the factors affecting it, with the goal of removing those factors in order to prevent reoccurrence. Failing to apply the above would only exacerbate the situation without adding value to the process.

Further, there are a lot of very good best practices coming from the activities in conventional oil and gas fields(Armacanqui, 2016). The first one is the OIL FIELD development, done with the participation of Conoco, of an oil field in Venezuelan Eastern in the late 90´s, whereby at the same time another company was founded that was in charge to plant eucalyptus trees surrounding the well clusters.

Another Sustainability Oriented Practice was taken further by a JV Company between Shell and Oman, Petroleum Development of Oman – PDO, in the Middle East, that treated the water coming from the oil wells to the point that it could be used for agricultural purposes. A third example comes also from PDO that incorporated the inhabitants of the villages located near the oil fields, in the development and operation of the fields. This occurred by means of founding small companies, called Local Community Companies, to carry out basic services, such as excavation, trenching, water transport, etc. Moreover, this gave opportunities for community members to be heard, enabling them to help set a local agenda.

3. RESULTS

3.1. Application of the Life Cycle Approach

Prior life cycle approaches presented by the author, have resulted in step change enhancements. Building on it and on the enabling factors, as described above, the "concept of the sustainable exploration, development and operation of unconventional fields" is presented in below schematic.

Figure 21. The life cycle concept (Own elaboration)

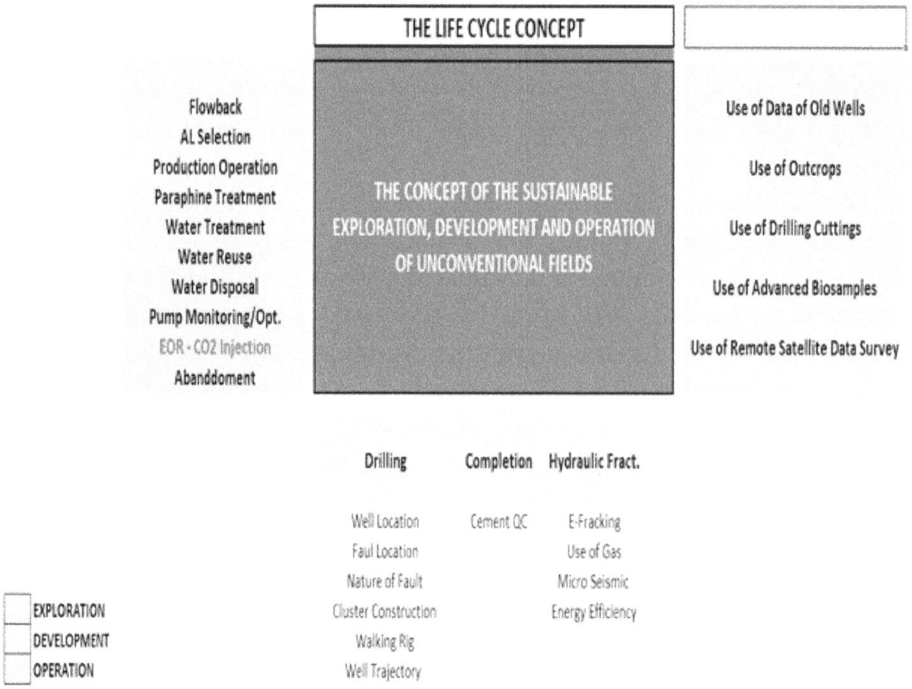

	THE LIFE CYCLE CONCEPT	
Flowback AL Selection Production Operation Paraphine Treatment Water Treatment Water Reuse Water Disposal Pump Monitoring/Opt. EOR - CO2 Injection Abanddoment	THE CONCEPT OF THE SUSTAINABLE EXPLORATION, DEVELOPMENT AND OPERATION OF UNCONVENTIONAL FIELDS	Use of Data of Old Wells Use of Outcrops Use of Drilling Cuttings Use of Advanced Biosamples Use of Remote Satellite Data Survey

	Drilling	Completion	Hydraulic Fract.
	Well Location	Cement QC	E-Fracking
	Faul Location		Use of Gas
	Nature of Fault		Micro Seismic
EXPLORATION	Cluster Construction		Energy Efficiency
DEVELOPMENT	Walking Rig		
OPERATION	Well Trajectory		

3.1.1. Important Technologies of Hydraulic Fracturing That Support the Oil and Gas Industry and CO2-Sequestration

Given the capital-intensive nature of the activities involved in the hf activities, several new technological advances have emerged. This is also justified by the significant achieved increase of the rate of produced oil or gas, that is manyfold as compared to conventional oil and gas wells.

The following Table 3, shows the potential impacts that can be generated by an inadequate hydraulic fracturing process, produced using excessive water consumption, possible contamination of water in aquifers, the use of chemicals that can be very toxic, the venting of flaring gas, induced seismicity, use of gas as a replacement for water-based drilling fluid.

The Concept of Sustainable Exploration, Development, and Operation

Table 3. Potential impacts in hydraulic fracturing and alternative solutions

Impacts	Causes	Existing Solution	Solution In Development
Environmental Impacts	Excessive water consumes	a. Using carbon dioxide instead of water (Ishida et al., 2012) b. Reutilization 'on site' of FPW (fracking produced water) (Mao et al., 2018a). c. Deep-well injection (Mao et al., 2018a). d. Wastewater disposal after pretreatment of FPW (Mao et al., 2018a) e. Using stranded gas in replace of the water-based fracturing fluid (Zheng & Sharma, 2021).	a. Domestic wastewater as well as seawater have great potential as water supplies for hydraulic fracturing. The development of high-efficiency equipment is essential, whether for the treatment of fracturing fluids or mine drainage, domestic wastewater, and seawater as a water resource for hydraulic fracturing. b. Gas Hydraulic Fracturing
Environmental Impacts/ Impacts on Human Health/ Impacts on wildlife	Water pollution	a. Well spacing and integrity (Gagnon et al., 2016). b. Establish water quality monitoring criteria (Gagnon et al., 2016). c. Submit chemical and its toxicological information (Gagnon et al., 2016). d. Fracturing using nitrogen foam (Todd et al., 2015). e. Best design in gas well development. f. Sealing abandoned wells g. improvement of leak control systems. h. Comply with occupational health and safety (HSE) guidelines. i. Constant monitoring of the wastewater storage area. j. Water body remediation k. Establish contingency plans in case of spills	
Impacts on Human Health	Chemicals used (carrier) can be toxic (carcinogen)	a. Discharge publicly owned municipal wastewater treatment plants. b. On-site reuse (Mao et al., 2018b).	

continued on following page

Table 3. Continued

Impacts	Causes	Existing Solution	Solution In Development
Consequences on wildlife / Environmental Impacts	Gas flaring and venting	a. Collect the gas produced through pipelines and sell it as liquefied natural gas (Zheng S. & Sharma M., 2021). b. Generate electrical energy with surplus natural gas (Zheng S. & Sharma M., 2021). c. Store the gas produced in suitable tanks (Zheng S. & Sharma M., 2021). d. Inject gas into reservoirs for oil recovery methods (Zheng S. & Sharma M., 2021). e. Use the produced natural gas for hydraulic fracturing, e.g., NG foam fracturing, natural gas fracturing (Zheng & Sharma, 2021).	
Geological impacts	Causing earthquakes	a. Determining how HF triggers earthquakes is important, as it represents the starting point for understanding the physical processes that control HF seismicity (Atkinson et al., 2020)	Further development of hazard forecasting and mitigation approaches is a critical future area of research.
Environmental Impacts	Release of greenhouse gases, mainly methane (25 times more powerful than CO_2)	a. Using stranded gas in replace of the water-based fracturing fluid can reduce the gas emission and the cost (Zheng & Sharma, 2021). b. Replacing diesel with cleaner-burning natural gas for engines (bifuel or dual-fuel) (Todd et al., 2015).	

3.1.2. Important Technologies of Hydraulic Fracturing That Support the Oil and Gas Industry and CO2-Sequestration

Given the capital-intensive nature of the activities involved in the HF Activities, several new technological advances emerged. This has also been justified by the achieved significant increase in the rate of produced oil or gas, that is manyfold as compared to conventional oil and gas wells.

3.1.3. HF Supports CO2 Sequestration

It is to note that the many technological advances associated with the hf of unconventional fields activities are real contributions to the whole oil and gas industry. one example of it is shown the above picture. It is about the use of co2 as an enhanced oil recovery technique applied to unconventional fields. Currently the oil recovery factor of this type of field is in the range of 6 to 8% - which is very low, as compared to the conventional fields, that can achieve the 50% range. The cycle co2 injection has demonstrated in lab conditions /xx/ important improvement of the oil recovery for unconventional oil fields, and could be used as an option for co2 sequestration.

3.2. Use of Gas instead of Water for Hydraulic Fracturing

Given that the use of water for hydraulic fracturing purposes requires the utilization of a high amount of chemicals to adapt the water flow properties to the requirements of the fracturing job, the use of Natural Gas (NG) as the fluid carrier of the propane has been proposed (Zheng & Sharma, 2021). In this paper the simulation results indicate that NG and foam fracturing fluids outperform water-base fracturing fluids with lower burst pressure, reduced leak off into the reservoir and higher pool efficiency. NG foams generate better propped fractures with shorter length and greater width due to their high viscosity. In addition, this improves the permeability of the stimulated rock volume, fracturing fluid flowback, significant reduction of water usage and increasing natural gas consumption.

4. CONCLUSION

In the present work a process centered Life Cycle Concept is described as part of a Method that enables the Sustainable Exploration, Development and Operation of Unconventional Oil and Gas Fields. Further, specific related examples of new technologies and innovations, and best practices are presented.

SPECIAL THANKS

Special thanks to the "Applied Research and Technological Development Project 2018-I" E041-2018- 01-BM, to the "World Bank", to "CONCYTEC" and "FONDECYT" as financing entity of the Subproject: 61096 and contract 139.

REFERENCES

Alameda García, D., Álvarez Álvarez, J. L., Hernández Zelaya, S. L., Fuertes Kronberg, C., García del Campo, L. A., Garcimartín Alférez, C., García Medina, J., Holguín Galarón, L. G., Llorente Valduvieco, J. I., Magdalena Miguel, L., Martín Martín, I., Matellán Pinilla, A., Reyes Reina, F. E., Rebollo Revesado, S., Rivas Herrero, L. A., & Sánchez Almaraz, J. (2021). Handbook of Sustainable Finance: A multidisciplinary approach. Academic Press.

Armacanqui, J. S. (2015). The Innovations Friendly Organization: Effective Introduction of New Technologies and Innovations in Oil and Gas Companies. *Society of Petroleum Engineers - SPE North Africa Technical Conference and Exhibition 2015*. *NATC*, 2015, 1575–1587. 10.2118/175876-MS

Armacanqui, J. S. (2016). The Concept of the Sustainable Oil Field Development Applied to Heavy and Extra Heavy Oil Fields. *Society of Petroleum Engineers - SPE Latin America and Caribbean Heavy and Extra Heavy Oil Conference 2016*, 510–516. 10.2118/181169-MS

Armacanqui, J. S., De Fátima Eyzaguirre, L., Flores, M. G., Zavaleta, D. E., Camacho, F. E., Grajeda, A. W., Alfaro, A. D., & Viera, M. R. (2016). Testing of Environmental Friendly Paraffin Removal Products. *Society of Petroleum Engineers - SPE Latin America and Caribbean Heavy and Extra Heavy Oil Conference 2016*, 171–179. 10.2118/181162-MS

Atkinson, G. M., Eaton, D. W., & Igonin, N. (2020). Developments in understanding seismicity triggered by hydraulic fracturing. *Nature Reviews. Earth & Environment*, 1(5), 264–277. 10.1038/s43017-020-0049-7

Campbell, C. J., & Laherrère, J. H. (1998). The End of Cheap Oil. *Scientific American*, 278(3), 78–83. 10.1038/scientificamerican0398-78

Chen, L., Dolado, J. J., Gonzalo, J., & Ramos, A. (2023). Heterogeneous predictive association of CO2 with global warming. *Economica*, 90(360), 1397–1421. 10.1111/ecca.12491

Clausthal, & Campbell, C. J. (2003, April 14). *Peak oil - A turning point for mankind*. Academic Press.

Colborn, T., Kwiatkowski, C., Schultz, K., & Bachran, M. (2011). Natural Gas Operations from a Public Health Perspective. *Human and Ecological Risk Assessment*, 17(5), 1039–1056. 10.1080/10807039.2011.605662

Columbia Industries. (2024). *Rig Walking Systems*. https://www.columbiacorp.com/oilfield-gas-industry-solutions/rig-walking-systems/

Deslances, N. (2022, September 27). *Almost 400 new mines needed to meet future EV battery demand, data finds.* https://techinformed.com/almost-400-new-mines-needed-to-meet-future-ev-battery-demand-data-finds/

Du, X., Carlson, K. H., & Tong, T. (2022). The water footprint of hydraulic fracturing under different hydroclimate conditions in the Central and Western United States. *The Science of the Total Environment*, 840, 156651. Advance online publication. 10.1016/j.scitotenv.2022.15665135700779

Energy Information Administration. (2009, June 29). *Annual U.S. Field Production of Crude Oil.* U.S. Field Production of Crude Oil(Thousand Barrels). https://web.archive.org/web/20091107181532/http://tonto.eia.doe.gov:80/dnav/pet/hist/LeafHandler.ashx?n=pet&s=mcrfpus1&f=a

Energy Information Administration. (2015). *World Shale Resource Assessments.* https://www.eia.gov/analysis/studies/worldshalegas/

Energy Information Administration. (2022, March 18). *EIA projects that renewable generation will supply 44% of U.S. electricity by 2050.* Author.

Energy Information Administration. (2023). *Annual Energy Outlook 2023.* https://www.eia.gov/outlooks/aeo/pdf/AEO2023_Narrative.pdf

Environmental Protection Agency. (2016). *Hydraulic Fracturing for Oil and Gas: Impacts from the Hydraulic Fracturing Water Cycle on Drinking Water Resources in the United States.* Author.

EPA. (2023). *Inventory of U.S. Greenhouse Gas Emissions and Sinks: 1990-2021 – Main Report.* https://www.epa.gov/system/files/documents/2023-04/US-GHG-Inventory-2023-Main-Text.pdf

Florides, G. A., & Christodoulides, P. (2009). Global warming and carbon dioxide through sciences. *Environment International*, 35(2), 390–401. 10.1016/j.envint.2008.07.00718760479

Gagnon, G. A., Krkosek, W., Anderson, L., McBean, E., Mohseni, M., Bazri, M., & Mauro, I. (2016). Impacts of hydraulic fracturing on water quality: A review of literature, regulatory frameworks and an analysis of information gaps. *Environmental Reviews*, 24(2), 122–131. 10.1139/er-2015-0043

Harde, H. (2019). What Humans Contribute to Atmospheric CO_2 Comparison of Carbon Cycle Models with Observations. *Earth Sciences (Paris)*, 8(3), 139. 10.11648/j.earth.20190803.13

Harde, H. (2023). *Science of Climate Change How Much CO2 and the Sun Contribute to Global Warming: Comparison of Simulated Temperature Trends with Last Century Observations*. 10.53234/scc202206/10

Harde, H., & Salby, M. L. (2021). What Controls the Atmospheric CO 2 Level? *Science of Climate Change*. 10.53234/scc202106/22

Hendricks, W. (2020). *CNX's E-Frac deal with evolution*. Academic Press.

Hong, Z., Moreno, H. A., & Hong, Y. (2018). Spatiotemporal Assessment of Induced Seismicity in Oklahoma: Foreseeable Fewer Earthquakes for Sustainable Oil and Gas Extraction? *Geosciences, 8*(12), 436. 10.3390/geosciences8120436

Howarth, R. W., Ingraffea, A., & Engelder, T. (2011). Should fracking stop? *Nature*, 477(7364), 271–275. 10.1038/477271a21921896

International Energy Agency. (2022). *The Role of Critical World Energy Outlook Special Report Minerals in Clean Energy Transitions*. https://iea.blob.core.windows.net/assets/ffd2a83b-8c30-4e9d-980a-52b6d9a86fdc/TheRoleofCriticalMineralsinCleanEnergyTransitions.pdf

IPCC. (2005). *Carbon Dioxide Capture and Storage*. https://www.ipcc.ch/report/carbon-dioxide-capture-and-storage/

IPCC. (2013). *The Physical Science Basis. Contribution of Working Group I to the Fifth Assessment Report of the Intergovernmental Panel on Climate Change*. IPCC.

IPCC. (2021). *Climate Change 2021: The Physical Science Basis. Contribution of Working Group I to the Sixth Assessment Report of the Intergovernmental Panel on Climate Change*. 10.1017/9781009157896

Ishida, T., Aoyagi, K., Niwa, T., Chen, Y., Murata, S., Chen, Q., & Nakayama, Y. (2012). Acoustic emission monitoring of hydraulic fracturing laboratory experiment with supercritical and liquid CO_2. *Geophysical Research Letters*, 39(16), 2012GL052788. Advance online publication. 10.1029/2012GL052788

Jackson, R. B., Vengosh, A., Darrah, T. H., Warner, N. R., Down, A., Poreda, R. J., Osborn, S. G., Zhao, K., & Karr, J. D. (2013). Increased stray gas abundance in a subset of drinking water wells near Marcellus shale gas extraction. *Proceedings of the National Academy of Sciences of the United States of America*, 110(28), 11250–11255. 10.1073/pnas.122163511023798404

Killgoar, B. A., Shutterstock, & Kleinberg, R. L. (2020). *The Global Warming Potential Misrepresents the Physics of Global Warming Thereby Misleading Policy Makers Abstract The Global Warming Potential Misrepresents the Physics of Global Warming Thereby Misleading Policy Makers*. Academic Press.

Liu, Y. (2023). The Impacts and Challenges of ESG Investing. *SHS Web of Conferences, 163*, 01015. 10.1051/shsconf/202316301015

Mao, J., Zhang, C., Yang, X., & Zhang, Z. (2018). Investigation on Problems of Wastewater from Hydraulic Fracturing and Their Solutions. *Water, Air, and Soil Pollution*, 229(8), 246. 10.1007/s11270-018-3847-5

Marc Airhart. (2007, January 26). *The Father of the Barnett Natural Gas Field George Mitchell*. Author.

Pimentel, D., & Burgess, M. (2017). World human population problems. *Encyclopedia of the Anthropocene, 1–5*, 313–317. 10.1016/B978-0-12-809665-9.09303-4

Potential Gas Committee. (2008, December 31). *Potential gas committee reports unprecedented increase in magnitude of U.S. natural gas resource base*. Author.

Ramírez-Orellana, A., Martínez-Victoria, M., García-Amate, A., & Rojo-Ramírez, A. A. (2023). Is the corporate financial strategy in the oil and gas sector affected by ESG dimensions? *Resources Policy*, 81, 103303. 10.1016/j.resourpol.2023.103303

Rasmussen, S. G., Ogburn, E. L., McCormack, M., Casey, J. A., Bandeen-Roche, K., Mercer, D. G., & Schwartz, B. S. (2016). Association Between Unconventional Natural Gas Development in the Marcellus Shale and Asthma Exacerbations. *JAMA Internal Medicine*, 176(9), 1334–1343. 10.1001/jamainternmed.2016.243627428612

Renee Choo. (2023, April 5). *The Energy Transition Will Need More Rare Earth Elements. Can We Secure Them Sustainably?* Academic Press.

Rogner, H., Dusseault, M. B., Rogner, H.-H., Authors Roberto Aguilera, L. F., & Archer, C. L. (2012). *Energy Resources and Potentials. In Global Energy Assessment-Toward a Sustainable Future Energy Resources and Potentials Convening Lead Author (CLA)*. 10.13140/RG.2.1.3049.8724

Salama, A., El Amin, M. F., Kumar, K., & Sun, S. (2017). Flow and transport in tight and shale formations: A review. *Geofluids*, 2017, 1–21. Advance online publication. 10.1155/2017/4251209

Samuel Armacanqui Tipacti, J. (2013). Arresting Unexpected Oil Production Decline in a Joint-Venture Environment. *Society of Petroleum Engineers - North Africa Technical Conference and Exhibition 2013. NATC*, 2013(2), 1496–1505. 10.2118/164785-MS

Soon, W., Connolly, R., & Connolly, M. (2015). Re-evaluating the role of solar variability on Northern Hemisphere temperature trends since the 19th century. In *Earth-Science Reviews* (Vol. 150, pp. 409–452). Elsevier B.V. 10.1016/j.earscirev.2015.08.010

Statista. (2023, August 16). *Which Country Consumes the Most Oil?* https://www.statista.com/chart/30609/countries-with-the-highest-oil-consumption-per-day/

Todd, B. M., Kuykendall, D. C., Peduzzi, M. B., & Hinton, J. (2015, March 16). Hydraulic Fracturing-Safe, Environmentally Responsible Energy Development. *All Days*. 10.2118/173515-MS

Toledano, P., Brauch, M. D., & Arnold, J. (2023). *Circularity in Mineral and Renewable Energy Value Chains: Overview of Technology, Policy, and Finance Aspects*. https://ccsi.columbia.edu/circular-economy-mining-energy

Viera Palacios, M. R. (2020). *Remoción de parafinas y mejoramiento del factor de recobro mediante uso de químicos multifuncionales y biodegradables para incrementar la productividad en pozos petroleros*. Academic Press.

Vtorov, I. P. (2017, January 17). *REE-table*. https://commons.wikimedia.org/wiki/File:REE-table.jpg

ZeroHedge. (2023, August 21). *U.S. And China Top The Chart In Global Oil Consumption*. U.S. And China Top The Chart In Global Oil Consumption.

Zheng, S., & Sharma, M. M. (2021). Modeling Hydraulic Fracturing Using Natural Gas Foam as Fracturing Fluids. *Energies*, 14(22), 7645. 10.3390/en14227645

Zou, C. (2017). Unconventional petroleum geology. In *Unconventional Petroleum Geology*. Elsevier. 10.1016/B978-0-12-812234-1.00002-9

Section 2
Exploration Module

Chapter 2
The Muerto Shale:
Stratigraphic Distribution of Carbonate Content and Organic Geochemistry in the Corcobado Outcrop

Jose Alfonso Rodriquez-Cruzado
National University of Engineering of Peru, Peru

Jorge Luis Ore-Rodriguez
National University of Engineering of Peru, Peru

ABSTRACT

The stratigraphic distribution of carbonate content (%CaO), total organic carbon (TOC), and pyrolysis parameters of a 277-meter section of the Muerto Formation, which was measured in the Corcobado outcrop, located in the Lancones Basin, Northwest Peru is characterized in the chapter. This is an organic-rich, calcareous sequence, with a thermal maturity in the oil window (Tmax = 455 °C and 1%Ro), which is also supported by the presence of bitumen in fractures. The authors identified an inverse relationship between %CaO and TOC, where it was observed that the highest organic enrichment (on average 2.42 wt.%) is associated with a carbonate content in the 30-40% range.

1. INTRODUCTION

The shale revolution (unconventional reservoirs in source rocks) has driven the increase in oil and gas production in the US. This has promoted the evaluation of important resources in shale reservoirs in the world (EIA, 2015).

DOI: 10.4018/979-8-3693-0740-3.ch002

The Muerto Shale

The current Peruvian oil context shows a growing demand for energy (IEA, 2020) and a sustained drop in oil production since the 1980s, this has caused the country to become a net oil importer (AIGLP, 2021; EIA, 2018; PERUPETRO, 2020), for this reason, unconventional sources in source rocks have attracted a growing interest in study (Morales-Paetán et al., 2018, 2020; Pairazamán et al., 2021; F. Palacios et al., 2015; Porlles et al., 2021).

In Peru, the northwestern basins (Figure 1) are one of the areas of interest for the evaluation of unconventional resources in source rocks, due to their proven oil tradition of almost 160 years, since the drilling of the first well in South America in 1863 (Bolaños Z., 2017). Some factors that could favor a source rock reservoir production project in northwestern Peru are the current presence of a hydrocarbon industry, which includes operating and service companies, extensive experience in fracturing low-permeability reservoirs, and the infrastructure such as the new Talara refinery that started operations in 2022, a gas treatment plant and ports.

In northwestern Peru, the Muerto Formation is one of the main source rocks (Fildani et al., 2005; Müller, 1993; PERUPETRO, 1999; Walters et al., 1992). This source rock was tested as a reservoir target in the 1980's, in more than 10 wells, however no commercial production was obtained, analogous to source rock production attempts made in other parts of the world at that time. The Muerto Formation is located in the Talara and Lancones basins (Castro-Ocampo, 1991; Tafur H., 1952) (Fig. 1). Of these areas, the Talara Basin has drilled more than 14,000 wells, produced more than 1.5 billion barrels of oil (BBO), and it has undiscovered recoverable resources in the range of 1.7 – 2.2 BBO and 4.8 – 5.8 TCFG (Infologic et al., 2006), which shows the potential of the area. On the other hand, the Lancones Basin is an exploratory area.

Figure 1. Left: Location of the Talara and Lancones basins, in northwestern Peru. Location of the outcrops of the Muerto Formation. Right: Location of the Corcobado creek (study area).

This study aims to characterize the organic rich sequence of the Muerto Formation, based on the stratigraphic distribution of carbonate content, total organic carbon (TOC) and pyrolysis parameters, for which intervals with distinguishable characteristics from adjacent intervals were defined.

The study area comprises the stratigraphic section of the Muerto Formation in the Corcobado creek outcrop (Figure 1). This outcrop is located on the eastern margin of the La Brea Mountains, in the Lancones Basin.

2. GEOLOGICAL FRAMEWORK

2.1 Muerto Formation

Since the identification of the Muerto Formation, exposed at surface (Bravo, 1921), in the outcrop that bears the same name, different studies have been carried out.

Different authors have called it a member of a larger sequence. Some authors had grouped the Muerto Formation with the underlying Pananga Formation (Bosworth et al., 1922; Chalco, 1954; Iddings & Olsson, 1928), or with the overlaying Copa Sombrero Group (Frizzell, 1944; Nauss, 1946; Olsson, 1934). In this study, the

The Muerto Shale

denomination of the Muerto Formation will be keep to the organic rich sequence, as used by International Petroleum Company (Nauss & Tafur H., 1946; Tafur H., 1952, 1954) and Petroperú (Castro-Ocampo, 1991; Pariguana M., 1979; Reyes & Vergara, 1985).

2.1.1 Lithology

The Muerto Formation is a fine-grained, calcareous sedimentary sequence, with a laminar stratification and sometimes with a concretional appearance, dark in color when freshly fractured, with a fetid odor and occasionally with traces of hydrocarbons in the outcrops (Figure 2). Occasionally with pyroclastic layers. Calcite-filled fractures are common.

Figure 2. Bitumen stains in the fractures of a Muerto Formation outcrop sample

2.1.2 Age

The Muerto Formation, of Albian Lower-Cretaceous age, is quite fossiliferous, containing Oxytropidoceras and Inoceramus in the lower part, fish teeth and scales and remains of thin shells in the middle part, and small ammonite casts filled with calcite. The fauna is made up of foraminifera, ammonites, pelecypods, radiolarians (Aldana A., 1994; Cruzado C. & Aliaga L., 1985; Fischer, 1956; Olsson, 1934; O. Palacios, 1994; Tafur H., 1952).

2.1.3 Sedimentary Context

This sedimentary sequence was deposited in the context of a regional transgressive event (Scotese, 2014), which is also evident in other areas of Peru, in the Pariatambo, Portachuelo and Raya Formations (Bianchi & Jacay, 2011). An oceanic anoxic event (OAE) is also associated, which coincides with the extensive deposition of source rocks during the Aptian-Albian in South America (Venezuela, Colombia and Peru) (Jenkyns, 1980), and other parts of the world (Alizadeh et al., 2012; Arthur et al., 1990; Dean et al., 1984; Flexer et al., 1986; Schlanger & Jenkyns, 1976; Sliter, 1989; Tarduno et al., 1985).

The Muerto Formation is a carbonate sedimentary sequence, deposited in a deep marine, platform and slope environment (Navarro-Ramirez et al., 2017).

2.1.4 Stratigraphic Relationship

The Muerto Formation lies concordantly on the carbonate bioclastic sequence of the Pananga Formation (Figure 3), which was deposited in a high-energy, shallow marine and shelf-reef environment (Séranne, 1987), during the Aptian-Albian. The top of the Muerto Formation shows a transitional sequence, which is not well defined, composed of siltstones, sandstones and carbonate layers, in the passage from the anoxic environment to the beginning of the turbiditic sedimentation, which continues during the Upper Cretaceous (Morris & Aleman, 1975; Tafur H., 1952, 1954; Uyen & Valencia, 2002).

Figure 3. Basal contact of the Muerto Formation in the Corcobado creek outcrop

2.1.5 Outcrops

The Muerto Formation outcrops are located on the margin of the La Brea and Amotape mountains, which divide the Talara and Lancones basins. The Muerto Formation is exposed in 4 main areas: (A) Pocitos Anticline, (B) La Brea Mountains, East Margin, (C) Amotape Mountains, East Margin, and (D) Amotape Mountains, Southwest Margin (Figure 1).

3. METHODOLOGY

3.1 Stratigraphic Section

The stratigraphic section was measured by Walac Research, from Universidad Nacional de Ingeniería (WR-UNI), during the field geology campaigns, between 2018 and 2020.

3.2 Laboratory Data

The laboratory data used in this study were obtained from declassified reports, stored by the Data Bank of Perupetro. The lab analyzes were carried out by the Petroperú laboratories, within the framework of the "Geochemical Study of the Cretaceous of the Lancones Basin", between 1984 and 1985. Laboratory data include carbonate content (55 samples), TOC (55 samples), and Rock-Eval pyrolysis (44 samples).

The stratigraphic position of the samples analyzed by Petroperú were estimated from the field notebook data (Reyes & Vergara, 1985).

3.2.1 Carbonate Content (%CaO)

Carbonate content was determined by acid treatment.

3.2.2 Total Organic Carbon (TOC)

The TOC was determined using a LECO EC-12 carbon analyzer, using 500 mg of rock, which was previously washed, dried, pulverized and treated with hydrochloric acid (HCl) to remove inorganic carbon (Alvarez et al., 1986; Alvarez & Garrido, 1987).

3.2.3 Open-System Programmed Pyrolysis (Rock-Eval)

Pyrolysis parameters were determined using a Rock-Eval 2 pyrolizer (RE2), using 100 mg of rock, following the standard operating parameters (Espitalie, 1986; Espitalie et al., 1985).

4. RESULTS AND DISCUSSION

Carbonate content, total organic carbon (TOC), and Rock-Eval pyrolysis data are summarized in Table 1 and were plotted according to their stratigraphic position in Figure 9.

4.1 Carbonate Content (%CaO)

The %CaO is highly variable, varying between 10-83%, with an average of 51% (Figure 4). The statistical distribution indicates a higher concentration of samples in the 40-50% range, and the absence of samples in the 20-30% and 90-100% ranges. A slight increase in the concentration of samples with %CaO in the 80-90% range is also shown.

Figure 4. Box plot and histogram of the carbonate content of the Muerto Formation in the Corcobado creek outcrop

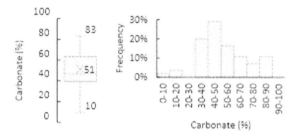

4.2 Organic Geochemistry

Thermal maturity, measured by Tmax, varies between 445 – 466 °C, with 441 and 472 °C as outliers, with a statistical distribution showing a notable higher frequency in the range 450 – 460 °C, and with an average Tmax of 455°C (Figure 5), which is consistent with the vitrinite reflectance (1%Ro) and with the Production

The Muerto Shale

Index (PI), mostly around 0.2, with the exception of some samples that exceed the threshold of 0.4 (Figure 6C). Thermal maturity suggests the oil generation window.

Given the advanced thermal maturity, the present-day TOC, generation potential (S2) and Hydrogen Index (HI), just shows the remaining organic richness and quality, which are reduced compared to original values. Therefore, they are an indicator of the present-day generation potential (Jarvie et al., 2007).

The TOC varies between 0.30 - 3.87 wt.%, on average 1.81 wt.% (Figure 5), which shows the organic enrichment of this section. The statistical distribution indicates a higher frequency of samples with TOC in the 2-3% range (Figure 5).

Figure 5. Box plot and histogram of the TOC and rock-eval Tmax of the Muerto Formation in the Corcobado creek outcrop

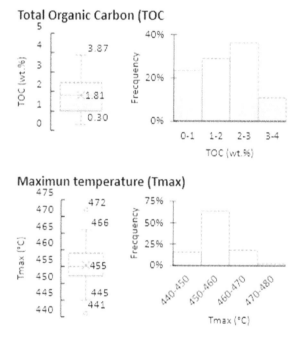

According to the OI versus HI and Tmax versus HI diagrams, which are used to classify a source rock according to the kerogen type (Banerjee et al., 1998; Espitalié et al., 1977; Peters, 1986), the data mainly follows the type I and type II pathways in an indetermined zone (Figure 6). It can be inferred that the original kerogen quality from the Muerto Formation corresponds to type II, because this type of kerogen is mostly associated with organic matter of marine origin (Jarvie, 2012), which is consistent with the sedimentary environment of the marine carbonate platform of the Muerto Formation and with the predominant amorphous organic matter, and with the absence of vitrinite particles on most of the samples.

Figure 6. (A) OI vs. HI diagram, (B) Tmax vs. HI diagram, and (C) Tmax vs. PI diagram

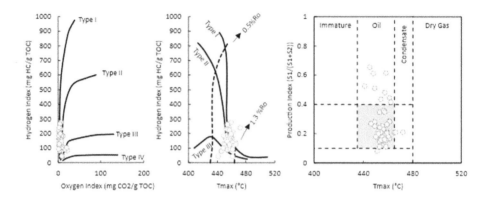

Table 1. Laboratory data of carbonate content, TOC and rock-eval pyrolysis

Sample	Carbonate content	TOC	S1	S2	S3	Tmax	HI	OI	Stratigraphic Position
	(%)	(wt.%)	(mg HC/ g rock)	(mg HC/ g rock)	(mg CO2/ g rock)	(°C)	(mg HC/ g TOC)	(mg CO2/ g TOC)	(m)
CB-1-CR	57.9	0.94	1.53	0.97	0.19	448	103	20	120
CB-2-CR	78.6	0.48							131
CB-3-CR	41.2	3.80	0.64	3.43	0.63	454	90	17	167
CB-4-CR	41.9	2.55	0.78	2.11	0.57	446	83	22	188
CB-5-CR	45.7	2.44	0.84	2.61	0.24	453	107	10	190
CB-6-CR	45.6	0.80	0.85	0.81	0.15	449	101	19	194
CB-7-CR	46.7	2.46	0.99	5.71	0.15	453	232	6	214
CB-8-CR	42.0	1.82	0.36	2.88	0.07	458	158	4	241
C-1	39.1	1.80	0.29	3.27	0.21	451	182	12	276.5
C-2	40.9	1.58	0.55	2.80	0.13	455	177	8	275.5
C-3	38.5	2.44	1.08	5.96	0.10	456	244	4	274.5
C-4	61.2	3.20	2.41	8.92	0.22	459	279	7	273.5
C-5	37.0	2.37	1.01	5.02	0.04	461	212	2	272.5
C-6	41.0	2.06	0.80	3.63	0.07	460	176	3	271.5
C-7	51.3	2.16	0.77	3.57	0.09	455	165	4	266

continued on following page

The Muerto Shale

Table 1. Continued

Sample	Carbonate content	TOC	S1	S2	S3	Tmax	HI	OI	Stratigraphic Position
	(%)	(wt.%)	(mg HC/g rock)	(mg HC/g rock)	(mg CO2/g rock)	(°C)	(mg HC/g TOC)	(mg CO2/g TOC)	(m)
C-8	48.6	2.25	1.03	3.98	0.11	459	177	5	263
C-9	36.5	2.41	1.12	6.40	0.09	454	266	4	262
C-10	53.5	0.62							249
C-11	42.3	2.30	0.54	4.32	0.19	453	188	8	245
C-12	34.9	1.49	0.76	1.94	0.05	449	130	3	242
C-13	63.2	0.43							240.5
C-14	34.4	2.60	1.63	4.84	0.05	449	186	2	235.75
C-15	47.8	0.52							226
C-16	58.1	1.41	0.90	1.61	0.06	456	114	4	234.5
C-17	55.2	2.52	1.18	4.21	0.04	452	167	2	232
C-18	37.8	3.87	2.36	6.93	0.08	455	179	2	222
C-19	39.3	2.64	1.13	6.01	0.05	455	228	2	214
C-20	32.9	2.11	0.74	2.79	0.05	454	132	2	208
C-21	42.2	2.86	0.77	3.50	0.23	450	122	8	199
C-22	42.0	2.79	0.96	5.19	0.04	458	186	1	195
C-23	81.8	0.74							188.5
C-24	18.4	1.49	0.56	2.13	0.02	465	143	1	186
C-25	43.1	2.54	1.00	3.97	0.18	452	156	7	188.5
C-26	54.6	3.83	1.63	8.57	0.06	457	224	2	156
C-27	16.5	2.05	2.25	2.80	0.11	461	137	5	136
C-28	51.4	0.44							118.5
C-29	69.1	1.15	1.30	1.68	0.09	454	146	8	107
C-30	63.1	1.33	1.45	1.56	0.06	457	117	5	90
C-31	55.7	1.88	0.64	2.41	0.05	466	128	3	70
C-32	35.7	3.27	2.11	7.89	0.07	472	241	2	78.5
C-34	75.9	0.31							53
C-35	74.0	0.46							38.5
C-36	80.0	1.01							22.5
C-37	82.4	1.13	0.42	1.06	0.01	461	94	1	17
C-38	77.2	0.59	0.30	0.27	0.05	441	46	8	3
COR-86-3	82.5	2.01	1.43	3.99	0.1	459	199	5	0.5
COR-86-3A	83.2	1.15							0.5
COR-86-4	69.9	1.41	1.18	1.65	0.02	462	117	1	91
COR-86-5	67.0	1.61	1.08	2.02	0.03	457	125	2	89
COR-86-6	9.7	1.23	1.92	1.2	0.03	458	98	2	109

continued on following page

Table 1. Continued

Sample	Carbonate content	TOC	S1	S2	S3	Tmax	HI	OI	Stratigraphic Position
	(%)	(wt.%)	(mg HC/ g rock)	(mg HC/ g rock)	(mg CO2/ g rock)	(°C)	(mg HC/ g TOC)	(mg CO2/ g TOC)	(m)
COR-86-7	80.1	0.30							115
COR-86-8	50.7	3.51	1.75	5.83	0.07	456	166	2	220
COR-86-9	43.3	2.10	1.4	2.61	0.08	453	124	4	235
COR-86-10	45.5	0.61	0.95	0.5	0.02	445	82	3	240
COR-86-11	36.0	1.62	0.3	0.75	0.17	461	46	10	250

4.3 Carbonate Content vs. TOC

The highest organic enrichment (2.42%) is related to samples with carbonate content in the range of 30-40% (Figure 7). An inverse relationship between TOC and carbonate content is observed, in the range 30 – 90% (Figure 7), where the higher the carbonate content, the lower the organic enrichment. This inverse relationship is analogous to that reported for other carbonate source rocks (Jones, 1984; Pratt, 1982).

The Muerto Shale

Figure 7. Plots of Carbonate content vs. TOC of the Muerto Formation in the Corcobado creek outcrop

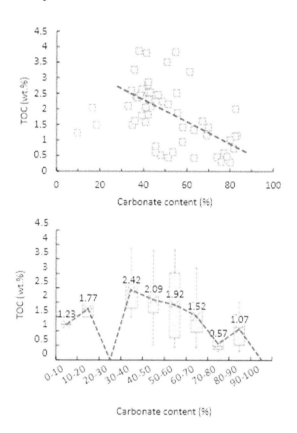

4.4 Stratigraphic Distribution

In this section, 3 intervals were identified within the sequence of the Muerto Formation, which should be understood as only valid for this area.

Compositionally, the %CaO is higher in the Lower Interval (79%, Figure 8) and although it decreases towards the top of the section, in the Upper Interval (45%, Figure 8), the %CaO still remains as an important component. In the Middle Interval, which coincides with the zone that presents intercalation with pyroclastic layers, the %CaO presents a greater heterogeneity, varying between 10-80% (Figure 8).

The organic richness, measured by TOC, is notably higher in the Upper Interval (2.13 wt.%, Figure 8), and it is lower in the Lower Interval (0.95 wt.%, Figure 8), which is logical, since both intervals of the section present different %CaO, and the organic enrichment, on average, is clearly related to the %CaO, as shown in Figure 7.

The HI, which is higher in the Upper Interval (Figure 9), suggests a better hydrocarbon generation potential. On the other hand, the PI, which is abnormally high in the Middle Interval (Figure 9), suggests a greater accumulation of hydrocarbons, possibly related to a greater storage capacity.

Table 2. Characteristics of the Muerto Formation on intervals. (…)=in the range of values, […]=outliers

	Lower Interval	Middle Interval	Upper Interval
Stratigraphic position (m)	0 – 55	55 – 140	140 – 278
thickness (m)	55	85	138
Carbonate content (%)	79 (74 - 83)	55 (10 – 80)	45 (33 – 63) [18; 82]
TOC (wt.%)	0.95 (0.31 – 2.01)	1.34 (0.30 – 3.27)	2.13 (0.43 – 3.87)
S1 (mg HC/g rock)	0.72 (0.30 – 1.43)	1.50 (0.64 – 2.25)	1.00 (0.29 – 1.63) [1.75; 2.36]
S2 (mg HC/g rock)	1.77 (0.27 – 3.99)	2.46 (0.97 – 2.80) [7.89]	3.96 (0.50 – 8.92)
S3 (mg CO2/g rock)	0.05 (0.01 – 0.10)	0.07 (0.02 – 0.19) [0.8; 1.4]	0.13 (0.02 – 0.24) [0.5; 0.6; 1.3; 1.6]
Tmax (°C)	454 (441 – 461)	459 (448 – 472)	454 (445 – 465)

The Muerto Shale

Figure 8. Box plots and frequency plots of carbonate content, TOC and rock-eval pyrolysis of the Muerto Formation intervals in the Corcobado creek outcrop

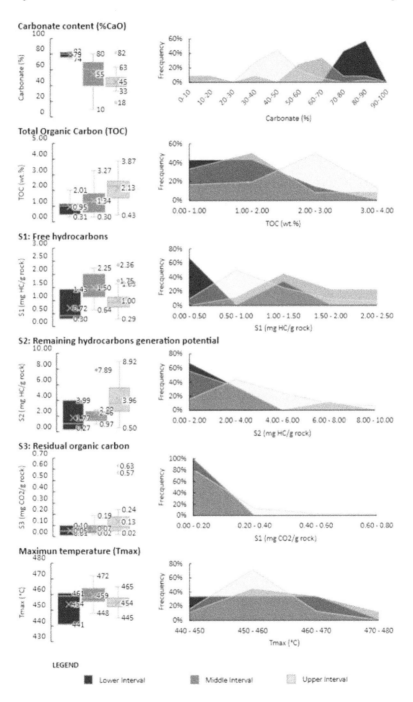

Figure 9. Stratigraphic profile of the Muerto Formation in the Corcobado creek outcrop, showing the stratigraphic distribution of carbonate content, TOC, Rock-Eval pyrolysis, and geochemical indices, which were grouped into 3 intervals

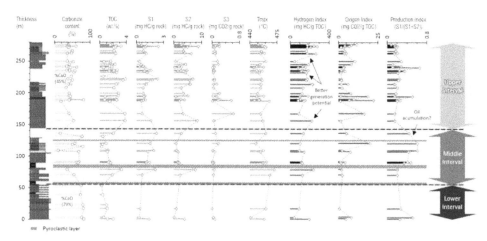

5. CONCLUSION

The Muerto Formation, in the Corcobado outcrop, is an organic-rich, calcareous sequence, with a thermal maturity in the oil window (Tmax = 455 °C and 1%Ro), which is also supported by the presence of bitumen in fractures.

An inverse relationship between %CaO and TOC was identified, where it was observed that the highest organic enrichment (on average 2.42 wt.%) is associated with a carbonate content in the 30-40% range.

Three intervals were defined in the Muerto Formation in the Corcobado outcrop, where:

- The Lower Interval (0 – 55 meters) is characterized by presenting the highest %CaO (on average, 79%) and the lowest TOC (on average, 0.95 wt.%).
- The Middle Interval (55 – 140 meters) presents intercalation with layers of volcanic tuffs, where the %CaO is highly heterogeneous (10 – 80%), and the TOC (on average, 1.34 wt.%), is slightly higher than in the Lower Interval. This interval presents a sector where there is a possible greater accumulation of hydrocarbons, as suggested by the production index close or higher than 0.4.

The Muerto Shale

- The Upper Interval (140 – 278 meters) is characterized by presenting a mostly homogeneous %CaO (on average, 45%), and the highest TOC (on average, 2.13 wt.%). This interval presents sectors with better source rock potential, as suggested by the higher hydrogen index.

ACKNOWLEDGMENT

A special thanks to the geologist Hugo Valdivia and Adrián Montoya for their guidance in carrying out the field geology studies. Special thanks to PERUPETRO SA for providing access to information from the Data Bank; and UNIPETRO ABC, ZEUS-OLYMPIC, and CONCYTEC-BANCO MUNDIAL for providing resources for this study. Our sincere thanks to the Peña family of the Corcobado farmhouse, for their hospitality and providing the facilities to enter the study area.

REFERENCES

AIGLP. (2021). *Perú: el GLP es el segundo combustible de mayor consumo y éste año crecerá la importación*. Author.

Aldana, A. M. (1994). Estudio de la Macrofauna de los cuadrángulos de Piura, Sullana, Quebrada Seca, Paita, Talara, Negritos, Lobitos, Tumbes, Zorritos y Zarumilla. In *Geología de los Cuadrángulos de Paita, Piura, Talara, Sullana, Lobitos, Quebrada Seca, Zorritos, Tumbes y Zarumilla* (pp. 145–190). Academic Press.

Alizadeh, B., Sarafdokht, H., Rajabi, M., Opera, A., & Janbaz, M. (2012). Organic geochemistry and petrography of Kazhdumi (Albian-Cenomanian) and Pabdeh (Paleogene) potential source rocks in southern part of the Dezful Embayment, Iran. *Organic Geochemistry*, 49, 36–46. 10.1016/j.orggeochem.2012.05.004

Alvarez, P., & Garrido, J. (1987). *Análisis de Roca Madre y Rocas Reservorio. Anexo 2*. Petroperu.

Alvarez, P., Garrido, J., & Aliaga, L. E. (1986). *Evaluación Geológica de la Cuenca Lancones. Perú Noroccidental. Estudios Geoquímicos del Cretáceo. Secciones Pan de Azucar, Cabrerías, Jaguay Negro, Huasimal, Corcobado, Gritón, Chilco*. Petroperu.

Arthur, M. A., Jenkyns, H. C., Brumsack, H.-J., & Schlanger, S. O. (1990). Stratigraphy, geochemistry, and paleoceanography of organic-carbon-rich Cretaceous sequences. In Ginsburg, R. N., & Beaudoin, B. (Eds.), *Cretaceous Resources* (Vol. 304). Events and Rhythms. 10.1007/978-94-015-6861-6_6

Banerjee, A., Sinha, A. K., Jain, A. K., Thomas, N. J., Misra, K. N., & Chandra, K. (1998). A mathematical representation of Rock-Eval hydrogen index vs Tmax profiles. *Organic Geochemistry*, 28(1–2), 43–55. 10.1016/S0146-6380(97)00119-8

Bianchi, C., & Jacay, J. (2011). *Distribución de Facies Anóxicas a través de la Margen Andina durante el Albiano*. VII Ingepet.

Bolaños, Z. R. (2017). Reseña histórica de la exploración por petróleo en las cuencas costeras del Perú. *Boletín de la Sociedad Geológica del Perú*, 112, 1–13.

Bosworth, T. O., Woods, H., Vaughan, W. T., Cushman, J. A., Brock, T. A., & Hawkins, H. L. (1922). Geology of the Tertiary and Quaternary Periods in the North-West Part of Peru. *Geological Magazine*, 60(1), 43–45. Advance online publication. 10.1017/S0016756800002272

Bravo, J. J. (1921). *Reconocimiento de la región costanera de los departamentos de Tumbes y Piura.* Asociación Peruana para el Progreso de la Ciencia.

Castro-Ocampo, R. (1991). *El Cretaceo en la Cuenca Talara del Noroeste del Perú.* Universidad Nacional de Ingeniería - Peru.

Chalco, A. (1954). *Informe Geológico Preliminar de la Region Sullana - Lancones.* Empresa Petrolera Fiscal.

Cruzado, C. J., & Aliaga, L. E. (1985). *Micropaleontología del Cretáceo del Área Pazul.* Petroperu.

Dean, W. E., Claypool, G. E., & Thide, J. (1984). Accumulation of organic matter in Cretaceous oxygen-deficient depositional environments in the central Pacific Ocean. *Organic Geochemistry*, 7(1), 39–51. 10.1016/0146-6380(84)90135-9

EIA. (2015). *World Shale Resource Assessments.* U.S. Energy Information Administration.

EIA. (2018). *Peru.* https://www.eia.gov/international/overview/country/PER

Espitalie, J. (1986). Use of Tmax as a Maturation Index for Different Types of Organic Matter. Comparison with Vitrinite Reflectance. *Thermal Modeling in Sedimentary Basins,* 475-496.

Espitalie, J., Deroo, G., & Marquis, F. (1985). La pyrolyse Rock-Eval et ses applications. Deuxième partie. *Revue de l'Institut Français du Pétrole*, 40(6), 755–784. 10.2516/ogst:1985045

Espitalié, J., Madec, M., Tissot, B. P., Mennig, J. J., & Leplat, P. (1977). Source rock characterization method for petroleum exploration. *Proceedings of the Annual Offshore Technology Conference,* 439–444. 10.4043/2935-MS

Fildani, A., Hanson, A. D., Chen, Z., Moldowan, J. M., Graham, S. A., & Arriola, P. R. (2005). Geochemical characteristics of oil and source rocks and implications for petroleum systems, Talara basin, northwest Peru. *AAPG Bulletin*, 89(11), 1519–1545. 10.1306/06300504094

Fischer, A. G. (1956). *Cretaceous of Northwest Peru. IPC Report WP-13*.

Flexer, A., & Rosenfeld, A. (1986). Relative Sea Level Changes During the Cretaceous in Israel. *AAPG Bulletin*, 70(11), 1685–1699. 10.1306/94886C9A-1704-11D7-8645000102C1865D

Frizzell, D. L. (1944). *Summary Report on the Stratigraphy and Paleontology of Northwestern Peru*.

Iddings, A., & Olsson, A. A. (1928). Geology of Northwest Peru. *AAPG Bulletin*, 12(1), 1–39. 10.1306/3D9327D7-16B1-11D7-8645000102C1865D

IEA. (2020). *Total final consumption (TFC) by source, Peru 1990-2019*. https://www.iea.org/countries/peru

Infologic, B.-R. (2006). *Geochemical-Solutions*. Petroleum Systems Evaluation – Talara/Tumbes Basins.

Jarvie, D. M. (2012). Shale resource systems for oil and gas: Part 1—shale-gas resource systems. In *M97: Shale Reservoirs—Giant Resources for the 21st Century* (pp. 69–87). American Association of Petroleum Geologists. 10.1306/13321446M973489

Jarvie, D. M., Hill, R. J., Ruble, T. E., & Pollastro, R. M. (2007). Unconventional shale-gas systems: The Mississippian Barnett Shale of north-central Texas as one model for thermogenic shale-gas assessment. *AAPG Bulletin*, 91(4), 475–499. 10.1306/12190606068

Jenkyns, H. C. (1980). Cretaceous anoxic events: From continents to oceans. *Journal of the Geological Society*, 137(2), 171–188. 10.1144/gsjgs.137.2.0171

Jones, R. W.R. W. Jones. (1984). Comparison of Carbonate and Shale Source Rocks: ABSTRACT. *AAPG Bulletin*, 68. Advance online publication. 10.1306/AD460EA4-16F7-11D7-8645000102C1865D

Morales-Paetán, W. J., Porlles-Hurtado, J., Rodriguez-Cruzado, J., Taipe-Acuña, H., & Arguedas-Valladolid, A. (2018). First Unconventional Play from Peruvian Northwest: Muerto Formation. *Unconventional Resources Technology Conference (URTeC)*. 10.15530/urtec-2018-2903064

Morales-Paetán, W. J., Rodriguez-Cruzado, J. A., Alvarez-Mendoza, B. G., Alarcón-Marcatoma, A. A., Oré-Rodriguez, J. L., Corrales-Hidalgo, R. W., Madge-Rodriguez, J. J., Porlles-Hurtado, J. W., Falla-Ruiz, J. L., & de Eyzaguirre-Gorvenia, L. (2020, July 27). Geochemical Heterogeneities Characterization of the Muerto Formation – Lancones Basin - Peru as a Source Rock Unconventional Reservoir, a Contribution for its Development. *SPE Latin American and Caribbean Petroleum Engineering Conference*. 10.2118/199088-MS

Morris, R., & Aleman, A. (1975). Sedimentation and Tectonics of Middle Cretaceous Copa Sombrero Formation in Northwest Peru. *SGP Bulletin, 48*(3).

Müller, D. (1993). *Geochemical Oil Correlations for UPPPL Oils and Offshore Oil Seep*. Talara Basin.

Nauss, A. W. (1946). *A Reconnaissance Geological Survey of the Pazul Area - Geological Memo #30*. Academic Press.

Nauss, A. W., & Tafur, H. I. A. (1946). *Geological Report of the Angostura district*. La Brea & Pariñas Estate.

Navarro-Ramirez, J. P., Bodin, S., Consorti, L., & Immenhauser, A. (2017). Response of western South American epeiric-neritic ecosystem to middle Cretaceous Oceanic Anoxic Events. *Cretaceous Research*, 75, 61–80. 10.1016/j.cretres.2017.03.009

Olsson, A. A. (1934). Contributions to the paleontology of northern Peru: the Cretaceous of the Amotape Region. In *Bulletins of American Paleontology* (Vol. 20, Issue 69, pp. 1–104). Academic Press.

Pairazamán, L., Palacios, F., & Timoteo, D. (2021). *Caracterización sedimentológica de alta-resolución de la Formación Muerto, Cuenca Lancones, NO Perú. ¿Un posible reservorio no convencional?* Academic Press.

Palacios, O. (1994). Geología de los cuadrángulos de Paita, Piura, Talara, Sullana, Lobitos, Quebrada Seca, Zorritos, Tumbes y Zarumilla. In *Carta Geológica Nacional* (1st ed.). Instituto Geológico, Minero y Metalúrgico.

Pariguana M., H. A. (1979). *Evaluación de las formaciones Muerto - Pananga - Área L.B.P*. Academic Press.

PERUPETRO. (1999). VOL 1. Generalidades. In *Estudios de Investigación Geoquímica del Potencial de Hidrocarburos. Lotes del Zócalo Continental y de Tierra*. PERUPETRO S.A.

PERUPETRO. (2020). *Estadística Anual de Hidrocarburos*. Academic Press.

Peters, K. E. (1986). Guidelines for Evaluating Petroleum Source Rock Using Programmed Pyrolysis. *AAPG Bulletin*, 70(3), 318–329. 10.1306/94885688-1704-11D7-8645000102C1865D

Porlles, J., Panja, P., Sorkhabi, R., & McLennan, J. (2021). Integrated porosity methods for estimation of gas-in-place in the Muerto Formation of Northwestern Peru. *Journal of Petroleum Science Engineering*, 202, 108558. Advance online publication. 10.1016/j.petrol.2021.108558

Pratt, L. M. (1982). *The paleo-oceanographic interpretation of the sedimentary structures, clay minerals, and organic matter in a core of the Middle Cretaceous Greenhorn Formation drilled near Pueblo Colorado*. Princeton University.

Reyes, L., & Vergara, J. (1985). *Libreta de campo 85-6. Geología de la Cuenca Lancones*. Petroperu.

Schlanger, S. O., & Jenkyns, H. C. (1976). Cretaceous Oceanic Anoxic Events: Causes and Consequences. *Netherlands Journal of Geosciences*, 55(3–4), 179–184.

Scotese, C. (2014). Atlas of Early Cretaceous Paleogeographic Maps, PALEOMAP Atlas for ArcGIS, volume 2, The Cretaceous, Maps 23 - 31, Mollweide Projection, PALEOMAP Project. 10.13140/2.1.4099.4560

Séranne, M. (1987). Evolution tectono-sédimentaire du bassin de Talara (nord-ouest du Pérou). *Bulletin de l'Institut Français d'Études Andines*, 16(3), 103–125. 10.3406/bifea.1987.952

Sliter, W. V. (1989). Aptian anoxia in the Pacific Basin. *Geology*, 17(10), 909. 10.1130/0091-7613(1989)017<0909:AAITPB>2.3.CO;2

Tafur, H. I. A. (1952). Cretaceous Geology of the East Front of the Amotape Mountains. IPC Report WP-12. *International Petroleum Company*.

Tafur, H. I. A. (1954). *Reconnaissance of Cretaceous Between Chira River and Amotape Mountains, Northwest Peru. IPC Report WP-14*.

Tarduno, J. A., McWilliams, M., Debiche, M. G., Sliter, W. V., & Blake, M. C.Jr. (1985). Franciscan Complex Calera limestones: Accreted remnants of Farallon Plate oceanic plateaus. *Nature*, 317(6035), 345–347. 10.1038/317345a0

Uyen, D., & Valencia, K. H. (2002). Anexo 16: North-Lancones Basin Surface Geology and Leads Evaluation. In *Informe Final. Segundo Periodo Exploratorio. Lote XII - Cuenca Lancones*. Pluspetrol Peru Corporation S.A.

Walters, C. C., Toon, M. B., Rae, D., Barrow, R., Flagg, E. M., & Hellyer, C. L. (1992). *Talara Basin, Peru: Source Rock Characterization. Academic Press*.

Chapter 3
Volumetric Estimation of the Gas in the Generator Rock Through an Evaluation of the Adsorption in the Northeast of Peru

Heraud Taipe- Acuña
National University of Engineering of Peru, Peru

Victor Huerta Quiñones
National University of Engineering of Peru, Peru

Ali Tinni
University of Oklahoma, USA

Hugo Valdivia Ampuero
National University of Engineering of Peru, Peru

Isabel Moromi Nakata
http://orcid.org/0000-0002-7298-565X
National University of Engineering of Peru, Peru

ABSTRACT

In Peru, energy security is based on gas. The decrease in natural gas reserves between 2016 and 2019 was from 16.1 to 10.1 TCF, which reflects a negative replacement

DOI: 10.4018/979-8-3693-0740-3.ch003

rate in this period and a decrease in exploration. However, the evaluation of shale gas reservoirs could influence the increase of gas resources in Peru. Technological advances in the exploitation of these resources made it possible to transform gas resources into economic reserves. The study of unconventional shale gas reservoirs in South America has been developing with greater interest, especially in the business of exploration, characterization, and pilot tests in these reservoirs. This research is the first to evaluate the volume of gas adsorbed in the source rock (Murder Formation) in the Lancones basin in Northeastern Peru, through isotherms obtained in the laboratory, and aims to volumetrically quantify the total gas content (scf/ton) in this formation.

1. INTRODUCTION

In the hydrocarbon industry, unconventional resources were classified for the first time by Jhon Master (Masters, 1979), who represented them within a "resource triangle", where shale oil and shale gas resources are located, later this triangle was modified by Stephen Holditch (Holditch, 2006), who incorporated permeability values. Hydrocarbons produced from shale gas and shale oil resources come from fine-grained sedimentary rocks rich in organic matter (Lakatos & Lakatos-Szabo, 2009). Juan Glorioso & Jesus Rattia defined the term "shale" as a reference to a laminar and fissile structure present in certain rocks (Glorioso & Rattia, 2012). It should also be considered that the organic matter that is preserved in these rocks depends on the level of dissolved oxygen in the water (Potter et al., 1981).

The beginning of the production of shale gas resources began in 1978 in the United States, thanks to the *Unconventional Gas Research Program,* started in 1976 by the Morgantown Energy Research Center (MERC), thanks to this program production increased by approximately 450% by the end of 1998 (Mershon & Palucka, 2013); The development period of this program lasted 5 years and the Devonian shale gas formations in the Appalachians and sectors of Illinois and Michigan were inventoried, where the recoverable reserves of shale gas were determined, as well as the most effective technologies for its use. extraction at the lowest possible cost. The world production of shale gas for 2014 in countries like the United States was 14.6 million cubic feet per day (MMSCFD) (U.S. Energy Information Administration EIA/AIR, 2015); in Canada it went from 1.9 to 3.9 MMSCFD from 2011 to 2014 (U.S. Energy Information Administration EIA/AIR, 2015); in China the production was 0.163 MMSCFD, which represented 1.5% of the total production of natural gas (U.S. Energy Information Administration EIA/AIR, 2015), considering that in 2011 the production was nil and in Argentina the production of shale oil coming mainly from the Neuquén Basin was close to 20,000 barrels per day (bpd), coming from the Vaca Muerto formation (U.S. Energy Information Administration EIA/AIR, 2015). These four countries mentioned are the only ones that register production in 2022, however, the technically recoverable resources (TRR) in South America are 1,429 trillion cubic feet (TCF) and 59.8 billion barrels (U.S. Energy Information Administration EIA/AIR, 2015).

Peru is not among the countries that determined its shale gas and shale oil resources, however, the Muerto Formation located in northeastern Peru was evaluated. This formation is considered the most important source rock in the Talara basin (Fernández et al., 2005) and the Lancones basin (Andamayo, 2008), the total organic content (TOC) determined in the Lanconces basin was on average 2.33% for outcrop samples (W. Morales et al., 2018) and 1.23% TOC in drill cuttings (W. J. Morales et al., 2020), the gas in place (GIP) determined by Porlles was 660.2 MSCF/(acre-ft) applying a multiple porosity model (Porlles et al., 2021). It is important to consider that energy security in Peru is based on gas, where the proven reserves of natural gas for 2019 was 10,142 TCF and compared to 2018 it decreased by 0.462 TCF (Ministerio de Energía y Minas & Dirección General de Hidrocaburos, 2019). The proven reserves of natural gas liquids (LNG) in 2019 was 493,221 million barrels of oil controlled at standard conditions (MMSTB) and compared to 2018 it decreased by 21,168 MMSTB (Ministerio de Energía y Minas & Dirección General de Hidrocaburos, 2019). The natural gas reserves replacement index as of 2019 was +0.29 and the natural gas liquids reserves replacement index as of 2019 was +0.35.

The evaluation of unconventional shale gas reservoirs is based on studies on coal methane or also known as "CBM", where the highest percentage of gas is adsorbed and is approximately 90% (Wang & Reed, 2009). To determine the total shale gas, add the free gas, adsorbed gas and the gas dissolved in the water and oil (Ahmed & Meehan, 2016; Rezaee, 2015), however, for the industry, the calculation results only from the sum of free gas and adsorbed gas, thus the volume of free gas is quantified by modifying the standard method of evaluation of a conventional reservoir (Ambrose et al., 2012), this volume must be subtracted from the volume of gas adsorbed on the organic nanopores that is not available for free gas storage. Therefore, without correction, the free gas is overestimated by including the pore space that is already occupied by adsorbed gas. Thus, the estimated free gas is adjusted with a correction factor for the thickness of the adsorbed layer (Belyadi et al., 2017). On the other hand, the adsorbed gas content is determined by the Langmuir isotherms, which was developed by Irving Langmuir (Langmuir, 1916; Langmuir, 1917), where the pressure and adsorbed gas content are related. However, it should be considered that the storage capacity of adsorbed gas content is related to mineralogical composition, total organic content (TOC), thermal maturity, pore diameter, and reservoir conditions such as pressure, temperature, and temperature.

among others.(Bustin et al., 2008; Chalmers et al., 2012; Sondergeld et al., 2010; Zhang et al., 2012; Zhou et al., 2018).

The volume of gas adsorbed is often significantly greater than the free gas stored, averaging 65% of the total gas in situ (Ahmed & Meehan, 2016; Murillo-Martínez et al., 2015). It is generally accepted that the adsorbed gas has an appreciable effect on the economics of the producing well (Mengal & Wattenbarger, 2011). Due to this, it is necessary to know the economically extractable total gas volume (free and adsorbed) in order to assess the economic viability of the reservoir. This research was the first to evaluate the volume of gas adsorbed in the Muerto Formation and in Peru, this by generating adsorption isotherms obtained in the laboratory, this research aims to volumetrically quantify the total gas content (scf/ton) in the Muerto Formation. These results will provide important knowledge for decision making in the following phases of exploration and development of the Lancones basin located in northeastern Peru.

2. GEOLOGICAL FRAMEWORK

The Lancones basin is located in the northeast of Peru and has been studied since 1927, the first exploration campaign was carried out by the IPC company, drilling the Tamarindo 1X, Tamarindo 2X and Tamarindo 3X wells (W. Morales et al., 2018). The company Petroperú carried out the second campaign in 1980, where geochemical studies, construction of stratigraphic columns, geological maps and other surface studies were carried out (W. Morales et al., 2018). The third campaign was carried out in the year 2000, by the company Pluspetrol, where it developed a geological model, incorporating 16 2D seismic lines and drilling an exploratory well Abejas 1X (W. Morales et al., 2018). Finally, the fourth campaign was developed by the BPZ exploration company in 2006 and they developed 14 seismic lines, which improved the geological model (W. Morales et al., 2018); all these works carried out in this basin are shown in Fig.1. In all these evaluations, the Muerto Formation is considered as the source rock of the Lancones basin (W. Morales et al., 2018).

2.1 Geochemical Parameters

The geochemical evaluation in unconventional reservoirs such as shale oil & gas, are by concept more detailed, due to the process of identifying the horizons rich in organic matter (Liborius-Parada & Slatt, 2016).

The geochemical information of the Muerto Formation, obtained from the study area, was evaluated by Petroperú, where they analyzed 77 samples from six different streams, as well as Pluspetrol analyzed 12 samples of drilling cuttings from the Abejas

Volumetric Estimation of the Gas in the Generator Rock

1X well. The results obtained from this evaluation show an average of 2.33% and 1.18% TOC respectively, the type of kerogen varies between II and III (TOC x S2), it is also found in the oil and gas window (TOC x HI), while the organic matter could be classified mainly in the mature zone (Tmax x HI). These results also show that the analyzed samples are in a regular to good production zone in terms of quantity and quality of organic matter (TOC x S2) (Tissot & Welte, 1978).

The existing geochemical data was the basis and starting point for the study and selection of a type of ravine, where the Muerto Formation outcrops from the base to the top, the ravine is located in the Lancones basin, a column was made stratigraphic, geochemical and mineralogical tests on 26 samples, petrophysical tests were also carried out on 6 samples. Sampling was carried out systematically from the base to the top in a conventional manner and with a manual core-draw tool (Eberli et al., 2017).

Figure 1. Geological map of Lancones basin incorporating seismic lines, wells and surface samples

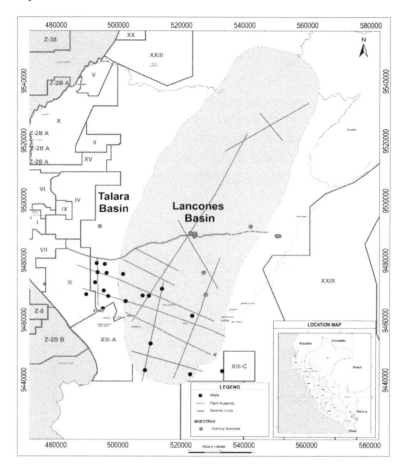

Volumetric Estimation of the Gas in the Generator Rock

Figure 2. TOC of samples evaluated

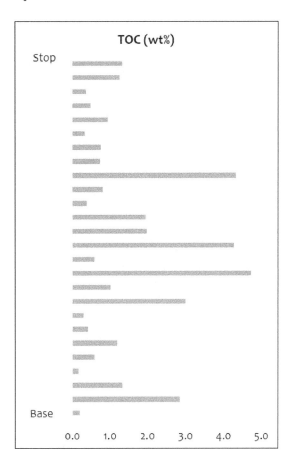

The new geochemical data were provided by the Core Laboratories laboratory where an average TOC (w%t) of 1.44% is shown as can be seen in Figure 2, however, the areas rich in organic matter of the unconventional reservoirs shale oil & gas type, varies by unit of stratigraphic facie and in our case facies can be observed with 4.73% TOC and in others 0.2% TOC. On the other hand, the quality of the kerogen (TOC x S2) can be seen in Figure 3, which indicates that the kerogen varies between type II and III and can generate oil and gas at the same time.

2.2 Mineralogy and Petrography

The laboratory evaluation was carried out by the company Laboratorio LCV del Perú, where petrography, diagenesis, and mineralogy by x-ray diffraction (XRD) were evaluated on 26 selected samples, subsequently, geochemical tests were carried out on these same 26 samples.

Table 1. Petrography and mineralogy of type samples

Sample	Petrographic Classification	Quartz, Feldspar, Mica	Carbonates	Clays
66-B	Mudstone/Siltstone	86	7	6
46-B	peloidal mudstone	57	40	TR
39-B	mudstone	46	48	3
37-B	silty mudstone	87	1	eleven
29-B	fossiliferous mudstone	48	38	eleven
25-B	silicified mudstone	76	eleven	10
7-B	Fossiliferous/radiolaritic wackestone	twenty-one	65	13

Figure 3. Ternary diagram of samples evaluated

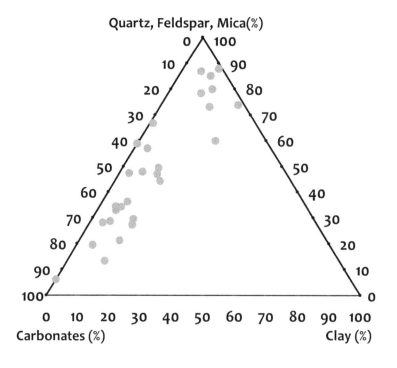

Volumetric Estimation of the Gas in the Generator Rock

These were washed with toluene to remove the hydrocarbon present, then impregnated with resin to study the pore network, and stained red with alizarin to differentiate calcite. All thin section cuts were microphotographed. The organic matter content was estimated visually, the carbonate rocks were classified following the terminology proposed by Rober Dunhan (Dunham, 1962). In Table I, the petrographic classification of the studied samples is presented, ordered following a stratigraphic criterion. It included the formation or group, sample name, petrographic classification and mineralogical composition of the main mineral species determined by XRD.

To synthesize the main characteristics, the samples were grouped into 6 microfacies, emphasizing all the analyzed samples corresponding to the Muerto Formation. On the other hand, regarding diagenetic studies, general sequences of diagenetic products and processes were established for mudstones, carbonate rocks and tuffs. They were listed in relative chronological order, not all events are present in all samples. The following were analyzed: calcite, dolomite, silica, pyrite, dissolution of grains and tuffs or tuffs.

Figure 4. Photograph of a thin section of sample 50-B: 73% carbonates, 5% clays, 19% silicoclatics and 3% others

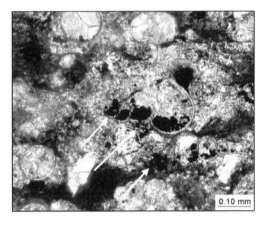

In Figure 4 you can see the photograph of a thin section of sample 50-B, where pyrite can be seen, which was identified as a replacement filled with skeletal grains in a subordinate way, as a replacement for plagioclase and disseminated in the matrix of some samples, Framboidal or massive habit is also present.

Finally, the results are presented, tabulated and graphed in a ternary diagram. Analyzing the XRD results and in association with the information from the petrographic study, some considerations can be mentioned, for example, regarding quartz, its proportion is high, with 41% on average, it is found in the skeletal grains (foraminifera and mainly radiolaria) and as part of the matrix. The clay content is low, reaching 9% on average for the samples from the Muerto Formation.

In the Fig.5, you can see one of the ternary diagrams of mineralogical variation made based on the mineralogical studies. The analysis of the samples, from the petrographic and mineralogical point of view (XRD), is a unit that is made up of fine-grained rocks with variable amounts of planktonic microfossils (foraminifera, radiolaria and pelecypods of the inoceramid type), which present a high content calcite and silica and low clay. These rocks would have been predominantly deposited by decantation in a relatively deep marine environment, in oxygen deficient conditions, from anoxic to disoxic bottoms. The analysis strongly suggests a biostratigraphic study, no ammonites were identified in the thin sections and echinoderms identified,

2.3. Porosity and Surface Area

Laboratory tests were carried out by the Mewbourne School of Petroleum & Geological Engineering, The University of Oklahoma. These tests determined the total porosity that resulted from the sum obtained by means of the Helium Porosimeter and the Nuclear Magnetic Resonance (NMR). NMR measured residual liquid volume, while Helium Porosity characterized the residual gas-filled pore space. To obtain the specific surface area, it was determined by the method Brunauer-Emmet Teller (BET), the results of these laboratories are shown in Table 2. Finally, Table 3 summarizes the petrophysical and geochemical properties of the samples evaluated.

Table 2. Total porosity and BET surface area

Sample	Total porosity (%)	BET surface area (m2/g)
46-B	2.9	1.79
39-B	8.2	3.41
37-B	4.6	1.50
29-B	1.8	2.68
25-B	4.9	1.25
7-B	2.2	0.58

continued on following page

Volumetric Estimation of the Gas in the Generator Rock

Table 3. Continued
Table 3. Petrophysical properties of samples analyzed

Sample	READ OCD (wt%)	Kerogen type	XRD Total Rock			Matrix	BET surface area (m2/g)	Overall porosity (%)
			Qz+Fk+Pl	Carbonates	Clays			
39B	4.28	II	46	48	3	siliceous	3.41	8.2
37B	1.98	II	87	1	11		1.50	4.6
25B	0.74	II	76	11	10	clayey-siliceous	1.25	4.9

3. METHODOLOGY

In the hydrocarbon industry, current techniques for evaluating storage in shale samples are designed for crushed rock (Tinni et al., 2017). However, measurements cannot be made under confining pressures, therefore these measurements may not be representative at reservoir pressure conditions. Due to this, the Mewbourne School of Petroleum & Geological Engineering laboratory, The University of Oklahoma, carried out storage measurements in plugs extracted from the samples acquired from the type of stream, with approximate lengths of 5.5 centimeters and diameters of 2.5 centimeters.

Figure 5. Experimental scheme of high-pressure saturation with methane

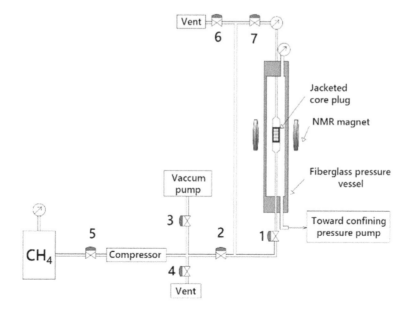

For methane adsorption was measured using a high-pressure gas (methane) adsorption manometer. Figure 5 shows a schematic of the experimental setup for methane saturation, including the Oxford Geospec2™ NMR that measured storage (ø). Before saturation, the plug was jacketed, and the confining pressure was increased to 5000 psi. Subsequently, the compressor was used to pressurize the methane to the desired pressure (in our case 500, 1000, 1500, 2000, 3000 and 4000 psi). Valves numbered 1 through 7 open and close allowing the vacuum and flow of methane due to the compressor. Once the desired pressure is reached, the interstitial (reservoir) pressure is controlled with a digital manometer with a resolution of ±1 psi. The sample is considered fully saturated if the pressure does not drop more than 5 psi over 8 hours, however if it drops more than 5 psi over an 8 hour interval, the sample is restored. For the analyzed samples, approximately 8 to 16 hours were required to saturate the samples with methane, at these conditions the temperature remained constant at 30°C. in the Table 4 the obtained values of pressure (psi) and adsorbed gas (scf/ton) are shown.

Table 4. Capacity of methane adsorbed at different pressures

Sample 25-B (T= 30 °C)		Sample 37-B (T= 30 °C)		Sample 39-B (T= 30 °C)	
P(psi)	**CH4 (scf/ton)**	**P(psi)**	**CH4 (scf/ton)**	**P(psi)**	**CH4 (scf/ton)**
0	0	0	0	0	0
500	31.39	500	32.45	500	42.15
1000	40.33	1000	42.31	1000	55.23
1500	42.96	1500	45.38	1500	58.96
2000	44.03	2000	47.74	2000	62.31
3000	47.32	3000	51.35	3000	67.05
4000	48.55	4000	52.76	4000	69.01

The values of pressure (psi) and adsorbed gas content (scf/ton) were statistically adjusted to the Langmuir equation (Langmuir, 1916), represented by the following equation:

$$G_{ads} = \frac{G_{SL} P}{(P_L + P)} \quad (1)$$

where, G_{ads} is the volume of gas adsorbed at the equilibrium pressure P, G_{SL} is the maximum storage capacity of the gas and is called the Langmuir volume constant, and P_L is the Langmuir pressure that corresponds to the pressure at 50% of the Langmuir volume.

Volumetric Estimation of the Gas in the Generator Rock

The adjustment made was through a non-linear regression, which minimizes the function of the average percentage error (APE) by means of the variation of the adjustable parameters of the model (Murillo-Martínez et al., 2015). On the other hand, to obtain the volume of free gas, it has been done using the Belyadi relation (Belyadi et al., 2017), who incorporated a correction factor ψ:

$$G_f = \frac{32.0368}{B_g}\left[\frac{\emptyset(1 - S_w - S_o)}{\rho_b} - \psi\right] \quad (2)$$

where G_f is the total free gas, $\emptyset$ is the total porosity of the sample, S_w is the water saturation, S_o is the oil saturation (in this case it is zero), B_g is the gas compressibility factor, M is the molecular weight of the single-component gas or the apparent molecular weight of the gas mixture, ρ_s is the density of the adsorbed gas, ρ_b is the bulk density of the rock, P is the equilibrium pressure, and P_L and G_{SL} are the pressure and volume of Langmuir, respectively.

4. RESULTS AND DISCUSSIONS

In this section, the values of free gas and adsorbed gas obtained by isotherm curves are shown, finally, the influence of petrophysical and geochemical parameters on adsorption is analyzed. To determine the total volume of shale gas resources, it is obtained from the sum of free gas and adsorbed gas (Ambrose et al., 2012):

$$GIP = G_f + G_{ads} \quad (4)$$

where GIP is the total gas in place (scf) per ton ($\approx$907.185 Kg) of rock.

4.1 High Pressure Methane Saturation

Figure 6, displays the generated isotherm curves, of samples 25-B (a), 37-B (b) and 39-B (c).

Figure 6. Adsorption isotherms at high pressures applying the Langmuir relation (blue curve is from sample 25B, red curve is from sample 37B and green curve is from sample 39B)

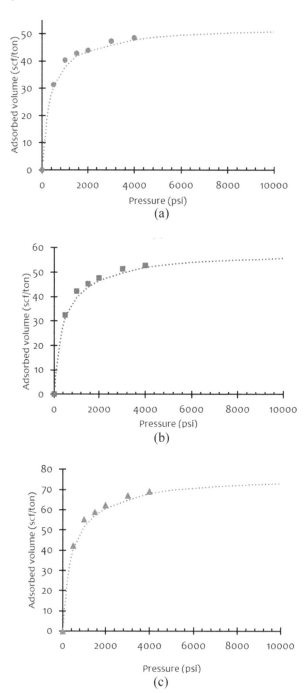

Volumetric Estimation of the Gas in the Generator Rock

Table 5 shows the values of and determined by the isotherm curves for the samples evaluated. P_L G_{SL}

Table 5. Capacity of methane adsorbed at different pressures

Sample	G_{SL} (scf/ton)	P_L (psi)
25-B	52.67	338.98
	$G_{ads} = \frac{(52.67)xP}{338.98 + P}$	
37-B	57.94	393.7
	$G_{ads} = \frac{(57.94)xP}{393.7 + P}$	
39-B	75.92	401.6
	$G_{ads} = \frac{(75.92)xP}{401.6 + P}$	

The values used to determine the volume of gas in equation (2) and (3) are shown in Table 6. The result obtained in this investigation is the first estimate developed in this basin, evaluated by the volume of adsorbed gas, total porosity, and surface area from the laboratory.

Table 6. Parameters obtained from the Muerto Formation (Porlles et al., 2021)

Parameters	Worth	Units
Formation Gas Volumetric Factor (βg)	0.00658	dimensionless
Water Saturation (S_w)	0.180	dimensionless
Oil Saturation (S_o)	0	dimensionless
Rock density (ρ_g)	2.71	g/cm3
Adsorbed gas density (ρ_s)	0.370	g/cm3
Reservoir pressure (P)	4280	psi

4.2 Determination of the Volume of Corrected Free Gas and Adsorbed Gas

The values of the parameters of Table VI and Table V, are replaced in equation (1), (2) and (4) to determine the adsorbed gas, free gas and total gas, respectively, which are shown in Table 7.

Figure 7 shows a comparison of the results, where the volume of adsorbed gas is lower and the volume of free gas is higher.

Volumetric Estimation of the Gas in the Generator Rock

Table 7. Summary of the parameters used to determine the total gas volume (scf/ton)

Parameter	Sample		
	25B	37B	39B
$\emptyset$	0.049	0.046	0.082
G_{SL}	52.67	57.94	75.92
P_L	338.98	393.7	401.6
G_{free}	72,188	67,768	120,804
ψ	13,543	14,724	19,260
$G_{free\ fixed}$	58,645 (54.6%)	53,044 (50.0%)	101,544 (59.4%)
G_{ads}	48,805 (45.4%)	53,059 (50.0%)	69,407 (40.6%)
G_{total} (scf/ton)	107,449	106,104	170,951

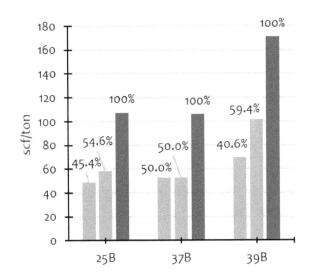

Figure 7. Bar diagram showing the summary of the volumes acquired through equations (1), (2), (3) and (4), the lead-colored bars represent the adsorption volumes, the light-blue bars represent the corrected free volumes, and the blue bars represent the sum of these volumes for each respective sample

The calculations carried out show a total gas storage capacity of 107,449 scf/ton for sample 25B, of which the largest contributor is the corrected free gas with a value of 58,645 scf/ton, which is equivalent to 54.6% of the total gas. For sample 37B, the total gas storage capacity is 106,104 scf/ton, of which the largest contributor

Volumetric Estimation of the Gas in the Generator Rock

is adsorbed gas with a value of 53,059 scf/ton, which is equivalent to 50.0% of the total gas. Finally, for sample 39B, the total gas storage capacity is 170,951 scf/ton, of which the largest contributor is corrected free gas with a value of 101,544 scf/ton, which is equivalent to 59.4% of the total gas.

In Figure 8, the influence of the percentage of TOC (a) and clay (b) on the surface area can be seen, where it is observed that the greatest contribution to the surface area was TOC.

Figure 8. Relationship of TOC (a) and the percentage of clay (b) with respect to the surface area, which shows a greater contribution of TOC

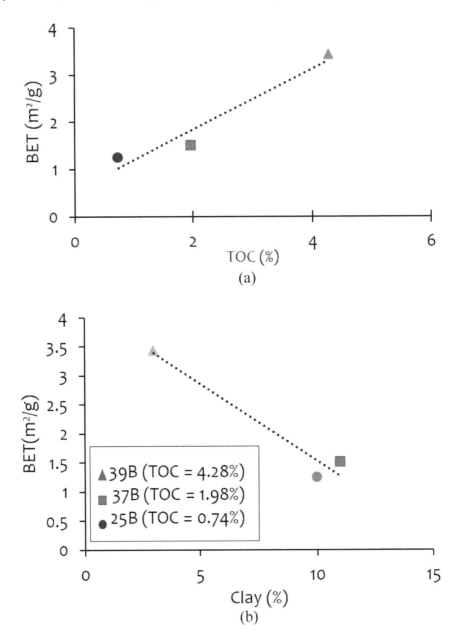

Volumetric Estimation of the Gas in the Generator Rock

Finally, Figure 9 shows the relationship between the adsorbed gas volume and the TOC of the three generated curves, which contributes to determine the relationship between these two properties.

Figure 9. Comparison of the 3 Adsorption isotherms generated (blue curve is from sample 25B, red curve is from sample 37B, and green curve is from sample 39B)

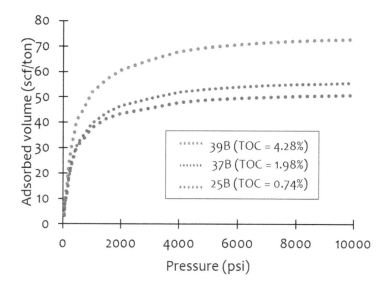

5. CONCLUSION

In the research areas, three zones rich in organic matter were obtained: 107,449 scf/ton, 106,104 scf/ton and 170,951 scf/ton, with a percentage of adsorbed gas of 45.4%, 50.00% and 40.6% respectively, which demonstrated the potential of gas in the Muerto Formation.

The adsorption results that were observed in the generated isotherms showed a direct relationship between the total organic content (TOC) and the adsorption capacity. Based on the petrophysical results, it could be inferred that the higher the TOC's capacity to generate hydrocarbons, the greater the porosity and surface area, where gas particles adhere to the latter, causing a higher content of adsorbed gas as shale rocks rich in organic matter mature.

On the other hand, the laboratory geochemical results showed a complicated classification of the type of kerogen, however, there is a possible difference that was evidenced in the maximum adsorption capacity according to the Langmuir isotherm, due to a higher content of aromatic molecules of the kerogen. product of greater maturity.

ACKNOWLEDGMENT

To the "Applied Research and Technological Development Project 2018-I" E041-2018-01-BM, to the "World Bank", to the National Council for Science, Technology and Technological Innovation (CONCYTEC) and the National Fund for Scientific, Technological and of Technological Innovation (FONDECYT), as the financing entity of the Sub-project: 61096 and contract 139 – 2018 UNI/FONDECYT that I was part of.

A special thanks to the Mewbourne School of Petroleum & Geological Engineering, The University of Oklahoma for the support provided in advising the high-pressure methane laboratories.

Also, a special thanks to the Soil Mechanics Laboratory (LBM-UNI) of the Faculty of Civil Engineering of the National University of Engineering for the support with the manual core-draw tool.

REFERENCES

Ahmed, U., & Meehan, D. N. (2016). *Unconventional Oil and Gas Resources Explotation and Development*. T. & F. Group.

Ambrose, R. J., Hartman, R. C., Diaz-Campos, M., Akkutlu, I. Y., & Sondergeld, C. H. (2012). Shale gas-in-place calculations Part I: New pore-scale considerations. *SPE Journal*, 17(1), 219–229. 10.2118/131772-PA

Andamayo, K. (2008). *Nuevo estilo estructural y probables sistemas petroleros de la cuenca Lancones*. The National University of San Marcos.

Belyadi, H., Fathi, E., & Belyadi, F. (2017). *Hydraulic Fracturing in Unconventional Reservoirs Theories, Operations, and Economic Analysis*. Elsevier Inc.

Bustin, R. M., Bustin, A. M. M., Cui, A., Ross, D., & Pathi, V. M. (2008). *Impact of Shale Properties on Pore Structure and Storage Characteristics*. All Days. 10.2118/119892-MS

Chalmers, G. R., Bustin, R. M., & Power, I. M. (2012). Characterization of gas shale pore systems by porosimetry, pycnometry, surface area, and field emission scanning electron microscopy/transmission electron microscopy image analyses: Examples from the Barnett, Woodford, Haynesville, Marcellus, and Doig uni. *AAPG Bulletin*, 96(6), 1099–1119. 10.1306/10171111052

Dunham, R. J. (1962). Classification of Carbonate Rocks According to Depositional Textures. *AAPG Bulletin,* 108–121.

Eberli, G. P., Weger, R. J., Tenaglia, M., Rueda, L., Rodriguez, L., Zeller, M., McNeill, D. F., Murray, S., & Swart, P. K. (2017). The unconventional play in the Neuquén Basin, Argentina - Insights from the outcrop for the subsurface. *SPE/AAPG/SEG Unconventional Resources Technology Conference 2017*, 1–12. 10.15530/urtec-2017-2687581

Fernández, J., Martínez, E., Calderón, Y., Hermoza, W., & Galdos, C. (2005). *Tumbes and Talara Basin; Hydrocarbon Evaluation PERUPETRO S.A*. Academic Press.

Glorioso, J. C., & Rattia, A. J. (2012). Unconventional Reservoirs: Basic Petrophysical Concepts for Shale Gas. In *SPE/EAGE European Unconventional Resources Conference and Exhibition* (p. 38). Society of Petroleum Engineers. 10.2118/153004-MS

Holditch, S. A. (2006). Tight Gas Sands. *Journal of Petroleum Technology*, 58(06), 86–93. 10.2118/103356-JPT

Lakatos, I. J., & Lakatos-Szabo, J. (2009). *Role of Conventional and Unconventional Hydrocarbons in the 21st Century: Comparison of Resources, Reserves, Recovery Factors and Technologies*. All Days.

Langmuir, I. (1916). The constitution and fundamental properties of solids and liquids. part i. Solids. *Journal of the American Chemical Society*, 38(11), 2221–2295. 10.1021/ja02268a002

Langmuir, I. (1917). The constitution and fundamental properties of solids and liquids. ii. liquids. *Journal of the American Chemical Society*, 39(9), 1848–1906. 10.1021/ja02254a006

Liborius-Parada, A., & Slatt, R. M. (2016). Geological characterization of La Luna formation as an unconventional resource in Lago de Maracaibo Basin, Venezuela. *SPE/AAPG/SEG Unconventional Resources Technology Conference 2016*, 3280–3299. 10.15530/urtec-2016-2461968

Masters, J. A. (1979). Deep Basin Gas Trap, Western Canada1. *AAPG Bulletin*, 63(2), 152–181.

Mengal, S. A., & Wattenbarger, R. A. (2011). Accounting for adsorbed gas in Shale gas reservoirs. *SPE Middle East Oil and Gas Show and Conference, MEOS. Proceedings*, 1(September), 643–657.

Mershon, S., & Palucka, T. (2013). *A Century of Innovation: From the U.S. Bureau of Mines to the National Energy Technology Laboratory*. Academic Press.

Ministerio de Energía y Minas. (2019). *Dirección General de Hidrocaburos*. Libro Anual de Recursos de Hidrocarburos. In Publicaciones Hidrocarburos.

Morales, W., Porlles, J., Rodriguez, J., Taipe, H., & Arguedas, A. (2018). First Unconventional Play From Peruvian Northwest: Muerto Formation. In *SPE/AAPG/SEG Unconventional Resources Technology Conference* (p. 14). Unconventional Resources Technology Conference.

Morales, W. J., Rodriguez-Cruzado, J. A., Alvarez-Mendoza, B. G., Marcatoma-Alarcón, A. A., Oré-Rodriguez, J. L., Corrales-Hidalgo, R. W., Madge-Rodriguez, J. J., Porlles-Hurtado, J. W., Falla-Ruiz, J. L., & Eyzaguirre-Gorvenia, L. de F. (2020). Geochemical heterogeneities characterization of the Muerto Formation - Lancones Basin - Peru as a source rock unconventional reservoir, a contribution for its development. *SPE Latin American and Caribbean Petroleum Engineering Conference Proceedings*, 1–15. 10.2118/199088-MS

Murillo-Martínez, C. A., Gómez-Rodríguez, O. A., Ortiz-Cancino, O. P., & Muñoz-navarro, S. F. (2015). Aplicación De Modelos Para La Generación De La Isoterma De Adsorción De Metano En Una Muestra De Shale Y Su Impacto En El Cálculo De Reservas. *Revista Fuentes El Reventón Energético*, 13(2), 131–140. 10.18273/revfue.v13n2-2015012

Porlles, J., Panja, P., Sorkhabi, R., & McLennan, J. (2021). Integrated porosity methods for estimation of gas-in-place in the Muerto Formation of Northwestern Peru. *Journal of Petroleum Science Engineering*, 202(January), 108558. 10.1016/j.petrol.2021.108558

Potter, P. E., Maynard, J. B., & Pryor, W. A. (1981). *Sedimentology of gas-bearing Devonian shales of the Appalachian Basin*. Academic Press.

Rezaee, R. (2015). *Fundamentals of Gas Shale Reservoirs*. John Wiley & Sons, Inc. 10.1002/9781119039228

Sondergeld, C. H., Newsham, K. E., Comisky, J. T., Rice, M. C., & Rai, C. S. (2010). *Petrophysical Considerations in Evaluating and Producing Shale Gas Resources*. All Days.

Tinni, A., Sondergeld, C., & Rai, C. (2017). *New Perspectives on the Effects of Gas Adsorption on Storage and Production of Natural Gas From Shale Formations*. Academic Press.

Tissot, B. P., & Welte, D. H. (1978). Sedimentary Processes and the Accumulation of Organic Matter. *Petroleum Formation and Occurrence*. 10.1007/978-3-642-96446-6_5

U.S. Energy Information Administration EIA/AIR. (2015). *World Shale Resource Assessments*. Author.

Wang, F. P., & Reed, R. M. (2009). Pore Networks and Fluid Flow in Gas Shales. In *SPE Annual Technical Conference and Exhibition* (p. 8). Society of Petroleum Engineers. 10.2118/124253-MS

Zhang, T., Ellis, G. S., Ruppel, S. C., Milliken, K., & Yang, R. (2012). Effect of organic-matter type and thermal maturity on methane adsorption in shale-gas systems. *Organic Geochemistry*, 47, 120–131. 10.1016/j.orggeochem.2012.03.012

Zhou, S., Ning, Y., Wang, H., Liu, H., & Xue, H. (2018). Investigation of methane adsorption mechanism on longmaxi shale by combining the micropore filling and monolayer coverage theories. *Advances in Geo-Energy Research*, 2(3), 269–281. 10.26804/ager.2018.03.05

Chapter 4
Characterization of the Structure of Mesopores in the Generating Rock of Northeastern Peru:
Muerto Formation

Kevin Chipana-Suasnabar
National University of Engineering of Peru, Peru

Heraud Taipe Acuna
National University of Engineering of Peru, Peru

Joseph Sinchitullo Gomez
http://orcid.org/0000-0003-3191-9055
National University of Engineering of Peru, Peru

Ali Tinni
University of Oklahoma, USA

Israel Chavez Sumarriva
National University of Engineering of Peru, Peru

Marco Tejada Silva
National University of Engineering of Peru, Peru

ABSTRACT

The contribution of organic matter to the total porosity in shale formations is mainly adsorbed in micropores and mesopores, which have a strong influence on the productive life of the reservoir, prolonging the productive life of the reservoir. An incorrect understanding of the total porosity and factors that influence it could lead to the rejection of potential productive zones during the initial stage of exploration. This research aims to characterize pores from 2 to 50 nm by low pressure N2 adsorption in conjunction with geochemical and mineralogical analysis of 5

DOI: 10.4018/979-8-3693-0740-3.ch004

samples from the Muerto Formation, acquired from the Lancones basin. The results indicated that the specific surface area is in the range of 0.58 to 3. 41 m2/ g values according to the oil and condensate generation window. In the pore size distribution, the presence of micropores was not observed; for the mesopores their distribution consists of two main ranges from 4 to 10 nm and from 20 to 50 nm, the porosities found were from 1.8 to 8.2%, being the average contribution of the mesopores to the total porosity is 35%.

1. INTRODUCTION

Shale is the term that has been used to describe a variety of fine-grained rocks, ie diameter less than 4 microns.

Research has been carried out in Shale formations to determine possible hydrocarbon reservoirs (Gao S., et al., 2021) (Solarin S. et al., 2020), the results show the positive impact of shales on hydrocarbon reserves, China is the country with the largest amount of technically recoverable resources (RTR) with 1115.2 tcf. (Energy Information Administration EIA, 2015).

The study of shale formations implies knowing their storage capacity, flow, and productivity, according to Mastalerz et al. (Mastalerz M. et al., 2013) are directly related to its pore structure. The types of pores that make up the shale structure can be classified in relation to their rock matrix as: (i) organic pores within the kerogen (ii) inter and intra pores in the minerals (Loucks R. et al., 2012.). Knowing the influence of these pore types on the total porosity is important due to their storage capacity as free gas, absorbed gas and, to a lesser extent, dissolved gas. (Curtis M. et al., 2012.). The adsorbed gas is mainly attributed to the nanopores in the kerogen and minerals due to their larger surface area favorable for adsorption. The free gas is attributed to pores in the matrix and microfractures. (Chalmers G. & Bustin R., 2008.), (Ross D. & Bustin R, 2007.). The development of nanopores in kerogen is frequently related to the level of maturity reached by the rock. (Loucks R. et al., 2009.). However, it has been documented that organic pores do not develop uniformly in shale formations with the same level of maturity; for RO> 0.9 it was found that the pores are poorly developed or there is no presence of these (Curtis M. et al., 2012.). This observation suggests that thermal maturity is an insufficient parameter to predict porosity development in kerogen, other factors such as organic matter composition complicate porosity development.

Currently, the methods of qualitative study in the pore structure are carried out using scanning electron microscope (SEM) and Broad/Focused Ion Beam (B/FIB) techniques, which provide 2D-3D images at the nanometric scale. (Qiu Z. et al., 2021.), (Guo H. et al., 2018.). While the quantitative determination of pore volume and pore size distribution in shale is mainly determined by low pressure N2 and CO2 gas adsorption techniques and Mercury Injection Capillary Pressures (MCIP) (Mastalerz M. et al., 2013.), (Ross D. & Marc Bustin R., 2009.). The choice of the most appropriate quantitative method is based on the study range, CO2 adsorption is used to characterize pores smaller than 2 nm, N2 adsorption for the study of pores from 2 to 50 nm and mercury intrusion capillary pressure (MICP) is used for pores larger than 50 nm. (Bustin R. et al., 2008.). For the present study, low pressure N2 adsorption was used to characterize the sample from 2 to 50 nm according to the scope of the investigation.

The Muerto formation in the Lancones basin, located in the department of Piura, Peru, is known for its high content of organic matter and for having reached the hydrocarbon generation window. (Morales W. et al, 2018). However, the pore characteristics in this formation are not known. The objective of this study is to characterize the pore structure by means of the low-pressure gas adsorption method with N2 and to know which factors influenced the development of micro and nano porosity with greater importance. Our results will provide important insights for decision-making for the next phases of exploration and development of the Muerto Formation in the Lancones basin.

2. METHODOLOGY

2.1 Geological Framework and Samples

The Lancones basin is in northwestern Peru, it is of the Forearc type where the Cretaceous formations are beveled towards the northwest on the Amotape Massif, the Lancones basin has been divided into two parts, defined by the observable deformation styles, and bounded by the Huaypirá normal displacement fault, the same one that separates the Cretaceous outcrops from the Cenozoic on the surface. (Andamayo K., 2008.). These two zones are known as the Lancones Folded Belt and the Cenozoic Coverage Zone.

The Muerto Formation belongs to the Albian age, it is made up of dark gray limestone in laminated and flimsy strata, interspersed with greenish gray calcareous sandstone in tabular strata with the presence of calcareous pads. (Morales W. et al, 2018). It presents an average total organic carbon (TOC) of 2.33%, which indicates its organic richness. The number of samples acquired from the Muerto Formation in outcrop was 26, the samples were systematically acquired from the base to the top, separated 5 meters vertically.

2.2 Geochemical Parameters

Organic geochemistry experiments were performed by the CoreLab laboratory, Houston. A total of 26 samples were used for the Total Organic Carbon (TOC) tests. The Rock-Eval Analysis (Tissot B. & Welte D., 1978.) was carried out using the Leco LC-632 carbon analyzer and the OCE-II pyrolizer. The results of hydrocarbon volatilization temperature (Tmax) were used to obtain maturity values.

2.3 Mineralogy

The tests were carried out by the LCV Laboratory, Argentina. X-ray diffraction analyzes were performed on 31 samples from the Muerto formation. Measurements of the diffraction of the incident radiation beam on the shale samples were recorded, and the Bruker D8 DISCOVER diffractometer was used. (BRAGG2D NWM-1, Bruker D8 Discover, Bruker Corporation, 2015) and for the processing of the diffractograms I use the CPSC procedure.

2.4 Total Porosity and N2 Adsorption at Low Pressure

Total porosity was calculated by integrating 2 techniques: (i) Helium Posimeter and (ii) Nuclear Magnetic Resonance (NMR). NMR measured residual liquid volume, while helium porosity characterized the residual gas-filled pore space. NMR measurements were performed with a 12 MHz Oxford Geospec analyzer, while helium porosity was performed with an Accupyc II 134.

Nitrogen adsorption analysis was performed with the TriStar II analyzer. Which was used to obtain the pore size distribution in the range of 2 to 50 nm and the specific surface area of the samples was calculated using the Brunauer-Emmet Teller (BET) method. (Brunauer S. et al., 1938.). The pore size and pore volume distribution were obtained from the sorption curves for a pore size range of 2-50 nm using the density functional theory (DFT) method. (Webb P. et al., 1997.).

3. ANALYSIS OF RESULTS

3.1 TOC Content and Pyrolysis Parameters

The organic content and pyrolysis parameters of the samples analyzed from the Muerto formation are shown in Table 1.

Table 1 shows that the TOC content of the samples is in the range of 0.74-4.32% with an average value of 2.59%, and the Tmax values are between 440 and 461 °C, indicating that the rock entered the window of hydrocarbon generation.

Table 1. Geochemical parameter results

Sample	OCD	Tmax	S1	S2
WLUC-46B	3	443	0.15	2.05
WLUC-39B	4.28	440	0.43	3.16
WLUC-29B	4.32	457	0.86	5.54
WLUC-25B	0.74	445	0.02	0.25
WLUC-7B	1.26	461	0.34	1.66

3.2 Mineralogical Composition

The mineralogy of 31 samples was determined using X-Ray Diffraction (XRD). The compositional triangle of the samples from the Muerto formation is presented in Figure 1. Quartz, carbonate are the main constituents and clays to a lesser extent with a range of 0% to 24%; the quartz content is between 5% to 80%; Calcite is the main carbonate mineral ranging from 2% to 93%. Five samples were taken for pore characterization by low pressure gas adsorption with N2 and their compositions are detailed in Figure 2.

Figure 1. Sample compositional triangle of the Quebrada Corcobado in the Lancones Basin

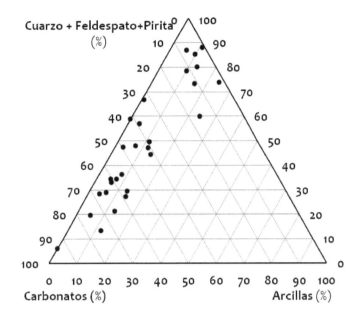

Figure 2. Mineralogical composition of samples from the Quebrada Corcobado in the Lancones basin

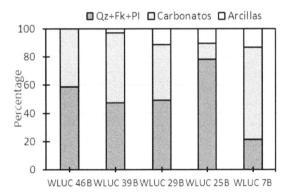

3.3 Specific Surface Area and Porosity

Nitrogen adsorption measurements were used to evaluate the pore structure in the range of 2 to 50 nm. It is observed in Table 2 that the surface area is in the range of 0.58 to 3.41 m2/g, with an average of 1.94 m2/g. In theFig.3surface area ranges of hydrocarbon-producing shale formations are shown (Sharadkumar A., 2017.), based on Tmax Table I and specific surface area and mineralogical composition, mainly carbonates, a certain comparison could be established with the Eagle Ford formation 1.

The pore volume is in the range of 0.248 to 0.548 cm3/100g with an average of 0.43 cm3/100g. The total porosity measurements of the samples from the Muerto formation are between 1.8 to 8.2%. In Table 3, total porosity and pore volume values are shown for the 5 samples analyzed in this study.

Table 2. BET surface area, DFT pore volume, pore diameter for samples from the Muerto formation

Sample	Superficial Area (m2/g)	Mesopore Volume (cm3/100g)	%	Volume Macropore (cm3/100g)	%
WLUC-46B	1.79	0.42	35	0.77	65
WLUC-39B	3.41	0.53	fifteen	2.94	75
WLUC-29B	2.68	0.55	76	0.17	24
WLUC-25B	1.25	0.46	23	1.51	77
WLUC-7B	0.58	0.25	30	0.59	70

Table 3. Total porosity and pore volume for samples from the Muerto formation, Lancones Basin

Sample	Total Porosity (%)	Pore Volume (cm3/100g)
WLUC-46B	2.9	1.19
WLUC-39B	8.2	3.47
WLUC-29B	1.8	0.72
WLUC-25B	4.9	1.97
WLUC-7B	2.2	0.84

Figure 3. BET surface area ranges for shale formations[twenty-one]

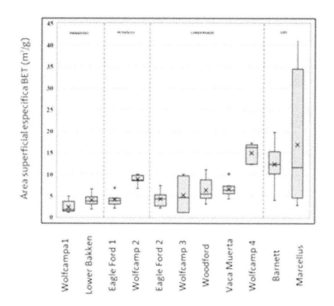

3.4 Maturity Level of Organic Matter

Maturity is the most important factor controlling the development of organic pores in shale, with increasing maturity level the number of nanopores increases and therefore the specific surface area also increases. (Chalmers G. & Bustin R., 2008.). The Tmax obtained from the Rock eval were used to determine the level of maturity.

For the analyzed samples the surface area values specify.

Table 2 shows that the organic pores have not fully developed, the lack of micropore structure in the Muerto formation is attributed to its level of thermal maturity and type of organic matter. (Curtis M. et al., 2012.), for the samples analyzed 443-461 °C and standard organic matter.

3.5 Adsorption Isotherm and Pore Structure

Isotherms of with nitrogen gas (N2) at low pressure and temperature (-196 °C <127kpa) are shown in *Figure 4*. According to the IUPAC classification of isotherms (Rouquerol J. et al., 1994.), the samples are classified as type IV and hysteresis type H3. These types of isotherms are generally associated with the presence of fractured pores in the shale, indicating the presence of slit-shaped pores. (Sing K. et al., 1985.). The amount adsorbed for p/p0 <0.01 is indicative of the presence of nanopores smaller than 2 nm (Kuila U. & Prasad M., 2013.). The samples analyzed show very little amount adsorbed for p/p0 <0.01, which would indicate the absence of nanopores at 2nm.

Samples WLUC-29B, 39B, 46B show a hysteresis loop for p/po of 0.45-0.50, while this loop is absent in samples WLUC 7-B and 25B. This phenomenon is attributed to the instability of the meniscus during capillary evaporation in pores smaller than 4nm in diameter. (Groen J. et al., 2003.). Therefore, the presence of pores smaller than 4 nm is demonstrated in samples WLUC 29-B, 39-B and 46-B from the Muerto formation.

Figure 4. Adsorption and desorption isotherms of samples from the Muerto Formation (a) WLUC-7B (b) WLUC-25B (c) WLUC-29 B (d) WLUC-39B (e) WLUC 46B

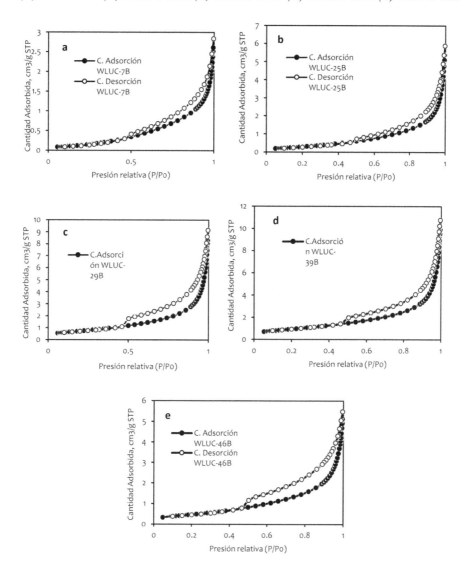

3.6 Porosity and Pore Size Distribution

To investigate the role of organic matter on the pore structure of samples from the Muerto formation, we chose five samples with different organic content. The pore size distribution was determined from the adsorption isotherms using the density functional theory (DFT) method. (Webb P. et al., 1997.).

The volumes of mesopores and macropores are summarized in Table II. It is observed that the volume of mesopores increases with TOC, except for sample WLUC 25B, this would indicate that the development of mesopores is related to the amount of organic matter. In Table I and Table III samples WLUC-29B and WLUC 39-B are observed with similar TOC, but with different porosities, both show a similar volume of micropores, with the volume of macropores being the main contributor.

The pore size distribution of the samples from the Muerto Formation shows different dominant ranges for our study of mesopores (2 to 50 nm), it can be observed in Figure 5. (a), ranges from 4 to 9 nm are observed, Figure 5(b) samples WLUC-29B and WLUC-39B are observed, which have a similar behavior with the presence of a dominant range between 20 to 50 nm, samples WLUC-46B and WLUC-25B in theFig.5(c) show a predominant range from 20 to 50 nm with a peak at 4 to 10 nm.

The presence of pores smaller than 10 nm can be attributed to the presence of organic pores within the kerogen and the presence of ranges greater than 50 nm is attributed to the presence of billable minerals such as quartz and calcite (Cao T. et al., 2015.).

Figure 5. Pore size distribution of samples from the Muerto Formation obtained by the DFT method[twenty]: (a) WLUC-7B, (b) WLUC 29 - WLUC 39B, (c) WLUC-46B - WLUC-25B

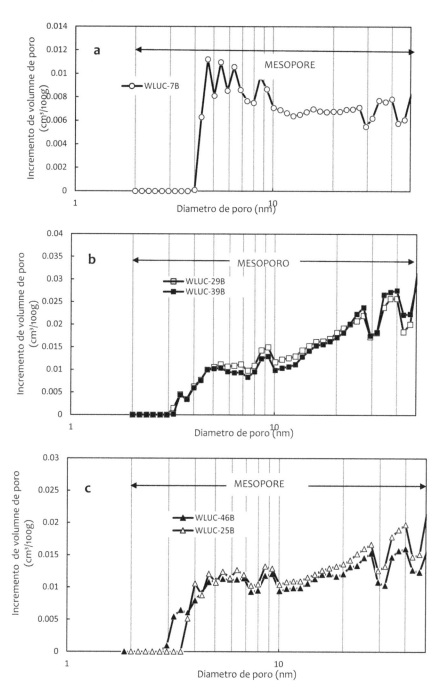

4. CONCLUSION

The use of low pressure N2 adsorption to characterize mesopores of the Muerto formation in combination with porosity measurements and their relationship with mineralogical and geochemical analyzes showed significant variations in their pore size distribution and specific surface area. Additionally, with the total porosity measurements, we were able to know the contributions to the pore volume of each pore size range to the total porosity of the sample.

The obtained isotherms were classified as type IV and with the presence of type H3 hysteresis, which are generally associated with the presence of fractured pores in the shale. In the pore size distribution obtained by DFT, no presence of micropores was observed, the volume of mesopores represents on average 35% of the total porosity of the rock.

The samples are characterized by a low amount of specific surface area which would confirm the level of oil and gas condensing generation found.

SPECIAL THANKS

To the "Applied Research and Technological Development Project 2018-I" E041-2018-01-BM, to the "World Bank", to the National Council for Science, Technology and Technological Innovation (CONCYTEC) and the National Fund for Scientific, Technological and of Technological Innovation (FONDECYT), as the financing entity of the Sub-project: 61096 and contract 139 – 2018 UNI/FONDECYT that I was part of.

A special thanks to the Mewbourne School of Petroleum & Geological Engineering, The University of Oklahoma for the support provided in advising the high-pressure methane laboratories.

Also, a special thanks to the Soil Mechanics Laboratory (LBM-UNI) of the Faculty of Civil Engineering of the National University of Engineering for the support with the manual core-draw tool.

REFERENCES

Brunauer, S., Emmett, P. H., & Teller, E. (1938). Adsorption of gases in multimolecular layers. *Journal of the American Chemical Society*, 60(2), 309–319. 10.1021/ja01269a023

Bustin, R. M., Bustin, A. M. M., Cui, A., Ross, D., & Pathi, V. M. (2008). Impact of Shale Properties on Pore Structure and Storage Characteristics. *SPE Shale Gas Production Conference*. 10.2118/119892-MS

Cao, T., Song, Z., Wang, S., Cao, X., Li, Y., & Xia, J. (2015). Characterizing the pore structure in the Silurian and Permian shales of the Sichuan Basin, China. *Marine and Petroleum Geology*, 61, 140–150. 10.1016/j.marpetgeo.2014.12.007

Chalmers, G. R. L., & Bustin, R. M. (2008, March). Lower Cretaceous gas shales in northeastern British Columbia, Part I: Geological controls on methane sorption capacity. *Bulletin of Canadian Petroleum Geology*, 56(1), 1–21. 10.2113/gscpgbull.56.1.1

Curtis, M. E., Cardott, B. J., Sondergeld, C. H., & Rai, C. S. (2012, December). Development of organic porosity in the Woodford Shale with increasing thermal maturity. *International Journal of Coal Geology*, 103, 26–31. 10.1016/j.coal.2012.08.004

Gao, S., Dong, D., Tao, K., Guo, W., Li, X., & Zhang, S. (2021). Experiences and lessons learned from China's shale gas development: 2005–2019. *Journal of Natural Gas Science and Engineering*, 85, 103648. 10.1016/j.jngse.2020.103648

Groen, J. C., Peffer, L. A. A., & Pérez-Ramírez, J. (2003). Pore size determination in modified micro- and mesoporous materials. Pitfalls and limitations in gas adsorption data analysis. *Microporous and Mesoporous Materials*, 60(1-3), 1–17. 10.1016/S1387-1811(03)00339-1

Guo, H., He, R., Jia, W., Peng, P., Lei, Y., Luo, X., Wang, X., Zhang, L., & Jiang, C. (2018, March). Pore characteristics of lacustrine shale within the oil window in the Upper Triassic Yanchang Formation, southeastern Ordos Basin, China. *Marine and Petroleum Geology*, 91, 279–296. 10.1016/j.marpetgeo.2018.01.013

Kuila, U., & Prasad, M. (2013, December). Application of nitrogen gas-adsorption technique for characterization of pore structure of mudrocks. *The Leading Edge*, 32(12), 1478–1485. 10.1190/tle32121478.1

Loucks, R., Reed, R., Ruppel, S., And, D., & Jarvie, D. (2009, November). Morphology, Genesis, and Distribution of Nanometer-Scale Pores in Siliceous Mudstones of the Mississippian Barnett Shale. *Journal of Sedimentary Research*, 79(12), 848–861. 10.2110/jsr.2009.092

Loucks, R. G., Reed, R. M., Ruppel, S. C., & Hammes, U. (2012, June). Spectrum of pore types and networks in mudrocks and a descriptive classification for matrix-related mudrock pores. *AAPG Bulletin*, 96(6), 1071–1098. 10.1306/08171111061

Mastalerz, M., Schimmelmann, A., Drobniak, A., & Chen, Y. (2013, October). Porosity of Devonian and Mississippian New Albany Shale across a maturation gradient: Insights from organic petrology, gas adsorption, and mercury intrusion. *AAPG Bulletin*, 97(10), 1621–1643. 10.1306/04011312194

Morales, W., Porlles, J., Rodriguez, J., Taipe, H., & Arguedas, A. (2018). First Unconventional Play From Peruvian Northwest: Muerto Formation. *SPE/AAPG/SEG Unconventional Resources Technology Conference*. 10.15530/urtec-2018-2903064

Passey, Q. R., Bohacs, K. M., Esch, W. L., Klimentidis, R., & Sinha, S. (2010). From Oil-Prone Source Rock to Gas-Producing Shale Reservoir – Geologic and Petrophysical Characterization of Unconventional Shale-Gas Reservoirs. *International Oil and Gas Conference and Exhibition in China*. 10.2118/131350-MS

Qiu, Z., Song, D., Zhang, L., Zhang, Q., Zhao, Q., Wang, Y., Liu, H., Liu, D., Li, S., & Li, X. (2021, November). The geochemical and pore characteristics of a typical marine–continentaltransitional gas shale: A case study of the Permian Shanxi Formation on the eastern margin of the Ordos Basin. *Energy Reports*, 7, 3726–3736. 10.1016/j.egyr.2021.06.056

Ross, D. J. K., & Bustin, R. M. (2007, March). Shale gas potential of the Lower Jurassic Gordondale Member, northeastern British Columbia, Canada. *Bulletin of Canadian Petroleum Geology*, 55(1), 51–75. 10.2113/gscpgbull.55.1.51

Ross, D. J. K., & Marc Bustin, R. (2009, June). The importance of shale composition and pore structure upon gas storage potential of shale gas reservoirs. *Marine and Petroleum Geology*, 26(6), 916–927. 10.1016/j.marpetgeo.2008.06.004

Rouquerol, J., Avnir, D., Fairbridge, C. W., Everett, D. H., Haynes, J. M., Pernicone, N., Ramsay, J. D. F., Sing, K. S. W., & Unger, K. K. (1994, January). Recommendations for the characterization of porous solids (Technical Report). *Pure and Applied Chemistry*, 66(8), 1739–1758. 10.1351/pac199466081739

Sharadkumar, A. (2017). *Surface area study in organic-rich shales using nitrogen adsorption*. University of Oklahoma.

Sing, K. S. W. (1985, January). Reporting Physisorption Data for Gas/Solid Systems with Special Reference to the Determination of Surface Area and Porosity. *Pure and Applied Chemistry*, 57(4), 603–619. 10.1351/pac198557040603

Solarin, Gil-Alana, & Lafuente. (2020). An investigation of long range reliance on shale oil and shale gas production in the US market. *Energy, 195*.

Chapter 5
The Shale of the Cabanillas Formation:
Integration Methodology and the Presence of the Gas Resource (Shale Gas) in the Marañón–Peru Basin

Walter Jacob Morales-Paetan
National University of Engineering of Peru, Peru

Pedro Zegarra Sanchez
National University of Engineering of Peru, Peru

ABSTRACT

Peru is economically dependent on hydrocarbons, so exploring gas shale in unconventional systems can be a new source of energy which has stimulated the interest of major oil companies and countries with greater energy dependence. The present study is based on the evaluation of the shales of the Cabanillas formation as a hydrocarbon generating rock and to be considered in the unconventional system and that encompasses the theoretical and conceptual framework, for which a collection and bibliographic review (digital data of exploratory well logs, seismic information, reports, and data analysis but not the state of the art) in the southern part of the Marañon Basin that exploits hydrocarbons with the objective of considering the shales as an unconventional reservoir was carried out.

DOI: 10.4018/979-8-3693-0740-3.ch005

The Shale of the Cabanillas Formation

1. INTRODUCTION

In the 1990s, George Mitchell made one of the most important discoveries that have been recorded in the energy industry in recent years. This American research is known in the world as the father of unconventional gas, a new source of energy that has aroused the interest of the main oil companies and countries with greater energy dependence (Cohen, 2016). When Mitchell began exploring unconventional gas in the United States, the costs were so high that the overwhelming results that are currently being obtained were not expected. The United States is the largest gas producer in the world, surpassing Russia in 2015, thanks to the strong investments made in recent years to extract unconventional gas. These investments have occurred largely because costs have been drastically reduced thanks to important technological advances. In the end, production costs are competitive by improving drilling and well stimulation techniques (Cohen, 2016).

Currently there are four types of unconventional gas, and the best known of all is shale gas, which is located between the layers of the sedimentary rock shale and shale. In addition, there is Tight gas (gas from compact reservoirs), stored in low permeability sandstones or compact sands and methane trapped in the coal beds CBM (Coal Bed Methane) (coal gas). The fourth option is methane hydrates, but experts rule it out due to its high costs.

The accelerated development of unconventional gas from shale has allowed a revolution in the international hydrocarbon market. Technological advances in recent years have allowed shale gas production to increase rapidly, causing a reduction in the price of natural gas worldwide National Society of Mining, Petroleum and Energy (SNMPE, 2012). As hydraulic fracturing technology has evolved, shale gas and oil production in the United States has increased considerably. To have an idea of this change in trend, its application increased crude oil production by 780,000 barrels/day in 2012, resulting in the greatest growth in its history (Morjandin, 2013), (Morales, 2013), and the tests Drilling for shale gas is underway in Europe. The geological setting is fundamentally different from that of the United States, complexity being the rule and not the exception, opportunities appearing abundant. In Poland it is mainly in rocks of Silurian age, in England, it is of Namurian and Wealden age, in France in rocks of the Jurassic, in Sweden of the Cambrian and in Germany in rocks of the Carboniferous, Jurassic and Wealden that are in focus. . Gash is the first major research initiative in Europe (Horsfield, Schulz, & Kapp, 2012).

According to the Energy Information Agency (EIA, 2013), the total global recoverable resources of conventional natural gas as of 2008 amount to 470.6 trillion cubic meters (TMC), of which 66.1 TMC have already been produced., and leaving estimated reserves of 404.5 TMC. At 2008 production, these resources give a reserve-production ratio of 136 years. According to the IEA's estimate, the global amount of recoverable unconventional gas, including Tight gas, CBM and shale gas, would reach 380 TMC, which would add another 120 years to the global reserve-production relationship. The estimates would be produced in the event that the behavior of the basins is similar to those of the United States and that the same technology is applied. It should be noted that of the total unconventional resources, shale gas is the most abundant of all with almost 50% of the total, according to the EIA (2013). In Latin America, Mexico, Colombia and Peru are also potential sources of shale gas (Honty, 2018).

Shale gas is an unconventional fossil fuel, meaning that additional procedures are required to extract beyond regular drilling. Many of these unconventional sources of oil and gas were previously too difficult (or unprofitable) to extract until recent advances in drilling and technology. A combination of directional drilling and the hydraulic fracturing process have produced large quantities of accessible natural gas locked in the tight pores of shale formations at depths of 2 km or more. Recent successes in the United States have boosted prospecting throughout Europe. In 2010, Cuadrilla Resources Holdings Limited began drilling near Blackpool in the Bowland shale (extending from Preston to the Irish Sea) (Honty, 2018).

Current estimates from the British Geological Society suggest that the UK's gas resources are equivalent to approximately 1.5 years of current gas consumption or 15 years of imported liquefied natural gas (LNG) from the UK. More recent figures from the US Energy Information Administration (EIA, 2013) estimate that the UK technically has recoverable gas shale resources equivalent to 5.6 years of consumption penalty or 56 years worth of imports. LNG. The EIA report (2013) estimates that shale gas adds 40% to the world's technically recoverable natural gas resources, particularly in China and the United States.

Accelerated gas development in the unconventional system from shale is being exploited commercially in the US, Canada and Argentina. Of special relevance is the case of the United States, which has achieved energy independence in terms of oil and gas, to the point that it is currently an exporter of li gas.

Liquid Technological advancement is generating interest in the international hydrocarbon market. Technological advances such as horizontal drilling and hydraulic fracturing fracturing or fracking) in recent years have allowed for a drastic reduction in the cost of drilling and completing unconventional shale wells. Oil and shale Gas. The only disadvantage is the high percentage of production decline per well. This forces new wells to be drilled continuously.

Gas shales have been little studied in Peru. Its exploration and development, if economical, would open another source of energy (gas) in which oil imports would be eliminated or reduced. The gas shale boom has completely altered the energy future in many countries around the world.

The Shale of the Cabanillas Formation

It is widely recognized that petroleum source rocks can have significant spatial variability in lithology, and the quality, quantity and thermal maturity of organic matter, which impacts resource potential. However, most studies focus on only a few selected rock samples that may not be representative of the entire source rock interval. In practice, most source rock analyzes use outcrop cuttings and/or samples, thus creating interpretive traps. Successful exploration of shale oil and shale gas shifts the focus of research from source rock and raises the importance of recognizing geochemical and lithological heterogeneity. Furthermore, this sparked scientific interest to identify new unconventional hydrocarbon facies within the source rock and to better understand the distributions of their reservoir and source rock properties for more accurate resource assessment (Jarvie, 2012; Schneider et al. ., 2013). Source rock cores are now an essential part of the precise unconventional shale resource exploration procedure (Jarvie, 2012; Schneider et al., 2013).

1.1. Study Area

The study area is located in the southern part of the Marañón basin as shown in figure 1 and figure 2, which includes the Sub-Andean morphostructural zone and the Amazon Plain where the meanders of the Ucayali and Huallaga rivers and their tributaries.

Figure 1. Location map of the research area

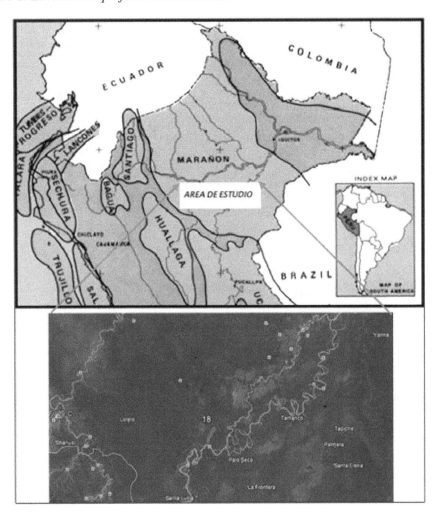

Figure 2. The wells studied in the southern part of the Marañón basin

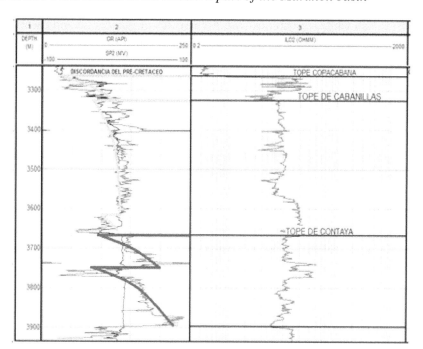

2. GEOLOGICAL FRAMEWORK

The Marañón basin is one of the most geologically studied in the exploration and exploitation of hydrocarbons as non-renewable energy resources in Peru. Currently in exploration and exploitation of heavy and extra-heavy crude oil, it is considered a great potential generator of hydrocarbons, with formations ranging in age from the Paleozoic to the Cenozoic as source rocks.

The Basement forms the base of this part of the Marañón basin, being identified in the Yarina well, which is made up of a gray to pale pink, dense granite, containing quartz, feldspar, mica and hornblende of Pre-Cambrian age.

Contaya Formation that overlies the Basement is made up at its base of white to red sandstones of medium to coarse grain, conglomeratic and partially with silicon cement and layers of dark gray shale of marine origin, having its greatest thickness to the east of the area between the Tapiche well and Yarina, and is attributed to Ordovician age

Cabanillas Formation have been reported in the Palmera, Tapiche and Yarina well that lithologically are dark gray to black, micaceous, fissile and fine sandstones, grayish white, calcareous with accessories of glauconite and carbonaceous matter. This interpretation served to identify and correlate the seismic lines with the Seismic-Stratigraphy method of the ARC-3 seismic line as shown in figure 4, which sedimentologically has been deposited during the marine transgression of the Devonian and platform environment according to the studies carried out on surface outcrops.

The Copacabana Formation was identified in the Tamanco well, made up of white, slightly gray, hard, very pyritic limestone, partly marly and clayey, occasionally with intercalations of clean white fine-grained, massive, fossiliferous and glauconitic sandstones. Deposited in a marine platform environment with intercalation of thin layers of shale and sandstone whose age is from the Lower Permian-Carboniferous, ending with continental sediments product of the Late-Hercynian tectonic phase, which can be considered as the regression of the sea from Copacabana time to give lead to sedimentation of the Ene formation in some places in the area.

Mitu Formation was identified seismically as filling the depressions produced by the Late-Hercynian Tectonics on the Copacabana Formation. This formation is considered of Triassic age and continental environment and is made up of slightly grayish white, orange-red, very fine-grained, rounded to sub-rounded, quartzite, very hard to friable, argillaceous, rarely calcareous sandstones.

Pucará Formation was found in the Loreto well and is made up of light gray to cream dolomites, very fine to fine grain, graded calcarenite and pinkish brown siltstone and pinkish brown claystones, it also presents limestones of the type of sparite, micrite, osparite with white sandstones, pinkish, considered to be from a Sabka environment, it represents tidal flats controlled by the wind, consisting of limestone, evaporites, and reddish gray shales in the western part of the area whose age is from the early Jurassic, which differs in lithology from that described in the area. Central Peru.

Sarayaquillo Formation consists of light brick red, orange red sandstones, fine to medium grained, occasionally coarse grained, mostly well rounded, frequent intercalations of red siltstones and claystones, conglomerates and volcanic tufos that are of Middle and Upper Jurassic age. This formation overlies the Pucara formation in the western part of the study area.

The Cretaceous is made up of a mainly silicon-clastic sequence, such as the fluvio-deltaic sands of the Cushabatay formation that rests indistinctly on the discordance with the Pre-Cretaceous formations.

The Shale of the Cabanillas Formation

2.1. Stratigraphic Column

The stratigraphic column of the area, as seen in figure 3. It was prepared based on the analysis of the exploratory wells with the drilling report and then its stratigraphic-structural correlations were made with the existing wells in the area and supported with the interpretation of the seismic lines that allowed us to recognize the tectonic events in this part of the basin. Greater emphasis was placed on Paleozoic formations.

Figure 3. Generalized stratigraphic column of the study area. Source: Own elaboration of the formations

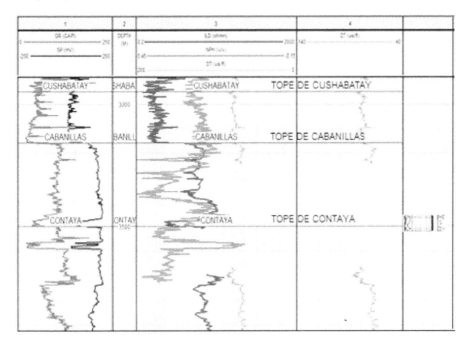

As can be seen in the stratigraphic column that at the end of the Devonian another tectonic phase called Neo-Hercynian that regionally affected the basin, in some places the efforts were of distension forming graben and horst structures, as well as in others slight foldings occurred in the Lower Paleozoic sediments (B, 1986) that were associated with a type of rift as shown in the seismic-geological interpretation in (Figure 5).

From the stratigraphic point of view, this tectonics is translated by the absence of the Upper Devonian and Carboniferous on folded terrains of the Lower Paleozoic in this area, to then continue with the sedimentation of the carbonates of the

Copacabana Formation of Lower Permian age. As can be seen in (Figure 3), they consist mainly of open sea platform limestone, from a shallow marine environment and moderately far from the coast.

Then the Late-Hercynian Tectonic phase of Compressional or epirogenic stresses took place that shaped the basin for the sedimentation of the Mitú Formation made up of red molasses that rest indistinctly on the limestones of the Copacabana Formation; In the study area, the folded Copacabana Formation can be seen (Figure 5) and the depressions filled by sediments of the Mitú formation according to the seismic-stratigraphic interpretation.

During the middle Permian, a folding phase affected southern Peru (Laubacher, 1970), which is also seen in the seismic interpretation of the study area, as can be seen in (Figure 5) in the central part. In Peru, this phase translates only into epurogenic movements accompanied by important volcanic phases. The denudation of these reliefs, the formation of important accumulations of red molasses is deposited throughout the Upper Permian and possibly reaching the Lower Triassic (B, 1986) is the Mitu formation.

The presence during the Upper Permian of an important volcanism of acidic composition associated with spilitic flows (E & N., 1974) is interpreted as the index of the functioning of an intercontinental "rifting" zone.

The study area was affected by Tectonic phases that influenced the different formations during their sedimentation, erosion, non-deposition and deformation. Some formations were not crossed by the exploratory wells in the area. The main tectonic events were Taconic, Neo-Hercynian, Late-Hercynian, Nevadiana, Mochica and Peruvian.

Based on the stratigraphic-structural correlation from West to East, as seen in (Figure 4), in the area from the Shanusi well to the Yarina well, the Paleozoic, Mesozoic and Cenozoic formations are shown.

Figure 4 shows the stratigraphic section carried out between the exploratory wells in the area, from the Shanusi well to Yarina, and the formations found and identified during drilling were correlated, which were from the Paleozoic, Mesozoic and Cenozoic, approximately the distribution is shown on the west side. of the Jurassic formations and the Cretaceous and Tertiary formations, while to the east, Paleozoic formations are observed, beginning with the Basamento and the Contaya, Cabanillas, Copacabana and Ene formations.

Figure 4. Stratigraphic correlation of the exploratory wells from Shanusi to Yarina. Source: Own elaboration

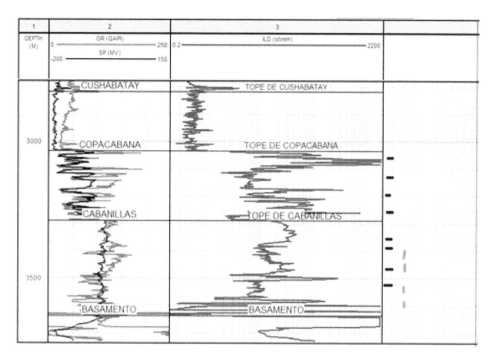

Figure 5. The interpreted and reduced ARC-3 seismic line showing the Paleozoic, Cretaceous and Tertiary seismic sequences

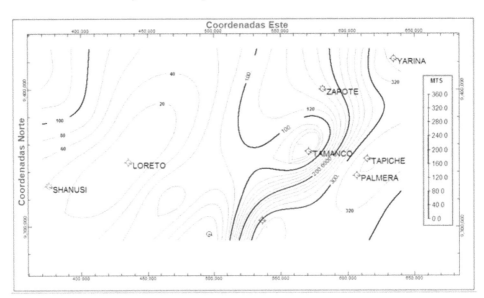

Figure 6 was prepared based on the interpretation of the ARC-3 seismic line, using the seismic stratigraphy method and its correlation with the electrical logs of the Tamanco and Tapiche wells, which allowed identifying and correlating the seismic sequences of the Formations, Cabanillas, Copacabana, Ene and Mitú of Paleozoic age; On the west side of the section we can see the wedging of the Pucara Formation of Jurassic age and its possible sedimentation limit; It also allows us to observe the Pre-Cretaceous unconformity; where it allowed us to group the Cretaceous formations in green and those from the Tertiary in yellow and the records from the Tamanco and Tapiche wells that allowed us to identify the formations in the subsurface.

The Shale of the Cabanillas Formation

Figure 6. Stratigraphic correlation of the exploratory wells from Shanusi to Yarina based on the ARC-3 seismic line. Source: Own elaboration

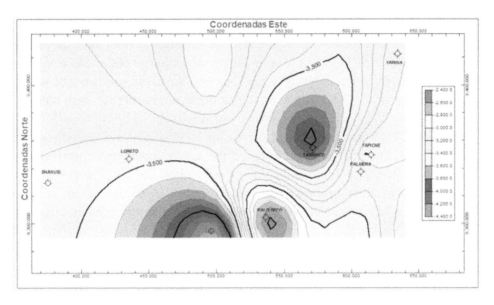

3. METHODOLOGY

This research was developed in two main phases:

The first phase involved the collection, analysis and bibliographic review of all the existing information about the topic to be investigated, for which the information was requested from the Data Bank department of the company PeruPetro S.A. Once the information was obtained, the stratigraphy and regional geology of the Marañon basin were analyzed based on the geology and petroleum engineering files; as well as all the information from the exploratory wells carried out in the study area, with their respective lithological, paleontological, electrical records, laboratory analysis and field tests to then carry out stratigraphic and structural correlations to delimit the study area.

The second phase as the area has 2D seismic lines which allowed an Interpretation of the lines using the stratigraphic seismic method to know the subsoil formations and with the electrical logs of the exploratory wells that allowed the determination of the sequences. seismic measurements of the formations in the study area, the tectonic-structural evolution was also carried out to determine its distribution and extension of the formation.

3.1. Seismic-Stratigraphic Interpretation and Tectonic Evolution

Seismic interpretation of unconventional resources has very different objectives from those of interpretation for conventional resources, although they have many points in common. The characterization of the unconventional reservoir is the greatest interest pursued as an objective to optimize the development of the unconventional reservoir. This interpretation was carried out with the 2D seismic lines of seismic reflection, the ARC-3 line with a West-East orientation being the most complete and covers the study area, these seismic lines were acquired by Western Geophysical for the Compañía Arco - Perú Corporation in 1975, and its location is shown in (Figure 6). The interpretation followed the seismo-stratigraphic methodology where the sequences are limited by the unconformities and the seismic reflectors indicate units of strata, geometry and unconformities or unconformities, or strata. Observation is key to seismic reflectors generally representing timelines based on these principles, seismic sequences from the Tertiary, Cretaceous and Paleozoic were determined, the latter being developed in more detail in order to carry out the tectonic evolution of the area and the sequences were determined. of the Cabanillas, Copacabana, Ene, Mitú, Pucara formation and the pre-Cretaceous unconformity on which the entire Cretaceous sequence and the Tertiary sequence rest without the assistance of commercial software for interpretation.

The stratigraphic-structural correlation between the Shanusi, Loreto, Tamanco, Palmera, Tapiche and Yarina wells that allowed determining the distribution, thickness and limits of the Paleozoic, Cretaceous and Tertiary formations as shown in figure 4.

Thus, the seismic sequences were grouped, identifying reflectors of similar frequencies along the seismic profile. Which allowed establishing the seismic sequences of the Cabanillas, Copacabana, Ene, Mitú Formations of the Paleozoic and the Cretaceous and Tertiary formations, the respective seismic sequences formed by reflectors both at the base and at the top of said formations; while for the Contaya Formation the base reflectors could not be determined, so the seismic sequence of said Formation could not be defined. For the Pucara formation, it was identified to the west of the area in the Loreto and Shanusi well, which allowed it to be seismically identified, as seen in Figure 5.

The seismic interpretation was integrated with the stratigraphic-structural correlation of the wells in the area as seen in (Figure 4), to determine the distribution, extension, thickness and limits of the formations and the Paleozoic seismic sequences, especially the Formation Cabanillas as seen in (Figure 5) and (Figure 6) which was used to carry out the Tectonic evolution and the unconventional petroleum system of the study area.

The Shale of the Cabanillas Formation

Figure 7 shows the horizontality carried out on the top of the Cabanillas Formation, showing that at the end of its sedimentation the Neo-Hercynian Tectonic phase acted with distension efforts generating normal faults that gave rise to graben and horst structures; as well as slight folding. In this part of the basin, according to regional field studies, a Rift type developed that conditioned an erosive phase, which is why the thickness of the Cabanillas formation is not homogeneous.

Figure 7. Interpretation of the ARC-3 seismic line correlation of horizontality of the Cabanillas Formation. Source: Own elaboration.

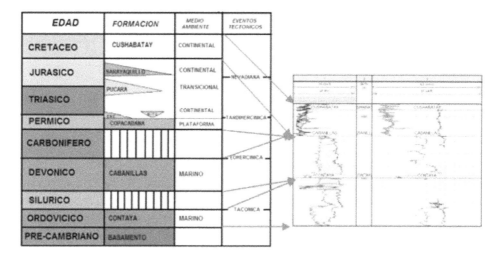

Figure 8 shows the horizontality carried out on the top of the Copacabana Formation, showing that after the Neo-Hercynian tectonics, carbonate sedimentation continued the eroded and irregular Top of the Cabanillas Formation, hence the heterogeneous thickness of the Formation Copacabana. At the end of this sedimentation, the Late-Hercynian tectonics occurred, which gave rise to the regression of the Copacabana Sea to deposit the Ene Formation and the folding of the Copacabana Formation.

Figure 8. Interpretation of the ARC-3 seismic line correlation of horizontality of the Copacabana and Cabanillas Formation. Source: Own elaboration.

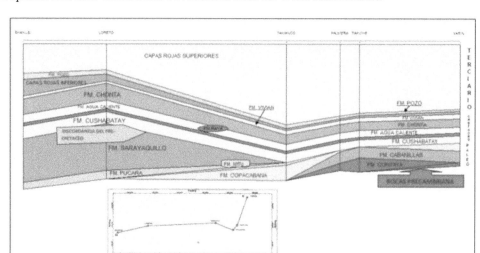

Figure 9 shows the horizontality in the Pre-Cretaceous Unconformity in the seismic-stratigraphic section, it shows the folding product of the Late-Hercynian tectonics that gave rise to depressions that were filled by continental sediments that correspond to the Mitú Formation and to the east to the Ene Formation. In addition, to the west a preserved wedge of the Pucara Formation is observed.

Figure 9. Interpretation of the ARC-3 seismic line correlation of horizontality of the Pucara, Mitu, Ene Copacabana and Cabanillas Formation. Source: Own elaboration.

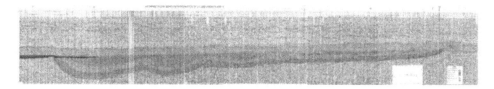

The Shale of the Cabanillas Formation

3.2. Analysis of Wells in the Study Area

This stage of the investigation of unconventional resources is much more detailed than that of conventional resources. Exhaustive petrophysical analyzes are required, which implies having many cores, lateral cores and good frequency in obtaining drilling cuttings.

The information from electrical logs, lithological logs, paleontological logs and the gas chromatograph is essential, such as the presence of gas and/or oil in drilling cuttings, especially in shale.

3.3. Analysis of Wells Containing Gas in the Shales

The Geology and Petroleum Engineering well files were reviewed, resulting in the following for each exploratory well in the area:

Palmera Well Within the Contaya Formation, traces of 10% fluorescence were observed, and 14 units of Total Gas were detected in the shales intercalated with sandstones in the interval 3774 m - 3786.5 m. As well as good indications of gas in the "neutron-density record" of the sandstones of the intervals 3756 m - 3760 m and 3762 m - 3763 m; Interpretation of the sonic log in these intervals shows low porosities of 7-9%. On the other hand, in the Cabanillas Formation, 3-9 units of Total Gas and complete Chromatography were detected C_1: 254 ppm, C_2: 60 ppm, C_3: 48 ppm, C_4: 41 ppm and C_5: 100 ppm in the interbedded dark gray shales with very fine sandstones, interval 3590.5 m – 3610 m, as can be seen in (Figure. 10).

Figure 10. Electrical records of SP, GR and Induction of the palmeras exploratory well. Source: Own elaboration.

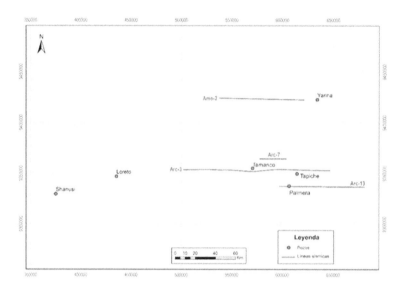

Tapiche Well (Corporation & Company, 1975) In the Contaya Formation, pale yellow to white fluorescence was observed, with instantaneous cutoff and Complete Chromatography was detected C1: 23.4 ppm, C2: 5 ppm, C3: 4.6 ppm, C4: 2.7 ppm and C5: 1.3 ppm in light-dark gray shales, interval 3605.7 m – 3612 m. Only fluorescence was also reported in black shales in the intervals 3508 m - 3509 m and 3688 m - 3691 m. In the basal sandstones of the Cabanillas Formation, oil stains, good fluorescence and shear were seen in the wall cores in the interval 3469.8m – 3496.0 m. Likewise, the Mud Logging unit detected methane, ethane, propane and butane gas, as seen in Figure 11.

Figure 11. Electrical records of SP, GR, ILD, DT and gas chromatography of the Tapiche well. Arco Perú Company. Source: Own elaboration.

The Shale of the Cabanillas Formation

Yarina Well (Peru, 1974). The Pre-Cretaceous rocks make up the Copacabana formation, its top was found at 3035m. Its base at 3287m. These are intercalations of limestone with shale with non-visible porosity, oil spots in the fractures 3057m -3065.3m and in the intervals 3115m-3294m, and were also recorded at 3057m. Gas C1, C2 & C3. From the depth of 3045mts. to 3508mts. the shales, limestones such as sandstones indicate traces of methane gas, as well as fluorescence, Traces of fluorescence and bluish white cut. In core N°2 3063.2m - 3069.9 m They recovered 14' of dense microcrystalline limestone with some fracture porosity. Fluorescen|ce and bluish-white medium section in fractured limestones in the interval 3067.8m - 3068.4 m. Hydrocarbon Indication: traces of fluorescence and shear in the interval 3064.7m - 3065.6 m Hydrocarbon Indication: Good fluorescence and shear in the interval 3045.5 m @ 3046.4 m (9992'-9995') as well as 100 units of methane gas. Good fluorescence and cutoff in the interval 3057.1m @ 3060.1 m (10030'-10040'). Fluorescence and shear traces in the interval 3060.1 @ 3064.7 m (10040'-10055'), also C1,-C2,-C3 gas. In the black shales of the Cabanillas Formation, "traces of methane gas" were recorded, as seen in Figure 12.

Figure 12. Electrical records of SP, GR, and ILD from the analysis of the channel samples from the Yarina-Amoco Perú Petroleum Company well. Source: Own elaboration.

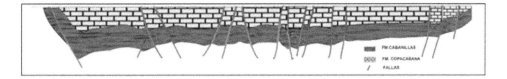

3.4. Tamanco Well

The result of the analysis of the electrical-radioactive, lithological, paleontological and chromatograph logs and the wall and conventional cores did not reveal hydrocarbon occurrences as during drilling, having penetrated 1819' feet into the pre-Cretaceous section. Both in the main objective, the pre-Cretaceous section, and in the Cretaceous, it was decided to abandon the well without completing. The pre-Cretaceous section of Upper Paleozoic age, composed mostly of dense apparent dolomitic limestone, does not constitute a reservoir rock, some sandy intervals present in the section are saturated with water. The palynological study carried out on the pre-Cretaceous rocks assigns a Permian age.

3.5. Loreto Well

The well was drilled in the largest structure in the Marañón Basin at the top of the Cretaceous and has a horizontal closure of 593,000 acres and a vertical closure of 1,300 meters. In the lower seismic horizon (Pucara Formation) the horizontal closure is 988,000 acres and the vertical closure is 1450 m. Some fluorescence was found in the lateral samples of the Raya and Agua Caliente formations and gas manifestations in Pucara (Lower Jurassic) indicating the possibility that there was an accumulation of hydrocarbon that was subsequently washed by fresh waters leaving only these traces. Low salinity indicates influence of meteoric water.

3.6. Shanusi Well

The well evaluated the potential of the Cretaceous reservoir rocks and the Pre-Cretaceous sequence of the Pucara group. The Cretaceous formations, also the Jurassic Sarayaquillo formation, but the rocks of the Pucara Group were not evaluated due to drilling problems.

3.7. Area Maps

Based on the seismic interpretation of the lines and the analysis of the electrical logs of the exploratory wells in the area, the following were prepared:

An **Isopaque map** of the Cabanillas formation of the area and the distribution of the thickness of the formation is shown as shown in Figure 13 to know the thickness of the formation and that can be used to quantify and estimate the volume of the formation in the area.

A **Structural map** at the top of the Cabanillas formation, as shown in figure 14, to know the depth of the formation in the area.

These two maps will serve to estimate the volume and size of the unconventional and challenging reservoir.

Figure 13. Isopaco Map of the Cabanillas Formation. Source: Own elaboration.

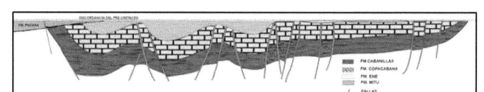

Figure 14. Structural Map at the top of the Cabanillas Formation. Source: Own elaboration.

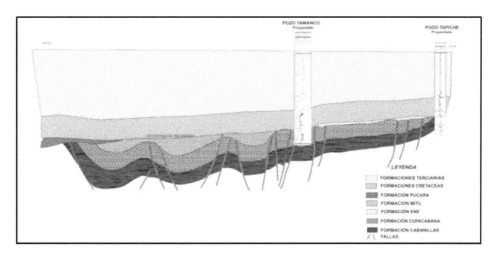

4. CONCLUSION

- The presence of gas and oil is concluded in the Shales of the Cabanillas formations and in the limestones of the Copacabana formation.
- The Cabanillas formation is distributed in the area and is made up of a sequence of black shales of Devonian age from the Paleozoic era in this part of the Marañón basin.
- The seismic stratigraphic interpretation of the seismic lines allowed the seismic sequence of the Cabanillas and Copacabana formation to be compared with the electrical and lithological records of the Tapiche and Tamanco exploratory wells, which determined its distribution and thickness in the area, having its greatest thickness to the east.
- The Paleozoic, Cretaceous and Tertiary formations were determined based on the seismic interpretation.
- The tectonic-structural evolution was carried out with the seismic interpretation of the sequences of the Cabanillas, Copacabana, Ene, Mitú and Pucara formations until the Pre-Cretaceous unconformity in which the Taconic, Neo-Hercynian, Late-Hercynian tectonic events can be seen, who performed in the area.

- It is suggested to continue with the petrographic analysis of the cores of the existing exploratory wells in the litho library of the company Perupetro S.A. in order to determine more accurately that the Cabanillas formation is an unconventional reservoir.

ACKNOWLEDGMENT

Special thanks to the "Applied research and technological development project 2018-I" E041-2018-01-BM", the World Bank, the National Council of Science, Technology and Technological Innovation (CONCYTEC) and the National Scientific Development Fund, Technological and Technological Innovation (FONDECYT), as the financing entity of Sub-project: 61096 and contract 139 – 2018 UNI/FONDECYT of which I was part.

To Doctor Luz Eyzaguirre G. for her recommendations to finish my Master's thesis. To Dr. Walter Barrutia for advice, guidance and corrections for editing the thesis. To my friends and colleagues Pedro Zegarra S., Cesar Montes A. Federico Diaz M., María Viera.

To the National University of Engineering, to the Faculty of Petroleum, Natural Gas and Petrochemical Engineering, which have allowed me to complete my master's degree studies. To the master's professors and to all the people who in some way supported and helped me.

To my sister Marlene and my nieces Lynne and Kelly for their great support and unconditional encouragement, which were essential to achieve all this and more. To all those who have been with me in the graduate classroom of the Faculty of Petroleum, Natural Gas and Petrochemical Engineering.

REFERENCES

B, D. (1986). *ORSTOM-INGEMMET Estudios Especiales*. Academic Press.

Corporation & Company. (1975)., *Well Completion Report 14-36-2X Tapiche*. Academic Press.

E, A., & N., V. I.-P. (1974). *The volcanism of the Northern part of Peruvian Altiplano and of the Oriental Cordillera*. Academic Press.

EIA. (2013). *Technically Recoverable Shale Oil and Shale Gas Resources: An Assessment of 137 Shale Formations in 41 Countries Outside the United States*. U.S. Energy Information Administration. Obtenido de https://www.eia.gov/analysis/studies/worldshalegas/archive/2013/pdf/fullreport_2013.pdf

Honty, G. (2018). Nuevo extractivismo energético en América Latina. *Ecuador Debate, 105*, 48-67. Obtenido de http://hdl.handle.net/10469/15261

Horsfield, B., Schulz, H.-M., & Kapp, I. (2012). Shale Gas in Europe. Search and Discovery Article #10380. Obtenido de https://www.searchanddiscovery.com/documents/2012/10380horsfield/ndx_horsfield.pdf

Jarvie, D., Burgess, J., Morelos, A., Mariotti, P. A., & Lindsey, R. (2001). Permian Basin petroleum systems investigations; inferences from oil geochemistry and source rocks. *AAPG Bulletin*, 85(9), 1693–1694.

Jarvie, D. M. (2012). Shale Resource Systems for Oil and Gas Part 1—Shale-gas Resource. In *Shale Reservoirs—Giant Resources for the 21st Century*. American Association of Petroleum Geologists. 10.1306/13321446M973489

Laubacher, M. (1970). La tectonica tardi-hercinica en la Cordillera oriental de los Andes del Sur del Peru. Academic Press.

Morales, I. (2013). La revolución energética en América del Norte y las opciones de política energética en México. LaReforma Energética, 109. Obtenido de http://www.foroconsultivo.org.mx/libros_editados/reforma_energetica.pdf#page=110

Morjandin, J. (2013). *La revolución energética en marcha en Estados Unidos*. Expansión.

Perú, A. P. C. (1974). YARINA 10-19-2X, Informe Diario Geológico.

Schneider, F., Laigle, J.M., Kuhfuss-Monval, L., & Lemouzy, P. (2013). Basin Modeling - the Key for Unconventional Play Assessment. Academic Press.

SNMPE. (2012). Shale Gas/ Gas de Lutitas. Informe quincenal de la SNMPE. Obtenido de https://issuu.com/sociedadmineroenergetica/docs/snmpe-informe-quincenal-hidrocarbur_63d9b410cfd751

Chapter 6
First Insights in the Estimation of the Petroleum Generation Potential of the Muerto Source Rock From el Cortado Outcrop, Northwestern Peru

Jose Alfonso Rodriguez-Cruzado
National University of Engineering of Peru, Peru

Jorge Oré-Rodriguez
National University of Engineering of Peru, Peru

ABSTRACT

The Muerto Formation is generally defined as an oil prone source rock in the Peruvian Northwest. As other worldwide source rocks, it could also act as a self-contained source-reservoir system; however, there are not quantitative estimations about the generated petroleum volume and its oil/gas composition. The data for this study includes TOC and Rock-Eval pyrolysis analyzes, which were performed by Petroperu in 1986 and Perupetro in 1999, on rock samples from the El Cortado outcrop, located in the Lancones Basin. Although both datasets contain lab analyzes performed more than 20 years ago, modern calculation techniques were applied to assess the petroleum generation potential. It was calculated that this sequence generates predominantly oil (45-73%) and cogenerates gas (27-55%), in the oil window; however,

DOI: 10.4018/979-8-3693-0740-3.ch006

the Muerto Formation is most probably a shale gas resource, due to its advanced thermal maturity in the Northern Lancones Basin.

1. INTRODUCTION

Shale resource systems have contributed to the dramatic increase on the supply of oil and especially gas from the United States, in fact, making it energy independent in natural gas production (Jarvie, 2012). These shale resource systems are typically organic-rich, source rocks, that act as a self-contained source-reservoir system and are a class of continuous petroleum accumulations (Schmoker, 1995). Success in producing oil and gas from these ultra-low-permeability (nanodarcys) and low-porosity (<15%) reservoirs has boosted the exploration efforts to locate these resource systems worldwide (U.S. Energy Information Administration, 2013). Despite the shale resources potential assessment from several South American countries, there is a lack of public information about the Peruvian shale resources potential (Morales-Paetán et al., 2018, 2020; Pairazamán et al., 2021; F. Palacios et al., 2015; Porlles et al., 2021). If unlocking the Peruvian shale resources potential, the development of shale-gas resource systems can potentially provide a long-term energy supply, with the cleanest and lowest carbon dioxide-emitting carbon-based energy source. Among the Peruvian productive areas, the Peruvian Northwest is the area with the greatest petroleum tradition since there was drilled the first oil well in South America in 1863 (Bolaños Z., 2017).

Since its identification in outcrops (Bravo, 1921), the Muerto formation has attracted attention in northwestern Peru due to its particular fetid odor and presence of bitumen in fractures. It is located in both the Talara and Lancones basins (Castro-Ocampo, 1991; Tafur H., 1952). The Muerto formation is an organic-rich calcareous source rock (Alvarez et al., 1986; Alvarez & Garrido, 1987), deposited on a carbonate platform and slope (Navarro-Ramirez et al., 2017), that occurred in Albian, at the end of the Lower Cretaceous (Aldana A., 1994; Cruzado C. & Aliaga L., 1985; Fischer, 1956; Olsson, 1934; O. Palacios, 1994; Tafur H., 1952), in the context of an Oceanic Anoxic Event (Bianchi & Jacay, 2011; Jenkyns, 1980).

This source rock overlies the shallow water carbonate sequence of the Pananga Formation, of Aptian age, Lower Cretaceous (Séranne, 1987), and transitionally underlies the siltstones and limestones of the Huasimal Formation, which represent the transition to a turbiditic environment that prevails during the Upper Cretaceous (Morris & Aleman, 1975; Tafur H., 1952, 1954; Uyen & Valencia, 2002). This organic rich sequence is considered one of the main rocks that have generated the oil that, after migration and trapping in clastic reservoirs, has produced the prolific

First Insights in the Estimation of the Petroleum Generation Potential

Talara Basin (Fildani et al., 2005; Müller, 1993; PERUPETRO, 1999; Walters et al., 1992), and amounting to more than 1,500 MMBO (Infologic et al., 2006).

Although the Muerto Formation is an organic rich sequence, considered as oil-prone, its generation potential is not evident. The objective of this research is to estimate the generation potential of hydrocarbons (in volume), and its components (proportion of oil and/or gas), which will provide quantitative evidence of its promising potential. The data for this study includes Total Organic Carbon (TOC) and Rock-Eval pyrolysis analyzes, which were performed by Petroperu in 1986 and Perupetro in 1999, on rock samples from the El Cortado outcrop (yellow star in Figure 1), located in the Northern Lancones Basin. Since the study area is restricted to one outcrop, it should be noted that this calculation is just locally valid and is not intended to be representative for the entire Muerto Formation.

Figure 1. Location of the Lancones and Talara basins in Northwest Peru. Location of the study area: El Cortado creek (yellow star)

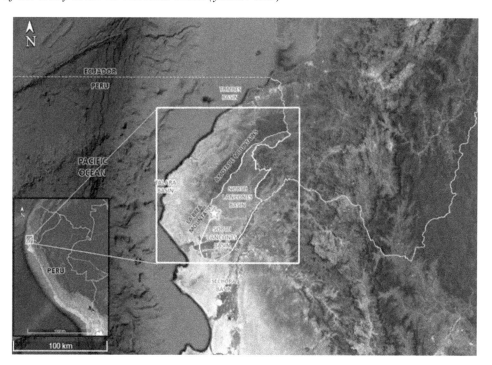

2. METHODOLOGY

The laboratory data used in this study were obtained from declassified reports, stored by the Data Bank of Perupetro. The data comprises TOC and Rock-Eval analysis carried out by Petroperu in 14 samples (Alvarez et al., 1986; Alvarez & Garrido, 1987), and Perupetro in 7 samples (DGSI, 1999).

Data of TOC, Rock-Eval pyrolysis and geochemical indices are detailed in Table 1. The geochemical indices were calculated according to their definition by Espitalié et al. (1977).

3. RESULTS AND DISCUSSION

Table 1. TOC and rock-eval data from the Muerto Formation at El Cortado outcrop

Sample	T0C (wt.%)	S1 (mg HC/ g rock)	S2 (mg HC/ g rock)	S3 (mg CO2/ g rock)	Tmax (°C)	HI (mg HC/ g TOC)	OI (mg CO2/ g TOC)	PI
EC-02	1.93	0.27	2.34	0.15	457	121	8	0.10
EC-03	2.67	0.90	4.31	0.07	455	161	3	0.17
EC-04	3.45	1.19	6.69	0.06	457	194	2	0.15
EC-05	2.20	1.33	4.32	0.04	464	196	2	0.24
EC-06	1.18	0.46	1.61	0.02	462	136	2	0.22
EC-07	4.75	2.08	11.32	0.18	454	238	4	0.16
EC-08	3.88	1.43	6.88	0.07	460	177	2	0.17
EC-09	1.50	1.21	1.89	0.02	464	126	1	0.39
EC-10	2.25	0.85	2.77	0.05	457	123	2	0.23
EC-11	1.95	0.52	2.94	0.07	455	151	4	0.15
EC-12	1.21	0.38	1.17	0.03	453	97	2	0.25
EC-13	0.46	-	-	-	-	-	-	-
EC-14	0.74	0.25	0.47	0.05	425*	64	7	0.35
EC-15	1.29	1.01	1.58	0.05	459	122	4	0.39
EC-98-01	1.89	0.70	2.68	0.29	452	142	15	0.21
EC-98-02	4.59	1.25	8.85	0.50	445	193	11	0.12
EC-98-03	4.19	0.71	7.95	0.47	444	190	11	0.08
EC-98-04	2.64	0.17	10.56	0.43	416*	400*	16	0.02

continued on following page

Table 1. Continued

Sample	TOC (wt.%)	S1 (mg HC/ g rock)	S2 (mg HC/ g rock)	S3 (mg CO2/ g rock)	Tmax (°C)	HI (mg HC/ g TOC)	OI (mg CO2/ g TOC)	PI
EC-98-05	3.07	0.68	4.89	0.49	446	159	16	0.12
EC-98-06	4.15	1.58	8.69	0.50	445	209	12	0.15
EC-98-07	2.02	0.45	2.56	0.35	444	127	17	0.15
MEAN	2.48	0.87	4.72	0.19	454	159	7	0.19
Range	0.46-4.75	0.17-2.08	0.47-11.32	0.02-0.5	444-464	64-238	1-17	0.02-0.39
Outliers					416; 425	400		

* Outlier values

3.1 Thermal Maturity

In the El Cortado outcrop, the thermal maturity of the Muerto Formation, measured by Tmax, varies between 444 – 464 °C, with 416 and 425 °C as outliers, and with an average Tmax of 454 °C (Table 1), which is consistent with the vitrinite reflectance of 1.0%Ro, reported by Petroperú (Garrido & Aliaga L., 1986).

In the Northern Lancones Basin, the Muerto formation outcrops have a thermal maturity of 1.0%Ro (Figure 2), which suggests that the outcrops reached the oil window, however, the thermal maturity in the Huasimal Formation outcrops (1.1, 1.6, 1.9 and 2.1%Ro, Fig. 2), that is a sequence overlying the Muerto Formation, suggests that the Muerto Formation is mostly in the gas window (>1.5%Ro).

Figure 2. Thermal maturity of the Muerto and Huasimal formations in the outcrops of the Northern Lancones Basin

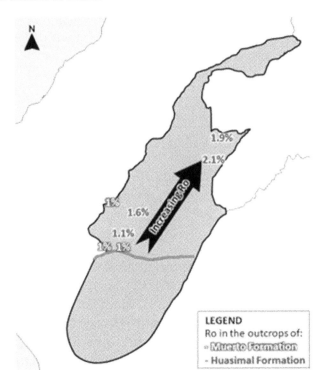

3.2 Organic Quality and Richness

The TOC varies between 0.46 – 4.75 wt.%, on average 2.48 wt.% (Table 1). On the other hand, S1 varies between 0.17 – 2.08 mg HC/g rock, on average 0.87 mg HC/g rock, and S2 between 0.47 – 11.32 mg HC/g rock, on average 4.72 mg HC/g rock.

Given the advanced thermal maturity, the TOC, generation potential (S2) and the Hydrogen Index (HI), just show the remaining organic richness and quality, which is reduced compared to the original values. Therefore, they are only an indicator of current generation potential (Jarvie et al., 2007).

According to the modified Van Krevelen (HI vs OI) and HI vs Tmax diagrams, which are used to classify a source rock according to the kerogen type (Banerjee et al., 1998; Espitalié et al., 1977; Peters, 1986), the data is mainly positioned on an indeterminate trend between type I and type II (Figure 3). It can be inferred that the original quality of the kerogen from the Muerto formation corresponds to type II, because this kerogen type is mostly associated with organic matter of marine origin

First Insights in the Estimation of the Petroleum Generation Potential

(Jarvie, 2012), which is consistent with the marine carbonate platform environment of the Muerto formation and with the predominance of amorphous organic matter, and the absence of vitrinite particles in most samples.

Figure 3. (a) Modified Van Krevelen (Espitalié et al., 1977) kerogen type diagram. (b) Espitalié et al. (1984) kerogen type and thermal maturity plot.

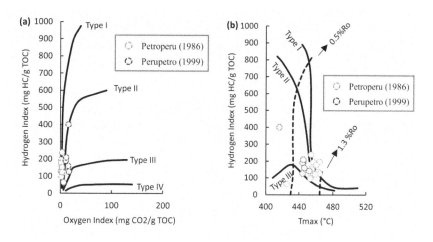

3.2.1 Live Hydrogen Index (HI_{LIVE})

The hydrogen index (HI) is used to assess a source rock quality (Pepper & Corvi, 1995). The HI, as conventionally calculated (Espitalié et al., 1977), is between 64-238 mg HC/g TOC, and 400 mg HC/g TOC as an outlier (Table 1).

However, a better method to determine the source rock quality is the HI the live kerogen (HI_{LIVE}), where "live" kerogen refers to the portion of organic matter from which hydrocarbons are obtained during pyrolysis (Dahl et al., 2004). The HI_{LIVE} is not intended to estimate gas or oil potential of individual samples, but rather an average value for a source rock section. The HI was calculated based on cross-plots of S2 versus TOC as described by Clayton et al. (1984) and Langford & Blanc-Valleron (1990), who regards the TOC to be a linear function of S2 with the HI of the live kerogen as the slope of the curve. This method allows to eliminate the problems related to S3, in the IH vs IO diagrams (Katz, 1983), and corrects the HI due to the effects of matrix and inert kerogen, which cause a reduction in the HI and a lower apparent quality (Espitalié et al., 1980; Peters, 1986).

Despite the differences on sources of data used in this research (Table 1), most of the data follows a same trend on a TOC vs S2 plot (Figure 4). Therefore, the datasets were considered as complementary. The HI_{LIVE} of 276 mg HC/g TOC was determined from the regression equation of the S2-TOC graph (Fig. 4), discarding the low values in S2 (<2 mg HC/g rock), since they are values where shows the retention effect of the matrix, and cause a deviation from the main trend.

First Insights in the Estimation of the Petroleum Generation Potential

Figure 4. TOC vs S2 graphs showing (a) the match between both datasets and (b) the HI_{LIVE} calculation from the slope of the main trend

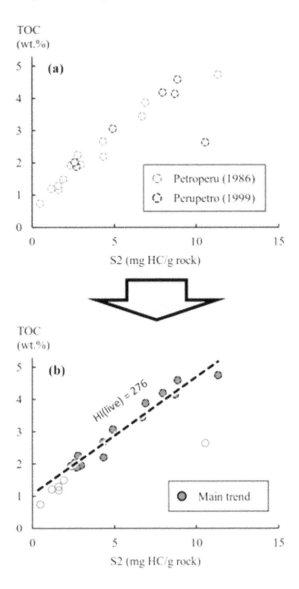

3.2.2 Original Hydrogen Index (HIo)

The HIo can be directly estimated from immature rock samples; however, no places have been found for the Muerto Formation on immature state. In fact, it is difficult to assign an HIo to a source rock in the absence of a collection of immature rocks samples. Given the advanced thermal maturity of the Muerto Formation in the outcrops (Figure 2), the present-day HI is less than HIo, and HI_{LIVE} is not indicative of the original kerogen quality.

The HIo of 451 mg HC/g TOC, as calculated using the Cornford et al. (2001) equation (Table 2), fits well into the range of oil-prone, immature marine shales, that has a predominant distribution of HIo between 300-700 mg HC/g TOC (Jarvie, 2012; Jones, 1984; Peters & Cassa, 1994).

3.2.3 Kerogen Transformation Ratio (TR)

Although maturity parameters such as vitrinite reflectance are empirically related to the oil and gas windows, it is feasible to evaluate conversion directly by measuring changes in organic matter yields, i.e., the extent of kerogen conversion by calculation of the kerogen transformation ratio (TR). In order to restore the original source rock potential, it is necessary to determine the kerogen transformation ratio (TR).

Transformation ratios can be found empirically by regional geochemical studies (e.g. Dahl & Augustson, 1993; Espitalié et al., 1987) or from basin simulation runs using software with a kinetic subroutine. TR can also be derived by determining HIo from maceral compositions as shown in this case. Using the average HIo value of 451 mg HC/g TOC (Table 2), two different equations were used to calculate the TR:

- The transformation ratio (TR) was calculated as the fraction of original organic matter (HIo) converted to hydrocarbons (Banerjee et al., 1998), resulting.
- $TR_{(1)}$ of 0.39 (Table 2).
- Using the equations of Claypool (Peters et al., 2004), the transformation ratio (TR), was derived from the change in HIo to present-day values (HI_{LIVE}), including a correction for early free oil content from the original production index (PIo) (Espitalié et al., 1984; Pelet, 1985; Peters et al., 2004), where PI is the production index (S1/(S1+S2)) as PIo = 0.02 to PIpd, resulting $TR_{(2)}$ of 0.52 (Table 2).

3.3 Gas/Oil Composition

The relative amounts of oil- and gas-producing organic material can be determined by combined pyrolysis-gas chromatography (pyrolysis GC) of source rock samples (Horsfield, 1989; Pepper & Corvi, 1995). Quantitative visual kerogen determinations can also be used to assess the approximate quantity of oil- and gas-producing components (Mukhopadhyay et al., 1985).

Due to the ease of the Rock-Eval method relative to pyrolysis GC analysis and visual kerogen typing, most geochemical databases contain more Rock-Eval data compared to pyrolysis GC and quantitative petrographic kerogen typing. Therefore, the division of source rock hydrocarbon potential into oil- and gas- producing constituents was performed using the workflow suggested by Dahl et al. (2004) that uses Rock-Eval and TOC data.

3.3.1 End-Members

The end-members were defined based on the "modified versions" of organic facies B and C (Cooper & Barnard, 1984; Jones, 1987), as suggested by Dahl et al. (2004).

The organo-facies, as defined in the literature, are usually qualitative with respect to oil and gas generation because only average total organic carbon (TOC), petroleum potential (S2), or the dominant kerogen type are presented (Cooper & Barnard, 1984; Demaison & Moore, 1980; Isaksen & Ledje, 2001). In order to assess the petroleum composition, petroleum potential was divided into their proportions of oil and gas (primary) producing constituents as proposed by (Cooles et al., 1986; Dahl & Meisingset, 1996; Dahl & Yükler, 1991).

- The oil-prone end-member (HI_{OIL}) was defined in the upper range of organic facies B, 700 mg HC/g TOC, where the vitrinite and other inertinite components are hypothetically "absent".
- The gas-prone end-member (HI_{GAS}) was defined in the upper range of organic facies C, 250 mg HC/g TOC, that contains subordinate herbaceous material and include the liquid component.

3.3.2 Overlays

To elaborate the overlays, the Hydrogen Indices with different gas/oil composition are calculated, by inserting the ideal values of the end-members, previously adjusted by the kerogen transformation (Table 2), and with an increase of 20% between the various oil and gas prone kerogen mixtures. Finally, the templates are plotted by converting the resulting hydrogen numbers to lines on S2 versus TOC plots. Because the overlays change with different increasing maturity, overlays for the different calculated TR (0.39 and 0.52) were prepared (Figure 5).

Figure 5. Overlays constructed for the two different calculated TR of 0.39 and 0.52

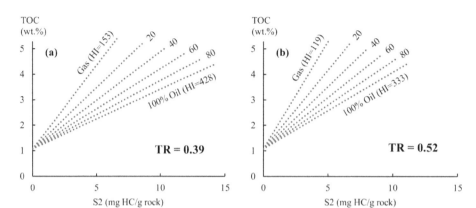

3.3.3 Gas/Oil Ratio Potential (GORP)

In order to split the kerogen into oil-prone and gas-prone organic carbon, GORP (gas-oil-ratio-potential) concept is used. To get around the determination of the GORP, the transparent overlays were superimposed on the main trend line and the kerogen mixtures can be graphically read (Figure 6). To be more accurate, the GORP was calculated as shown in Table 2.

The GORP suggest a source rock from slightly gas-prone to highly oil-prone kerogen (45-73%, Figure 6). These values agree with kerogen analysis studies reported by Petroperu (Gamarra, 1987), which suggest that this is a section dominated by amorphous organic matter, with potential for oil generation.

Figure 6. Graphical representation of the GORP calculation by the superposition of the HI_{LIVE} and overlays adjusted for TR, resulting in (a) GORP of 0.55 and (b) GORP of 0.27

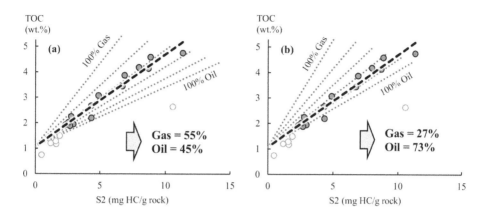

3.4 Original Generation Potential (S2o)

Determination of the original generation potential (S2o) of a source rock provides a quantitative means to estimate the total volume of hydrocarbons that it can generate. Given that this source rock has been subjected to thermal degradation, it is not straightforward to determine original values. As an alternative, there were explored two different approaches to calculate original S2 (S2o) from TOC and pyrolysis Rock-Eval data.

3.4.1 Approach #1 (Dahl et al., 2004)

The calculation of S2o by this approach, using both kerogen conversion factors, indicates that the original generation potential was between 9.02-11.59 mg HC/g rock (Table 2).

continued on following page

Table 2. Continued

Table 2. Calculation of the original Hydrogen Index (HIo), kerogen transformation ratio (TR), Gas/Oil Ratio Potential (GORP), and original generation potential (S2o)

Description	Value		Units	Derivation	Source
HIo	451		mg HC/g TOC	$HI + HI(T\max - 30)$	(Cornford et al., 2001)
$TR_{(1)}$	0.39		dimensionless	$\frac{HIo - HI}{HIo}$	(Banerjee et al., 1998)
$TR_{(2)}$	0.52		dimensionless	$\frac{HI[1177 - HIo(1 - PIo)]}{HIo[1177 - HI(1 - PI)]}$	(Jarvie, 2012)
	Cálculos basados en				
	$TR_{(1)}$	$TR_{(2)}$			
$HI_{OIL;TR}$	428	333	mg HC/g TOC	$HI_{OIL}(1 - TR)$	(Dahl et al., 2004)
$HI_{GAS;TR}$	153	119	mg HC/g TOC	$HI_{GAS}(1 - TR)$	(Dahl et al., 2004)
GORP	0.55	0.27	dimensionless	$\frac{HI_{OIL;TR} - HI_{LIVE}}{HI_{OIL;TR} - HI_{GAS;TR}}$	(Dahl et al., 2004)
S2o	9.02	11.59	mg HC/g rock	$\frac{S2o}{1 - TR}$	(Dahl et al., 2004)

3.4.2 Approach #2 (Jarvie et al., 2012)

For this approach, it is taken into consideration that the TOC, as measured by any geochemical laboratory, do not provide an indication of the original hydrocarbon generation potentials because they largely represent the present-day non-generative organic carbon (NGOC), especially when they are at gas window thermal maturity. Given that it is desired to show the hydrocarbons generation potential, the original generative organic carbon (GOCo) with derivation of original generation potential (S2o) were necessary (Jarvie et al., 2007).

Assuming that a source rock generates hydrocarbons that are approximately 85% carbon, which is a reasonable mean value for marine shales (Jarvie, 2012). Bitumen- and/or oil- and kerogen-free TOC (TOC_{bkfree}) is calculated by the subtraction of carbon in S1 and S2 from TOC. The removal of this residual organic carbon (OC) yields a TOC_{bkfree} of 1.92 wt.% (Table 3). Then, the non-generative organic carbon (NGOC) was corrected for the increased char formation ($TOC_{NGOCadjusted}$), subtracted from base TOC values, resulting a $TOC_{NGOCadjusted}$ of 1.74 wt.% (Table 3). Given that the HIo was already estimated (Table 3), the percent GOC in TOCo (%GOCo) can readily be determined. The maximum HIo can be estimated by the reciprocal of carbon content in hydrocarbons, that is, 1/0.085 or 1177 mg HC/g TOC. Using 1177 mg HC/g TOC as the maximum HIo, the %GOCo can be calculated by the relation HIo/1177. The percentage of reactive carbon in the immature shale yields 38% of the TOCo that could be converted to petroleum (Table 3). Using the measured present-day TOC with the correction for bitumen and/or oil and kerogen in the rock and any increase in NGOC caused by hydrogen shortage, an original TOC of 2.83 wt.% is calculated. This means that the original generation potential (S2o) is 12.76 mg HC/g rock. Data for this calculation are summarized in Table 3.

Table 3. Calculation of the original generation potential S2o using the equations proposed by (Jarvie, 2012)

Description	Value	Units	Derivation
$OC_{in}S1 + S2$	0.55	wt. %	$0.085(S1 + S2)$
TOC_{bkfree}	1.92	wt. %	$TOC - (OC_{in}S1 + S2)$
$NGOC_{correction}$	0.18	wt. %	$HI_o \times 0.0004$
$TOC_{NGOCadjusted}$	1.74	wt. %	$TOC_{bkfree} - NGOC_{correction}$
%GOCo	38%		$HI_o/1177$
TOCo	2.83	wt. %	$TOC_{NGOCadjusted}/(1 - \%GOCo)$
GOCo	1.08	wt. %	$TOCo - TOC_{NGOCadjusted}$
$S2o_{(3)}$	12.76	mg HC/g rock	$GOCo/0.085$

3.5 Generation Potential vs. Storage Capacity

During primary cracking, the equivalent, immature Muerto Formation original generation potential (S2o), that was converted to oil and gas, resulted in a yield of 88-205 bbl/ac-ft and 405-926 mcf/ac-ft respectively (Table 4).

The high termal maturity (>1.5%Ro) on most of the Northern Lancones Basin suggest that secondary cracking has occurred. The cracking of oil to gas is limited by the amount of available hydrogen in the system needed to form wet and dry gas. Gases in the shale play are typically less than 100% methane, and a reasonable average is about 90% methane across the entire productive area (Jarvie et al., 2007). At 90% methane, the atomic H/C ratio requirement for condensate wet-gas formation is about 3.8, so hydrogen deficiency is approximately 53% (Jarvie et al., 2007). Taking hydrogen deficiencies into account, the total gas-generation potential for the Muerto Formation is about 903-1278 mcf/ac-ft (Table 4).

If a gas storage capacity corresponding to porosity between 3-6% is considered, which is a typical range in shale reservoirs (Jarvie et al., 2007), it results in a storage capacity that varies between 270-540 mcf/ac-ft for typical reservoir conditions (70°C, 3800psi), and between 430-850 mcf/ac-ft for higher burial conditions (180°C, 8000psi). In both cases, the generation potential exceeds the storage capacity, as shown in Figure 7, so the Muerto Formation in the depth would be expected to be saturated by gas.

Figure 7. Comparison of the total gas generation potential with the storage capacities of typical shale reservoirs based on 3-6% porosity and different reservoir pressure and temperature conditions

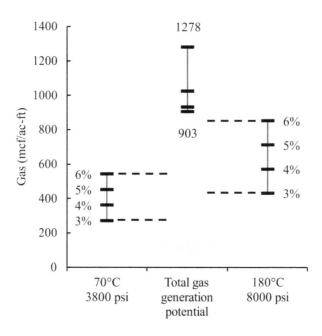

Table 4. Computation of the total gas generation potential of the Muerto Formation in the El Cortado outcrop

	$GORP_{(1)} = 0.55$			$GORP_{(2)} = 0.27$			Units
	$S2o_{(1)}$	$S2o_{(2)}$	$S2o_{(3)}$	$S2o_{(1)}$	$S2o_{(2)}$	$S2o_{(3)}$	
	9.02	11.59	12.76	9.02	11.59	12.76	
Primary cracking							
Oil generated from kerogen (*)	88	-	125	-	186	205	bbl/ac-ft
Gas generated from kerogen (**)	654	-	926	-	405	445	mcf/ac-ft
Secondary cracking							

continued on following page

Table 4. Continued

	$GORP_{(1)} = 0.55$			$GORP_{(2)} = 0.27$			Units
	$S2o_{(1)}$	$S2o_{(2)}$	$S2o_{(3)}$	$S2o_{(1)}$	$S2o_{(2)}$	$S2o_{(3)}$	
Correction factor for insufficient Hydrogen in oil	47%	-	47%	-	47%	47%	
Gas generado a partir del craqueo secundario del petróleo	249	-	352	-	525	578	mcf/ac-ft
Total gas generation potential (gas from primary cracking + gas from secondary cracking of oil)	903	-	1278	-	930	1023	mcf/ac-ft

*Conversion of Rock-Eval S2 in mg HC/g rock to bbl oil/ac-ft, multiply by 21.89.
**Conversion of Rock-Eval S2 in mg HC/g rock to mcf/ac-ft, multiply by 131.34.

4. CONCLUSION

The Muerto Formation at El Cortado outcrop is an oil prone source rock, as indicated by the calculated GORP. Organic matter from the Muerto Formation generates about 45-73% oil and 27-55% gas in the oil window from primary cracking. Despite the Muerto Formation is an oil prone source rock, it most probably has potential for gas production due to the elevated thermal maturity (>1.50%Ro) in the Northern Lancones Basin.

This source rock has a total gas-generation potential for about 903-1278 mcf/ac-ft. That is larger than the storage capacity (270-540 mcf/ac-ft) of a common shale with 3-6% porosity at typical reservoir conditions (70°C, 3800psi). Therefore, the Muerto Formation is expected to be saturated by gas in the depth.

ACKNOWLEDGMENT

Special thanks to PERUPETRO SA for providing access to information from the Data Bank; and to CONCYTEC-BANCO MUNDIAL for providing resources for this study.

REFERENCES

Aldana, A. M. (1994). Estudio de la Macrofauna de los cuadrángulos de Piura, Sullana, Quebrada Seca, Paita, Talara, Negritos, Lobitos, Tumbes, Zorritos y Zarumilla. In *Geología de los Cuadrángulos de Paita, Piura, Talara, Sullana, Lobitos, Quebrada Seca, Zorritos, Tumbes y Zarumilla* (pp. 145–190). Academic Press.

Alvarez, P., & Garrido, J. (1987). *Análisis de Roca Madre y Rocas Reservorio. Anexo 2*. Petroperu.

Alvarez, P., Garrido, J., & Aliaga, L. E. (1986). *Evaluación Geológica de la Cuenca Lancones. Perú Noroccidental. Estudios Geoquímicos del Cretáceo. Secciones Pan de Azucar, Cabrerías, Jaguay Negro, Huasimal, Corcobado, Gritón, Chilco*. Petroperu.

Banerjee, A., Sinha, A. K., Jain, A. K., Thomas, N. J., Misra, K. N., & Chandra, K. (1998). A mathematical representation of Rock-Eval hydrogen index vs Tmax profiles. *Organic Geochemistry*, 28(1–2), 43–55. 10.1016/S0146-6380(97)00119-8

Bianchi, C., & Jacay, J. (2011). *Distribución de Facies Anóxicas a través de la Margen Andina durante el Albiano*. VII Ingepet.

Bolaños, Z. R. (2017). Reseña histórica de la exploración por petróleo en las cuencas costeras del Perú. *Boletín de la Sociedad Geológica del Perú*, 112, 1–13.

Bravo, J. J. (1921). *Reconocimiento de la región costanera de los departamentos de Tumbes y Piura*. Asociación Peruana para el Progreso de la Ciencia.

Castro-Ocampo, R. (1991). *El Cretaceo en la Cuenca Talara del Noroeste del Perú*. Universidad Nacional de Ingeniería - Peru.

Clayton, J., Ryder, R. T., & Meissner, F. (1984). Geochemistry of Black Shales in Minnelusa Formation and Desmoinesian Age Rocks (Permian-Pennsylvanian), and Associated Oils, Powder River Basin and Northern DJ Basin, Wyoming and Colorado: ABSTRACT. *AAPG Bulletin*, 68(7), 935–935. 10.1306/AD4614C1-16F7-11D7-8645000102C1865D

Cooles, G. P., Mackenzie, A. S., & Quigley, T. M. (1986). Calculation of petroleum masses generated and expelled from source rocks. *Organic Geochemistry*, 10(1–3), 235–245. 10.1016/0146-6380(86)90026-4

Cooper, B. S., & Barnard, P. C. (1984). Source Rocks and Oils of the Central and Northern North Sea. In Demaison, G. J., & Murris, R. J. (Eds.), *Petroleum Geochemistry and Basin Evaluation* (Vol. 35). American Association of Petroleum Geologists. 10.1306/M35439C17

Cruzado, C. J., & Aliaga, L. E. (1985). *Micropaleontología del Cretáceo del Área Pazul*. Petroperu.

Dahl, B., & Augustson, J. H. (1993). The influence of Tertiary and Quaternary sedimentation and erosion on the hydrocarbon generation in Norwegian offshore basins. In A. G. Dore (Ed.), *Basin Modelling: Advances and Applications* (pp. 419–431). Elsevier.

Dahl, B., Bojesen-Koefoed, J., Holm, A., Justwan, H., Rasmussen, E., & Thomsen, E. (2004). A new approach to interpreting Rock-Eval S2 and TOC data for kerogen quality assessment. *Organic Geochemistry*, 35(11–12), 1461–1477. 10.1016/j.orggeochem.2004.07.003

Dahl, B., & Meisingset, I. (1996). Prospect resource assessment using an integrated system of basin simulation and geological mapping software: examples from the North Sea. In Doré, A. G., & Sinding-Larsen, R. (Eds.), *Quantification and Prediction of Hydrocarbon Resources* (pp. 237–251). Elsevier. 10.1016/S0928-8937(07)80021-X

Dahl, B., & Yükler, A. (1991). The Role of Petroleum Geochemistry in Basin Modeling of the Oseberg Area, North Sea. In Merrill, R. K. (Ed.), *Source and migration processes and evaluation techniques*. American Association of Petroleum Geologists. 10.1306/TrHbk543C6

Demaison, G. J., & Moore, G. T. (1980). Anoxic environments and oil source bed genesis. *Organic Geochemistry*, 2(1), 9–31. 10.1016/0146-6380(80)90017-0

DGSI. (1999). VOL 3. Datos analíticos de rocas de afloramiento. In W. G. Dow (Ed.), *Estudios de Investigación Geoquímica del Potencial de Hidrocarburos. Lotes del Zócalo Continental y de Tierra*. PERUPETRO S.A.

Espitalié, J., Madec, M., & Tissot, B. P. (1980). Role of Mineral Matrix in Kerogen Pyrolysis: Influence on Petroleum Generation and Migration. *AAPG Bulletin*, 64(1), 59–66. 10.1306/2F918928-16CE-11D7-8645000102C1865D

Espitalié, J., Madec, M., Tissot, B. P., Mennig, J. J., & Leplat, P. (1977). Source rock characterization method for petroleum exploration. *Proceedings of the Annual Offshore Technology Conference,* 439–444. 10.4043/2935-MS

Espitalié, J., Marquis, F., & Barsony, I. (1984). Geochemical Logging. In Voorhees, K. J. (Ed.), *Analytical Pyrolysis—Techniques and Applications* (pp. 276–304)., 10.1016/B978-0-408-01417-5.50013-5

Espitalié, J., Marquis, F., & Sage, L. (1987). Organic geochemistry of the Paris Basin. In Brooks, J., & Glennie, K. (Eds.), *Petroleum geology of north West Europe* (pp. 71–86). Geological Society of London Graham & Trotman.

Fildani, A., Hanson, A. D., Chen, Z., Moldowan, J. M., Graham, S. A., & Arriola, P. R. (2005). Geochemical characteristics of oil and source rocks and implications for petroleum systems, Talara basin, northwest Peru. *AAPG Bulletin*, 89(11), 1519–1545. 10.1306/06300504094

Fischer, A. G. (1956). *Cretaceous of Northwest Peru*. IPC Report WP-13.

Gamarra, S. (1987). *Evaluación Geológica Cuenca Lancones. Estudio de Materia Orgánica Formaciones Huasimal y Muerto - Cretaceo*. Petroperu.

Garrido, J., & Aliaga, L. E. (1986). *Madurez de Roca Madre por Reflecancia de Vitrinita. Formaciones Muerto y Huasimal. Cretáceo Cuenca Lancones (Secciones: Corcobado, El Cortado, Angelitos, etc)*. Petroperu.

Horsfield, B. (1989). Practical criteria for classifying kerogens: Some observations from pyrolysis-gas chromatography. *Geochimica et Cosmochimica Acta*, 53(4), 891–901. 10.1016/0016-7037(89)90033-1

Infologic, B.-R. (2006). *Geochemical-Solutions*. Petroleum Systems Evaluation – Talara/Tumbes Basins. In Geochemical Database Development and Petroleum Systems Definition Project.

Isaksen, G. H., & Ledje, K. H. I. (2001). Source Rock Quality and Hydrocarbon Migration Pathways within the Greater Utsira High Area, Viking Graben, Norwegian North Sea. *AAPG Bulletin*, 85(5), 861–883. 10.1306/8626CA23-173B-11D7-8645000102C1865D

Jarvie, D. M. (2012). Shale Resource Systems for Oil and Gas: Part 1—shale-gas resource systems. In *M97: Shale Reservoirs—Giant Resources for the 21st Century* (pp. 69–87). American Association of Petroleum Geologists. 10.1306/13321446M973489

Jarvie, D. M., Hill, R. J., Ruble, T. E., & Pollastro, R. M. (2007). Unconventional shale-gas systems: The Mississippian Barnett Shale of north-central Texas as one model for thermogenic shale-gas assessment. *AAPG Bulletin*, 91(4), 475–499. 10.1306/12190606068

Jarvie, D. M., Jarvie, B. M., Weldon, D., & Maende, A. (2012). *Components and Processes Impacting Production Success from Unconventional Shale Resource Systems*. https://doi.org/10.3997/2214-4609-PDB.287.1226756

Jenkyns, H. C. (1980). Cretaceous anoxic events: From continents to oceans. *Journal of the Geological Society*, 137(2), 171–188. 10.1144/gsjgs.137.2.0171

Jones, R. W. (1987). Organic facies. In Brooks, J., & Welte, D. H. (Eds.), *Advances in Petroleum Geochemistry* (pp. 1–90). Academic Press.

Jones, R. W. (1984). Comparison of Carbonate and Shale Source Rocks: ABSTRACT. *AAPG Bulletin*, 68. Advance online publication. 10.1306/AD460EA4-16F7-11D7-8645000102C1865D

Katz, B. J. (1983). Limitations of 'Rock-Eval' pyrolysis for typing organic matter. *Organic Geochemistry*, 4(3), 195–199. 10.1016/0146-6380(83)90041-4

Langford, F. F., & Blanc-Valleron, M.-M. (1990). Interpreting Rock-Eval Pyrolysis Data Using Graphs of Pyrolizable Hydrocarbons vs. Total Organic Carbon. *AAPG Bulletin*, 74(6), 799–804.

Morales-Paetán, W. J., Porlles-Hurtado, J., Rodriguez-Cruzado, J., Taipe-Acuña, H., & Arguedas-Valladolid, A. (2018). First Unconventional Play from Peruvian Northwest: Muerto Formation. *Unconventional Resources Technology Conference (URTeC)*. 10.15530/urtec-2018-2903064

Morales-Paetán, W. J., Rodriguez-Cruzado, J. A., Alvarez-Mendoza, B. G., Alarcón-Marcatoma, A. A., Oré-Rodriguez, J. L., Corrales-Hidalgo, R. W., Madge-Rodriguez, J. J., Porlles-Hurtado, J. W., Falla-Ruiz, J. L., & de Eyzaguirre-Gorvenia, L. (2020, July 27). Geochemical Heterogeneities Characterization of the Muerto Formation – Lancones Basin - Peru as a Source Rock Unconventional Reservoir, a Contribution for its Development. *SPE Latin American and Caribbean Petroleum Engineering Conference*. 10.2118/199088-MS

Morris, R., & Aleman, A. (1975). Sedimentation and Tectonics of Middle Cretaceous Copa Sombrero Formation in Northwest Peru. *SGP Bulletin, 48*(3).

Mukhopadhyay, P. K., Hagemann, H. W., & Gormly, J. R. (1985). Characterization of kerogens as seen under the aspect of maturation and hydrocarbon generation. *Erdoel Kohle, Erdgas, Petrochem. Brennst. -Chem*, 38(1), 7–18.

Müller, D. (1993). *Geochemical Oil Correlations for UPPPL Oils and Offshore Oil Seep*. Talara Basin.

Navarro-Ramirez, J. P., Bodin, S., Consorti, L., & Immenhauser, A. (2017). Response of western South American epeiric-neritic ecosystem to middle Cretaceous Oceanic Anoxic Events. *Cretaceous Research*, 75, 61–80. 10.1016/j.cretres.2017.03.009

Olsson, A. A. (1934). Contributions to the paleontology of northern Peru: the Cretaceous of the Amotape Region. In *Bulletins of American Paleontology* (Vol. 20, Issue 69, pp. 1–104). Academic Press.

Pairazamán, L., Palacios, F., & Timoteo, D. (2021). *Caracterización sedimentológica de alta-resolución de la Formación Muerto, Cuenca Lancones, NO Perú. ¿Un posible reservorio no convencional?* Academic Press.

Palacios, O. (1994). Geología de los cuadrángulos de Paita, Piura, Talara, Sullana, Lobitos, Quebrada Seca, Zorritos, Tumbes y Zarumilla. In *Carta Geológica Nacional* (1st ed.). Instituto Geológico, Minero y Metalúrgico.

Pelet, R. (1985). Evaluation quantitative des produits formés lors de l'évolution géochimique de la matière organique. *Revue de l'Institut Français du Pétrole*, 40(5), 551–562. 10.2516/ogst:1985034

Pepper, A. S., & Corvi, P. J. (1995). Simple kinetic models of petroleum formation. Part I: Oil and gas generation from kerogen. *Marine and Petroleum Geology*, 12(3), 291–319. 10.1016/0264-8172(95)98381-E

PERUPETRO. (1999). VOL 1. Generalidades. In *Estudios de Investigación Geoquímica del Potencial de Hidrocarburos. Lotes del Zócalo Continental y de Tierra*. PERUPETRO S.A.

Peters, K. E., & Cassa, M. R. (1994). Applied Source Rock Geochemistry. *AAPG Memoir*, 60, 93–120. 10.1306/M60585C5

Peters, K. E. (1986). Guidelines for Evaluating Petroleum Source Rock Using Programmed Pyrolysis. *AAPG Bulletin*, 70(3), 318–329. 10.1306/94885688-1704-11D7-8645000102C1865D

Peters, K. E., Walters, C. C., & Moldowan, J. M. (2004). *The Biomarker Guide* (2nd ed., Vol. 1). Cambridge University Press. 10.1017/CBO9780511524868

Porlles, J., Panja, P., Sorkhabi, R., & McLennan, J. (2021). Integrated porosity methods for estimation of gas-in-place in the Muerto Formation of Northwestern Peru. *Journal of Petroleum Science Engineering*, 202, 108558. Advance online publication. 10.1016/j.petrol.2021.108558

Schmoker, J. W. (1995). Method for Assessing Continuous-Type (Unconventional) Hydrocarbon Accumulations. In Gautier, D. L., Dolton, G. L., Takahashi, K. I., & Varnes, K. L. (Eds.), *National Assessment of United States Oil and Gas Resources: Results, Methodology, and Supporting Data*. U.S. Geological Survey. 10.3133/ds30

Séranne, M. (1987). Evolution tectono-sédimentaire du bassin de Talara (nord-ouest du Pérou). *Bulletin de l'Institut Français d'Études Andines*, 16(3), 103–125. 10.3406/bifea.1987.952

Tafur, H. I. A. (1952). Cretaceous Geology of the East Front of the Amotape Mountains. IPC Report WP-12. *International Petroleum Company*.

Tafur, H. I. A. (1954). *Reconnaissance of Cretaceous Between Chira River and Amotape Mountains, Northwest Peru*. IPC Report WP-14.

U.S. Energy Information Administration. (2013). *World Shale Gas and Shale Oil Resource Assessment. Technically Recoverable Shale Gas and Shale Oil Resources: An Assessment of 137 Shale Formations in 41 Countries Outside the United States.* U.S. Energy Information Administration.

Uyen, D., & Valencia, K. H. (2002). Anexo 16: North-Lancones Basin Surface Geology and Leads Evaluation. In *Informe Final. Segundo Periodo Exploratorio. Lote XII - Cuenca Lancones*. Pluspetrol Peru Corporation S.A.

Walters, C. C., Toon, M. B., Rae, D., Barrow, R., Flagg, E. M., & Hellyer, C. L. (1992). *Talara Basin, Peru: Source Rock Characterization Academic Press.*.

Chapter 7
Lithological Facies Classification, Surface Gamma Ray, and Toc Analysis Using an Outcrop of Unconventional Rock Field:
Case in the Muerto Formation, Peru, for the Identification of New Areas With Hydrocarbon Potential

Brayan Nolasco Villacampa
http://orcid.org/0000-0002-1924-465X
National University of Engineering of Peru, Peru

Jorge Luis Oré- Rodriguez
National University of Engineering of Peru, Peru

José Alfonso Rodriguez-Cruzado
National University of Engineering of Peru, Peru

Israel J. Chavez-Sumarriva
National University of Engineering of Peru, Peru

Manuel Lopez Reale
LCV Group, USA

Humberto Chiriff-Rivera
National University of Engineering of Peru, Peru

DOI: 10.4018/979-8-3693-0740-3.ch007

Lithological Facies Classification, Surface Gamma Ray, and Toc Analysis

Jesus Samuel Armacanqui-Tipacti
National University of Engineering of Peru, Peru

Alfredo Vazquez-Barrios
National University of Engineering of Peru, Peru

Luz Eyzaguirre-Gorvenia
National University of Engineering of Peru, Peru

ABSTRACT

In the present work, the stratigraphic distribution of the Muerto Formation was evaluated, with the objective of identifying the lithological facies, using petrography studies through thin sections and gamma ray analysis for an elevation. The section under study crops out on the eastern margin of the Amotape Mountains. Stratigraphic, facies, and sedimentological analyzes are methods of vital importance to evaluate the potential of an unconventional reservoir. These studies make it possible to predict the spatial distribution of facies, as well as to identify the areas with the most favorable petrophysical properties for the prospecting and exploitation of hydrocarbons, in addition to being able to estimate with greater certainty the volume of hydrocarbons in situ and the optimal completion of a well.

1. INTRODUCTION

Stratigraphic, facies and sedimentological analyzes are methods of vital importance to evaluate the potential of an unconventional reservoir. These studies make it possible to predict the spatial distribution of facies, as well as to identify the areas with the most favorable petrophysical properties for the prospecting and exploitation of hydrocarbons, in addition to being able to estimate with greater certainty the volume of hydrocarbons in situ and the optimal completion of a well. (Kietzmann & Vennari, 2013). Unconventional deposits have been commercially exploited for approximately 20 years in various basins around the world, but in recent years they have sparked renewed interest. (Barreriro & Masarik, 2011) Because total gas production figures will likely increase as new unconventional gas reserves are discovered and quantified and extraction methods improved. Furthermore, it is very possible that unconventional gas reserves greatly exceed conventional gas reserves. The potential that is beginning to be glimpsed in unconventional hydrocarbons, and particularly in unconventional gas, is generating the conviction that the world is on the verge of a true energy revolution. (Luis & Moncayo, 2013)

In Peru there are 18 sedimentary basins with potential for hydrocarbon exploration. All these basins are related, to a greater or lesser degree, to the processes of plate tectonics and uplift of the Peruvian Andes.(Ministerio de Energia y Minas, 2001) 8 of the 18 basins are totally or partially offshore, they are: Tumbes Progreso, Talara, Sechura, Salaverry, Trujillo, Lima, Pisco and Mollendo. The remaining 10 basins are located on the mainland: Lancones, Moquegua, Santiago, Bagua, Huallaga, Ene, Titicaca, Marañón, Ucayali and Madre de Dios. (Ministerio de Energia y Minas, 2001) 4 of these basins concentrate 98% of Peru's exploratory wells, which are the Talara, Sechura, Marañón and Ucayali basins. (Perúpetro, 2017)

In the North of Peru, in the department of Piura, is the Lancones basin, which is the study basin for this project, where favorable conditions exist for a thick sequence of Cretaceous sediments, with diverse horizons rich in organic matter, potential reservoirs in the basal sandstones (Gigantal formation) and attractive folding structures for hydrocarbon exploration. However, it is estimated that the presence of nearby igneous bodies and associated hypoabyssal volcanism may be a risk factor for exploration.(Villar & Pardo, 2010a)

Within the Lancones Basin is the Muerto Formation, which is presumed to be the main source rock of the Lancones Basin, Northwestern Peru. (Álvarez P., 1987), and it is also considered the First Non-Conventional Play Reservoir in this country. (Morales et al., 2018b) The Muerto Formation is of Albian age (Cruzado, 1985), related to a regional marine transgression (Uyen & Valencia, 2002). The Muerto Formation is characterized by the deposition of black clayey fossiliferous limestones with a strong smell of petroleum in the fractures and with pelagic foraminifera. (Fernández et al., 2005)

Petroleum geochemistry is a fundamental factor in understanding the properties of the source rock, such as its productive and non-productive zones, oil migration and oil field development. For this, parameters such as Total Organic Carbon (TOC) were used, which indicates the total amount of organic matter; S1 (mgHC/g rock) records the free hydrocarbons that are released from the rock sample without separating the kerogen during the first heating stage at a temperature of 300°C, and S2 (mgHC/g rock) records the hydrocarbons that are released from the sample during the second stage of programmed heat application of the pyrolysis process that represents the milligrams of residual hydrocarbons contained in one gram of rock, therefore indicating the potential amount of hydrocarbons that the source rock could continue to produce if the thermal maturation process continued. (McCarthy et al., 2011) Previous studies have shown that the geochemistry of the Muerto Formation acts as an excellent source rock. This source rock has a TOC of 0.8 to 4.0 wt.% in calcareous lithology, S1 of 0.05 to 1.56 (mg/g) and S2 of 0.8 to 0.09 to 64.8 (mg /g) (Morales, Porlles, Rodriguez, Taipe, & Arguedas, 2018) which indicates that the organic matter is autochthonous, behaving as a good source rock with type

Lithological Facies Classification, Surface Gamma Ray, and Toc Analysis

III kerogen and capable of generating gas. By thermally maturing, the rock in the Muerto Formation is mature and located in the oil window. (Morales et al., 2018b)

The main objective of this project is the identification and classification of lithofacies in the Muerto Formation, through petrography studies (thin sections), correlating it with TOC analyzes and spectral Gamma Ray measurement carried out on the outcrop, in order to identify the areas with possible hydrocarbon potential within the stratigraphic column of the Muerto Formation.

Figure 1. (A) Location of the area of interest, Lancones Basin, Piura, Peru; and (b) geological outcrops in the Lancones Basin, highlighting the Muerto Formation

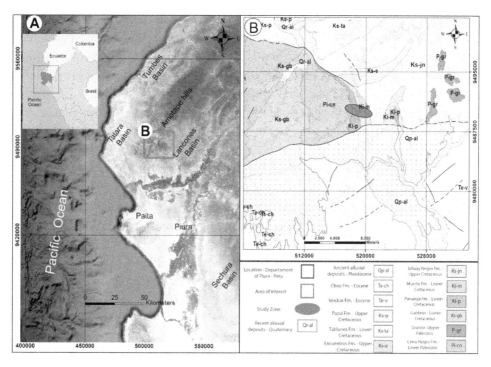

2. GEOLOGICAL SET UP

The Muerto formation is in the Talara Basin and the Lancones Basin. This study focuses on the Lancones Basin, which is located in northwest Peru, in the transition zone from the central Andes to the northern Andes, known as the Huancabamba deflection; and extends south of Ecuador, limited to the east by the Western Cordillera in the portion known as the Celica arc, to the west by the Amotapes Block and to the south by the Sechura basin.(García, Jaimes, Concha, Astete, & Chapilliquen, 2015) In Peru, it covers the region of Piura, province of Sullana. The Lancones Basin has a NNE-SSW direction.

The Muerto formation is attributed to an age between the lower Albian and the early upper Albian. (Jaillard et al., 1999) This era is characterized by a marine transgression that allowed rapid sedimentation of carbonate shelf levels on the Andean continental margin. (Jaillard & Soler, 1994) Furthermore, the middle to upper Albian is known for the manifestations of volcanic emissions and for local extensional tectonic events. (E Jaillard et al., 2000)(Winter, 2008). This formation overlies the Pananga Formation, which is formed by a basal unit of calcareous sandstone and conglomerate with quartzite and shale boulders that changes to neritic and fossiliferous reef-type crystalline limestone. The sequence culminates with the development of widespread anoxic euxinic conditions that characterize the Muerto Formation.

The limestones of the Muerto Formation were deposited in shallow marine or shallow carbonate platform environments, where the relative rise in sea level caused the deposition of fetid black limestones in thin layers of the Muerto Formation, because these sequences were developed in deep internal shelf environments with low energy, no oxygen and plenty of organic content.(García et al., 2015b) The anoxic environments where these limestones were deposited are probably due to the contemporary volcanic events of the San Lorenzo Formation, located in the eastern domain of the basin and which have an age of 104.4 ± 1.9 Ma. (Jaimes et al., 2012)

The upper part is characterized by dark gray limestones and claystones affected by landslide structures (Slump), and olistoliths can also be seen; covered by nodular and flaky micrites in strata of 10 to 15 cm interspersed with finely laminated claystones. (Andamayo, 2008)

Likewise, geochemical studies reveal that the Muerto formation is the unit with the highest organic content, with adequate kerogen and with a thermal maturity that passes from the oil window to that of wet gas and condensate. (Villar & Pardo, 2010b)

Figure 2. Contacto basal de la Formación Muerto, Quebrada Corcobado

3. METHODOLOGY

The area of interest focuses on the outcrops of the Quebrada Corcobado; where various stratigraphic sections were made at a scale of 1:50, following classic methodologies (Reineck y Singh 1973; Walker y James 1992), and likewise, the acquisition of Spectral Gamma Ray (SGR) profiles using the portable device "GMS310 Core Gamma Logger", with a separation between measurements of 50 centimeters. At each recording point, the value of K (%), Th (ppm), U (ppm) and Gamma Ray Total (API) were determined with a count stabilization time of two (2) minutes.

The outcrops of the Corcobado Gorge were studied through various field and laboratory methods, for which various stratigraphic cuts were made along the creek, where 2.5kg samples were extracted approximately every 5 meters vertically, to which Subsequently, X-ray diffraction (XRD), total organic carbon (TOC) and petrographic studies of thin sections were carried out.

For petrographic analysis, 34 thin sections were examined; the same ones that were washed with toluene to eliminate the hydrocarbon present. They are then impregnated with resin to study the pore network, and stained with Alizarin Red-S, for the differentiation of calcite. Clastic rocks were classified according to the terminology proposed by Dott (R. H. Dott, Jr., 1964) y Folk (Folk, 1974). For fine-grained rocks (pelites, shales and related rocks), the proportion of skeletal grains, non-skeletal grains, the percentage of siliciclastic grains or some other significant characteristic, such as diagenetic replacement or bioturbations, was described.

Carbonate rocks are classified following the terminology proposed by Dunham (1962), (Table 1) and pyroclastic rocks are named according to the classification of the IUGS Subcommittee (1980).

Table 1. Classification table of Depositional Textures developed by (Dunham (1962) and modified by Hallsworth & Knox (1999) and Embry and Klovan (1971))

Allochthonous limestone original components not organically bound during deposition						Autochthonous limestone original components organically bound during deposition		
Less than 10% >2mm components				Greater than 10% >2 mm components		Boundstone		
						By organisms which act as barriers	By organisms which encrust and bind	By organisms which build a rigid framework
Contains lime mud (<0.02 mm)			No lime mud	Matrix supported	>2 mm component supported			
Mud supported		Grain supported						
Less than 10% grains (>0.02mm to <2mm)	Greater than 10% grains							
Mudstone	Wackestone	Packstone	Grainstone	Floatstone	Rudstone	Bafflestone	Bindstone	Framestone

The X-ray diffraction analysis was carried out on 30 samples, through the analysis of total rock and clay fraction. To carry out the total rock analysis, a representative sample of the total rock was obtained, grinding the sample to 230 ASTM mesh. The resulting powder was placed in the stainless-steel sample holder and compacted uniformly, obtaining a smooth and regular surface to be exposed to X-rays. In this way all the mineral components present were identified.

In this work, the lower member of the Muerto Formation was described in detail, which has a thickness of 389m according to (Petroperú, 1987), where the geology was petrographically described to subsequently identify the lithofacies.

4. RESULTS

4.1. Lithostratigraphy

From the union of various stratigraphic columns raised in the field, the following stratigraphic column was obtained (Figure 3), which corresponds to the Muerto Formation in the Corcobado ravine with a total thickness of 276.50 meters.

Figure 3. Stratigraphic column of the Muerto Formation in Quebrada Corcobado with 276.50 meters thick, with lithology between limestone, siltstone, marl, and presence of volcanics

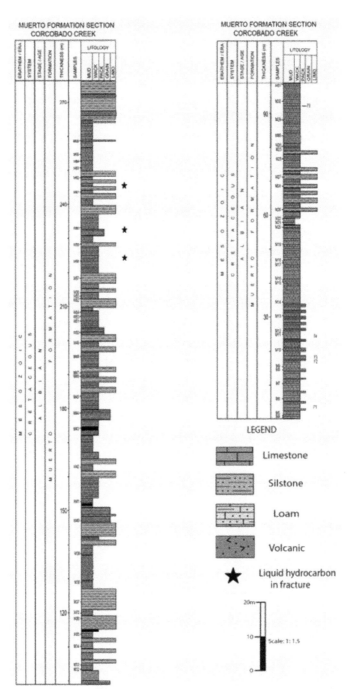

4.2. Geochemical Evaluation

A geochemical study of petroleum is essential to understand the properties of the source rock, being the 3 main characteristics to determine the hydrocarbon potential of the source rock, (1) the geochemical properties of the organic matter, (2) the thermal maturity and (3) the abundance of hydrocarbon. Thermal maturity is one of the most important parameters to know, which determines at what temperature the generating rock can produce oil, natural gas or condensate. ("Evaluación de La Materia Orgánica, Potencial de Hidrocarburos y Madurez Térmica En El Noreste Peruano: Formación Muerto," 2018)

To evaluate the source rock, the Rock-Eval pyrolysis method has been used for this project, where the S1 (mg HC/g) was evaluated, the amount of free hydrocarbons obtained at a temperature of 300°C; S2 (mg HC/g), amount of hydrocarbons that are generated from the cracking of heavy hydrocarbons and the thermal decomposition of kerogen; S3 CO2 (mg CO2/g), represents the CO2 that is released from the thermal cracking of kerogen during pyrolysis; hydrogen index (HI), which indicates the potential for oil generation; oxygen index (OI), which is related to the amount of oxygen contained in the kerogen; the production index (PI), to characterize the evolution of organic matter and Tmax °C, temperature at which the maximum amount of hydrocarbon is generated; which helped to evaluate the geochemical characteristics of the source rock and also determine its hydrocarbon potential.

From these parameters we can describe the quality and quantity of organic matter (S2 and %TOC), generative potential (S1, S2 and %TOC), type of hydrocarbon generated (S2, HI and %TOC) and thermal maturation (S1, S2, %TOC). For this purpose, 19 samples were analyzed that exhibit TOC values in a range of 0.15 and 4.73 wt.% (Table 2). A systematic variation was observed along the section (Figure 4), for which the stratigraphic column was divided into 3 horizons, lower horizon, in the interval of 0-78m (0-255.9 ft); middle horizon between 78-175m (255.9-574.2 ft) and upper horizon in the range of 175-276.5m (574.2-907.15 ft).

Table 2. Bulk geochemical data and parameters produced on 19 analyzed samples

SampleID	Height (m)	TOC wt%	S1 (mgHC/g rock)	S2 (mgHC/g rock)	S3 CO_2 (mgCO_2/g rock)	HI (mg HC/g TOC)	OI (mgCO_2/g TOC)	T_{max} °C
WLUC 64-B	250.5	2.84	0.77	4.81	0.1	169	4	458
WLUC 58-B	223	0.15	0.07	0.15	0.09	99	59	371
WLUC 52-B	215	1.2	1.27	1.99	0.09	166	8	453

continued on following page

Table 2. Continued

SampleID	Height (m)	TOC wt%	S1 (mgHC/g rock)	S2 (mgHC/g rock)	S3 CO_2 (mgCO_2/g rock)	HI (mg HC/g TOC)	OI (mgCO_2/g TOC)	T_{max} °C
WLUC 48-B	195	0.29	0.46	0.37	0.07	127	24	298
WLUC 44-B	178.5	1.01	0.21	0.85	0.2	84	20	447
WLUC 42-B	162.5	4.73	1.02	6.44	0.11	136	2	461
WLUC 40-B	147	0.59	0.02	0.21	0.47	35	79	439
WLUC 39-B	137.5	4.28	0.43	3.16	1.57	74	37	440
WLUC 37-B	123	1.98	0.03	0.48	1.47	24	74	442
WLUC 35-B	113.5	1.95	0.27	0.69	1.43	35	73	440
WLUC 31-B	98.5	0.8	0.01	0.1	0.84	12	104	505
WLUC 29-B	88	4.32	0.86	5.54	0.11	128	3	457
WLUC 25-B	76.5	0.74	0.02	0.25	0.4	34	54	445
WLUC 22-B	59	0.77	0.16	1.01	0.01	132	1	466
WLUC 18-B	46.5	0.32	0.01	0.09	0.15	28	46	505
WLUC 15-B	40.5	0.94	0.31	1.42	0.02	151	2	456
WLUC 13-B	30.5	0.49	0.08	0.38	0.01	77	2	453
WLUC 7-B	10.5	1.26	0.34	1.66	0.01	132	1	461
WLUC 4-B	0.5	1.33	0.44	1.85	0.02	139	2	459

Abbreviations: HC = hydrocarbons; HI = hydrogen index (S2/TOC · 100); OI = oxygen index (S3/TOC · 100); Tmax = the temperature at which peak S2 HC evolution occurs; TOC = total organic carbon

The lower horizon (LM) presents an initial decrease in TOC from an average of 1.29 wt.% to 0.49 wt.% at 30.5 m (100 ft), followed by values less than 1 wt.% of TOC until the beginning of the middle horizon (MM) with 4.32 wt.%. TOC values in the MM fluctuate more relative to the LM, showing minimal variability between 113.5 and 123 m (372.4-403.54 ft), followed by a strong peak of 4.28 wt.% at 137.5 m (451.1 ft), a pronounced drop to 0.59 wt.% at 147 m (482.3 ft) and again, a strong peak of 4.73 wt.% at 162.5 m (533.1 ft). The upper horizon (UM), with the fewest samples, exhibits TOC values that fluctuate less relative to the MM between 0.15 and 1.2 wt.%, followed by a strong peak of 2.84 wt.% at 250.5 m (821.9 ft).

4.2.1. Quantity and Quality of Organic Matter

Determining the quality and total quantity of organic matter in the samples (%TOC) is important for the evaluation of organic content, as well as being able to discover the generation potential that the rock has, using the parameters S1 and S2 for this purpose.

For the samples of this project corresponding to the Muerto Formation, an average TOC of 1.58 wt% is presented, which indicates that it has a regular richness of organic matter, regardless of whether it can produce hydrocarbons or not; In other words, it indicates the quantity of organic matter, but not the quality of the organic matter. For this reason, the parameter S2 was used, which records the hydrocarbons released in the second stage of heat application within the pyrolysis process, that is, it represents the milligrams of residual hydrocarbons contained in one gram of rock, therefore indicating the amount hydrocarbon potential that the source rock could continue to produce if the thermal maturation process continued. (McCarthy et al., 2011)

The graph of S2(mgHC/g) vs TOC (wt.%) shows the quality of the source rock. Where it is observed that the Muerto formation, for the most part, is in the zone from poor to fair, however, 3 samples from the intermediate horizon (MM) present a "good" quality, which are the WLUC29 samples. -B, WLUC39-B and WLUC42-B with an S2 value of 5.54, 3.16 and 6.44 (mgHC/g rock) and TOC of 4.32, 4.28 and 4.73 wt.% respectively, and 1 sample from the upper horizon (UM), WLUC64 -B that presents a "regular" quality with an S2 value of 4.81 and TOC of 2.84 wt.%.

To determine if the hydrocarbon is indigenous (in-situ) or non-indigenous (migrated), the graph of S1(mgHC/g) vs TOC (wt.%) (Figure 4) was used, where it was deduced that the migrated hydrocarbons should have a high S1 and low OCD. Therefore, it was mostly observed in the graph that the hydrocarbon of the Muerto Formation is indigenous (in-situ), without considering 2 samples corresponding to the upper horizon (UM); WLUC48-B and WLUC52-B, with S1 values of 0.46 and 1.27 (mgHC/g rock) and TOC of 0.29 and 1.2 wt.% respectively.

Figure 4. Geochemical analysis of the Muerto Formation, in TOC (wt%), HI and Tmax (°C), lower, middle, and upper horizon

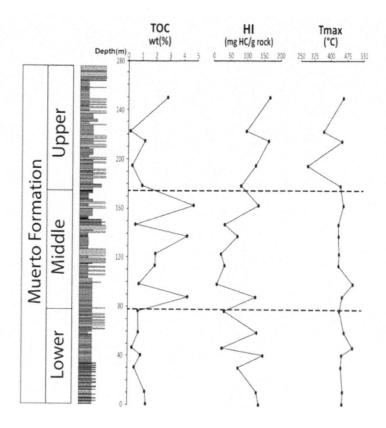

Figure 5. Quality and quantity of organic matter from the Muerto Formation, Lancones Basin, Peru. a)TOC vs S1. b)TOC vs S2

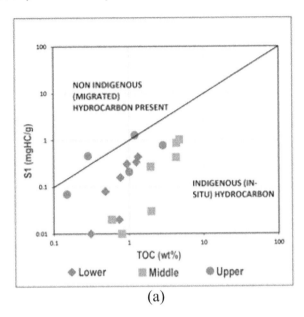

(a)

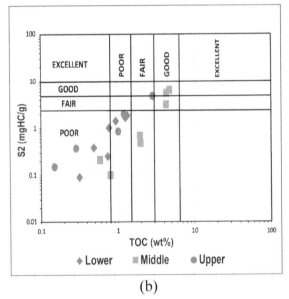

(b)

4.2.2. Generating Potential

The generating potential is the sum of S1 and S2, according to Hunt, source rocks with a GP less than 2, 2 to 5, 5 to 10 and greater than 10 are considered to have a poor, regular generation potential, good and very good respectively. The relationship between the generating potential S1+S2 (mg/g) and TOC (wt.%) is observed in Figure 5, where it is observed that the Muerto Formation presents a poor to good generating potential.

Also plotted, HI (mg/g) vs TOC (wt.%) showing the type of source or type of hydrocarbon that can be produced in Figure 6. For this work, it is predicted to be gas or oil source rock. mostly.

Figure 6. Generating potential of the Muerto Formation, Lancones Basin, Peru

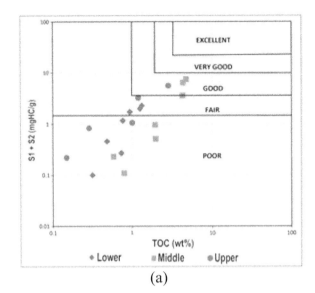

(a)

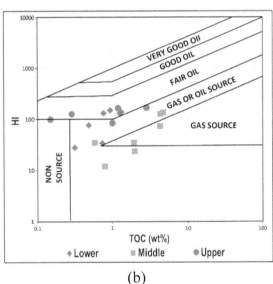

(b)

4.2.3. Genetic Type of Organic Matter

Knowing the genetic type of the organic matter of a source rock is essential for predicting the oil and gas potential it can produce. The hydrogen index (HI) is obtained from the relationship between hydrogen and TOC; It is defined as 100 x S2/

TOC. The HI is proportional to the amount of hydrogen contained in the kerogen; therefore, a high HI indicates a greater potential for oil generation. Likewise, it helps to differentiate the types of organic matter, that is, the type of kerogen.

An HI less than 150mg/g indicates that the source rock has the potential to generate gas (mainly type III kerogen). The HI between 150 to 300 mg/g contains more type III kerogen than type II kerogen and therefore they can generate a mixture of gas and oil, but mostly gas. An HI greater than 300 mg/g contains mostly type II kerogen and therefore is considered to have a good source to generate oil, but to a lesser extent gas.

For this work we will use the graph to identify the type of kerogen according to Langford and Blanc-Valleron, which we will represent in Fig. 6; This shows that the Muerto Formation kerogen represents Type III kerogen.

The oxygen index, OI, is obtained from the relationship between CO_2 and TOC; It is defined as 100 x S3/TOC. The OI is related to the amount of oxygen contained in the kerogen and is useful for tracking the maturation or type of kerogen. Therefore, the HI versus the OI was also graphed, which can be seen in Figure 7; Just as Van Krevelen did, this is used to determine the type of kerogen.

Figure 7. Genetic type of organic matter from the Muerto Formation, Lancones Basin, Peru

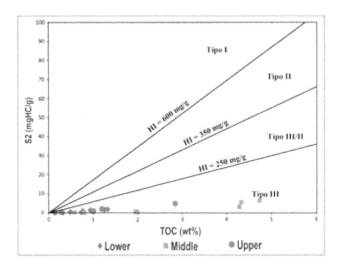

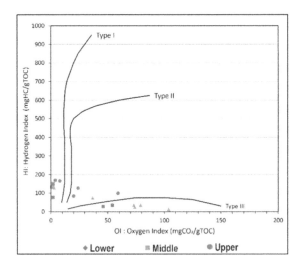

The production index (PI) is obtained from the relationship between the hydrocarbons generated during the first and second stages of the pyrolysis process; is defined as S1/(S1 + S2). This relationship is used to characterize the evolution of organic matter because the PI index tends to increase gradually with depth in a fine-grained rock. Furthermore, it tends to increase with the maturation of the source rock before the expulsion of the hydrocarbons, as the thermally degradable components of the kerogen are converted into free hydrocarbons. Abnormally high values of S1 and PI can also be used to identify oil accumulations or impregnated producing layers.

The PI (Production Index) and Tmax (°C) were also graphed, which can be seen in Figure 8; This is used to determine the production window in which the organic matter is found. The graph shows that the Muerto formation is in the range of oil window to mostly wet gas zone; It is also noted that the Muerto formation has points in the dry gas zone range.

Figure 8. Muerto Formation hydrocarbon production window, Lancones Basin, Peru

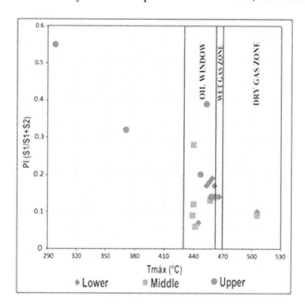

4.2.4. Thermal Ripening

The generation of petroleum from organic matter during its burial history is a part of the total process of thermal metamorphism of organic matter. The concentration and distribution of hydrocarbon contained in a particular source depends on the type of organic matter and its degree of thermal alteration. Peter and Espítale say that the generation of oil in the source rock begins at Tmax = 435-465°C.

For this work, it is shown that the Muerto formation is within the oil and gas generation window zone, likewise, that it is in the mature zone for the most part, with some samples in the post-mature zone.

Figure 9. Thermal maturation of the Muerto Formation, Lancones Basin, Peru

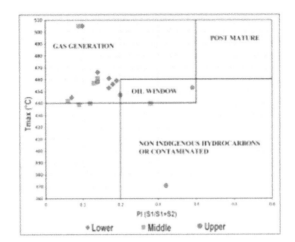

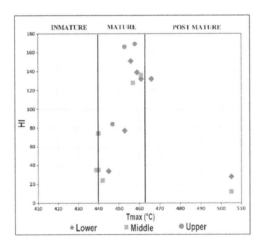

4.3. Gamma Ray

The acquisition of Spectral Gamma Ray (SGR) profiles was based on the use of the portable device "GMS310 Core Gamma Logger", with a vertical separation between measurements of 50 centimeters. At each recording point, the value of K (%), Th (ppm), U (ppm) and Gamma Ray Total (API) were determined with a count stabilization time of two (2) minutes. Based on this, the Muerto Formation was divided into several sections, in order to perform gamma ray analysis on surface outcrops. Showing below some that comprise the basis of the Muerto Formation. Section names are based on a related name in the field for better location of work equipment.

Figure 10. Amphitheater Section and Culebra Section, beginning of the Muerto Formation, Lancones Basin, Peru

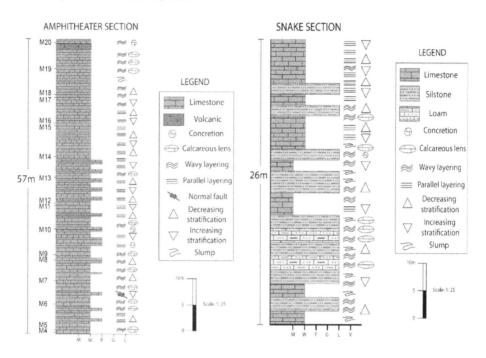

Figure 11. Amphitheater section correlated to gamma ray. The dispersion of the record may be due to the fact that it is the base of the formation, that is, the first meters of sedimentation

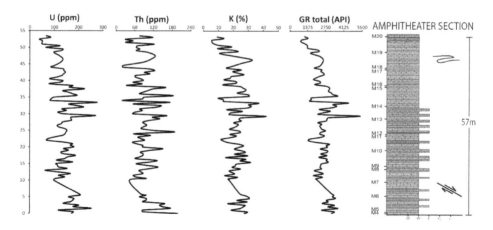

Figure 12. Snake section correlated to gamma ray. A trend of intercalation of the gamma ray record is observed due to the presence of siltstone and limestone

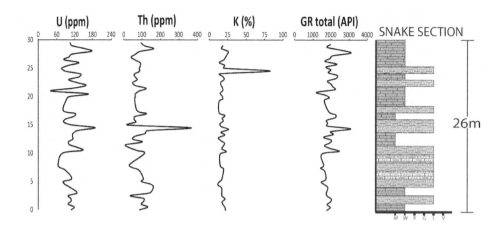

Figure 13. Nap section correlated to gamma ray. A correlation of the volcanic with a low radioactivity and with a more porous horizon compared to the others

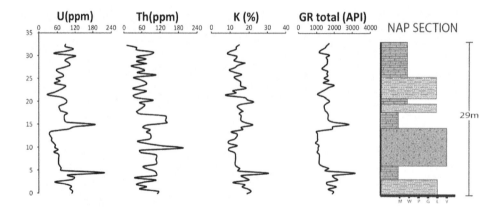

4.4. Lithofacies

To synthesize the main characteristics, the samples were grouped into 6 microfacies, emphasizing the 31 samples corresponding to the Muerto Formation.

Microfacie 1 (MF-1): wackestones, packstones and fossiliferous mudstones

This microfacies includes 16 samples, being the most abundant microfacies. Grouped here are those samples that present abundant calcite content either in the matrix (wackestones and packstones) or with a large amount of calcitic or calcitized skeletal elements (radiolaritic, foraminiferal or fossiliferous mudstones). The composition of the matrix is variable and there does not seem to be a predominance of any component. Silicoclastic grains are rare (7.6% on average) and skeletal grains are frequent on average (31%) and are mainly represented by foraminifera and radiolarians. The proportion of organic matter is variable and is scattered throughout the matrix.

Microfacie 2 (MF-2): mudstones, silty mudstones and silty claystone

This microfacies includes 5 samples. In general, they present a weak parallel and discontinuous horizontal lamination, occasionally affected by bioturbations. The composition of the matrix is siliceous-clayey. They are characterized by having a higher content of silicoclastic elements that reach 25% on average, while skeletal grains reach 16% on average and are mainly represented by foraminifera and radiolarians. The proportion of organic matter is moderate to abundant and is disseminated in the matrix.

Microfacie 3 (MF-3): tuffs/tuffites

This microfacies comprises 3 samples, of which 2 are rich in lithoclasts (samples M23 and M26) and one in plagioclase crystals. The 3 samples are intensely altered. The observed porosity is very low and is linked to the dissolution of plagioclase, that is, it presents intracrystalline texture (photo, arrows pointing to pores impregnated with blue epoxy resin).

Microfacie 4 (MF-4): lithoclastic rocks

Three samples are grouped in this lithofacies: M31, M40 and M44. Despite having different classifications, which vary depending on the percentage of lithoclasts and interstitial material, these 3 samples have the following features in common:
1. In all of them, intraclasts of similar composition are recognized, carbonatic, immersed in predominantly fine silicoclastic interstitial material. If clasts larger than 2 mm are not in contact with each other, and are more than 10%, the rock is called floatstone. If components larger than 2 mm are in contact, it is called rudstone (clasificación de Embry y Klovan, 1971).
2. The interstitial material corresponds to a mudstone with abundant microfossils: radiolarians and foraminifera. It presents few scattered plagioclase crystals. It is interpreted as accumulated in a low energy, possibly dysoxic environment.

3. The intraclasts reach a maximum size of 8 mm, and are fragments with a nodular appearance, clear edges and a composition comparable to a bindstone*: irregular textures, micropeloidal to pelletoidal, with a recrystallized matrix to micro and subsparite.

It is interpreted that the intraclasts come from shallower positions than the interstitial material, indicating erosion and transport of these, which accumulate in low energy positions.

Microfacie 5 (MF-5): dolomite

A single sample belongs to this microfacies (M18). It presents an intense replacement by dolomite. The dolomite appears as euhedral to subhedral crystals (arrows), equigranular in size and with planar S-type fabric. The size of the crystals is very fine (0.1 mm) and organic matter associated with the interstitial material is recognized, among the crystals. A subtle discontinuous and parallel horizontal lamination is preserved. The sample also presents replacement by silica in the form of chert. The composition of the matrix is somewhat clayey calcareous. Silicoclastic grains are very rare (3%) and skeletal grains are rare (10%).

Microfacie 6 (MF-6): peloidal mudstones

This microfacies comprises 3 samples. They present a marked planar, parallel, and discontinuous lamination given by the abundance of peloids and flattened lenses with a concentration of inoceramid remains. The matrix is siliceous with abundant disseminated organic matter. Peloids (non-skeletal grains) represent 25% while silicoclastic and skeletal grains are subordinate. The presence of dispersed altered plagioclase crystals is observed.

Figure 14. (MF-1) M64, mudstone with foraminifera: Abundant skeletal grains (arrows). (MF-2) M25, silicified mudstone: Planar, parallel, and discontinuous lamination, given by the abundance of peloids. (MF-3) M20, Crystalline tuff, altered: Subordinate intracrystalline pores in plagioclase. (MF-4) M31, lithoclastic wackestone: Carbonate lithoclasts of bindstone composition immersed in wackestone or mudstone. (MF-5) M18, Dolomite: Detail of euhedral to subhedral dolomite rhombuses (circle). Parallel Nicoles. (MF-6) M46, peloidal mudstone: Abundant silicoclastic grains (white dots).

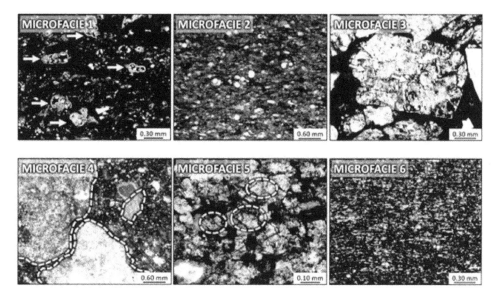

5. CONCLUSION

The Muerto Formation of the Lancones Basin is classified into 6 microfacies along the 276.50 meters.

Microfacies 1 (MF-1) is the most abundant throughout the entire formation and prevails especially in the lower and upper horizon of the Muerto Formation. Likewise, the greatest microfacies variation occurs in the middle horizon of the Muerto Formation.

The greatest fluctuation of TOC values occurs in the intermediate member, having the highest TOC values of 4.32 wt% at 88m (288,714 ft), 4.28 wt% at 137.5 m (451.1 ft), of 4.73 wt% at 162.5 m (533.1 ft).

The Muerto Formation is found in a greater percentage within the gas generation window, but with a poor generating potential.

The greatest hydrocarbon potential would occur in microfacie 1 with TOC values greater than 0.7wt% and with HI values greater than 80 (mg/g).

REFERENCES

Andamayo, K. (2008). Nuevo estilo estructural y probables sistemas petroleros de la cuenca Lancones [Thesis: Ingeniero Geólogo]. Universidad Nacional Mayor de San Marcos.

Barreriro, E., & Masarik, G. (2011). Los reservorios no convencionales, un "fenómeno global." Petrotecnia, 10–18.

Evaluación de la Materia Orgánica, Potencial de Hidrocarburos, & Madurez Térmica en el Noreste Peruano: Formación Muerto. (2018). LACCEI International Multi-Conference for Engineering, Education and Techonoly: "Innovatios in Education and Inclusion," 1–4.

Fernández, J., Martínez, E., Calderón, Y., Hermoza, W., & Galdos, C. (2005). Tumbes and Talara basins hydrocarbon evaluation. Basin Evaluations Group Exploration Department, internal r(PERUPETRO S.A.), 130.

García, B., Jaimes, F., Concha, R., Astete, I., & Chapilliquen, J. (2015). Evolución tectono-sedimentaria del dominio occidental de la Cuenca de Lancones. INGEMMET Revista Institucional, 12–21.

Jaillard, E., Laubacher, G., Bengtson, P., Dhondt, A. V, Bulot, L. G., Cuenca, L., & Suroeste, D. A. C. (1999). Estratigrafía y Evolución de la Cuenca Cretácica Ante-Arco Celica-Lancones en el Suroeste del Ecuador. Abstracto Resumen y Conclusiones : Discusión e Interpretación.

Jaillard, E., & Soler, P. (1994). Cretaceous to early Paleogene tectonic evolution of the northern Central Andes (0-18°S) and its relations to geodynamics. Tectonophysics, 259(1-3 SPEC. ISS.), 41–53. 10.1016/0040-1951(95)00107-7

Jaimes, F., Navarro, J., Alan, S., & Bellido, F. (2012). Geología del cuadrángulo de Las Lomas. INGEMMET Revista Institucional, 59(Boletin N°146), 23–26.

Kietzmann, D. A., & Vennari, V. V. (2013). Sedimentología y estratigrafía de la formación vaca muerta (Tithoniano-Berriasiano) en el área del cerro Domuyo, norte de Neuquén, Argentina. *Andean Geology*, 40(1), 41–65. 10.5027/andgeoV40n1-a02

Luis, F., & Moncayo, G. (2013). Hidrocarburos no convencionales. Tierra y Tecnología, 41(Redacción), 9.

McCarthy, K., Rojas, K., Niemann, M., Palrnowski, D., Peters, K., & Stankiewicz, A. (2011). Basic petroleum geochemistry for source rock evaluation. *Oilfield Review*, 23(2), 32–43.

Ministerio de Energia y Minas. (2001). Cuencas sedimentarias. Author.

Morales, W., Porlles, J., Rodriguez, J., Taipe, H., & Arguedas, A. (2018). First unconventional play from Peruvian northwest: Muerto Formation. SPE/AAPG/SEG Unconventional Resources Technology Conference 2018, URTC 2018, 1–14. 10.15530/urtec-2018-2903064

Perúpetro. (2017). Situacion Actual Y Potencial Hidrocarburifero Del Perú. Author.

Villar, H., & Pardo, A. (2010). Potencial de hidrocarburos y sistemas de petróleo en las cuencas costeras del Perú. Geología Integrada, 6–13.

Winter, L. S. (2008). the Genesis of 'Giant' Copper-Zinc-Gold-Silver Volcanogenic Massive Sulphide Deposits At Tambogrande, Perú: Age, Tectonic Setting, Paleomorphology. *Lithogeochemistry and Radiogenic Isotopes.*, 6(11), 951–952.

Chapter 8
Evaluation of Geological Boundary Conditions of a Water Well for Aquifer Integrity Conformance by Pressure Transient Analysis:
Presence of Hydraulic Fracturing, the Operation of Wastewater, CO2 Storage, and Geothermal Wells

Gustavo Enrique Rodriguez-Robles
National University of Engineering of Peru, Peru

Jesus Samuel Armacanqui-Tipacti
National University of Engineering of Peru, Peru

ABSTRACT

It is very important to determine the characteristics of the nature of geological faults in the stage of production and exploration of hydrocarbons. However, there is no cost-effective methodology to determine the sealing or non-sealing character of a geological fault. There are methods to determine these characteristics, which have some drawbacks, e.g., it is required to maintain the production conditions of an oil well at a constant flow rate or to shut down the production of the well, which entails operating expenses for the period of time that the well is not in production. The

DOI: 10.4018/979-8-3693-0740-3.ch008

objective of this study is to develop a cost-effective and environmentally sustainable methodology in order to determine the effect of geological faults (i.e., boundary effect). Therefore, a water well was located near the geological fault and the pressures that were measured during transient pressure tests were recorded, the benefit obtained by performing a water well was the speed of the pressure response, which It is transmitted more quickly, due to the characteristic of the fluid.

1. INTRODUCTION

In the first months of 2009, an increase in well drilling permits was observed in the United States, which lasted until the beginning of 2015(Vera R., 2016), In consequence, debate grew on the benefits and environmental consequences of hydraulic fracturing (Raimi D., 2017). Besides of the benefits that the increase in hydrocarbon extraction entails through this process, such as: Increased income and employability(Bartik A. et al., 2019), there are many negative consequences of hydraulic fracturing; 1) Consequences to Human Health: Effects on human health for workers and people living near well platforms, new cases of asthma with the development of unconventional reservoirs(Rasmussen S. et al., 2016) ; 2) Consequences for the fauna: Release of gases in the aquifers that would have a polluting effect on farmed animals.(Bamberger M. & Oswald R., 2012) 3) Environmental Consequences: Changes in air quality because these chemicals can have long-term effects(Colborn T., 2011);Methane gas leaks and toxic chemicals seeping from the well(Osborn S. et al.,2011);High concentration values Methane, Ethane and propane in drinking water wells near the hydraulic fracturing zone (Jackson R. et al., 2013); Surface-level spills of hydraulic fracturing fluids in significant volumes and concentrations as a result of operational accidents that reach aquifers or groundwater (Environmental Protection Agency, 2016). Besides, exist Circumstances (causes) where hydraulic fracturing would affect drinking water sources a) Fluid seepage through activation of geological faults(Reagan M. et al., 2015), b)Poor design or operational failure of the casing cementation would result in a leak of hydraulic fracturing fluids(Darrah T et al.,2014), c) Terrible stimulation techniques(Reagan M. et al., 2015) d) Water consumption in areas or times of the year with low water availability.(Environmental Protection Agency, 2016) then it shows in Figure 1 the potential routes that would lead to the contamination of an aquifer: poor operation in the cementation of the casing pipe, migrations of drilling fluids through geological faults to an aquifer. 4) geological consequences: induced seismicity (Hong Z. et al., 2018) in a space-time study of the seismic activity that occurred in the water injection zones for hydraulic fracturing (wastewater disposal) related to the hydraulic fracturing industry in Oklahoma, correlated the incidence of earthquakes with the water injection zones between 2006 – 2015 and 2016 – 2017(Hong Z. et al., 2018). Also, I know presented three factors causing seismicity: (i) Production water injection (disposal) wells (Kendall J. et al., 2019). (ii) fracturing water injection wells next to producing wells (Hornbach M. et al, 2015). (iii) Occurrence of poro-elastic stresses (Buijze L. et al, 2017).

Figure 1. Contamination due to failure in the production casing

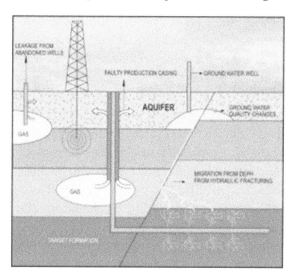

A quantitative and qualitative study of the nature of geological faults would allow increasing the number of future reservoirs for exploration and drilling, due to the fact that in Peru new alternative sources are required that allow increasing the production of oil fields to achieve the development of the hydrocarbon sector (Morelo C., 2015). Peru is a country with hydrocarbon potential because it has 18 sedimentary basins with potential for hydrocarbon exploration, however, only 4 basins are explored (Ministry of Energy and Mines, 2001). Consequently, it is required to explore and develop unconventional reservoirs to increase the number of hydrocarbon reservoirs. The Muerto Formation, located in Piura, Peru, is currently being studied, which has the conditions to be the first unconventional reservoir in Peru due to its thermal maturity and its high content of total organic carbon (TOC). (Morales W. et al., 2018), this would mean a great advance for energy sustainability. Peru is an energy-dependent country because of every 100 barrels that are consumed, 75 of them are imported (Bandach A.,2018). In the year 2020 due to the COVID -19 pandemic and the abrupt drop in the price of oil, oil fields were closed because they are not economically profitable, the closure of the North- Peruvian pipeline caused a decrease of 61.2 MBPD in the month of February 2020 to 31.5 MBPD in the month of May 2020. (Yesquen S., 2020)

In the exploration and production stage it is important to know the characteristics of the geological faults, studies carried out on the nature of the geological faults in aquifers, show the sealing or non- sealing character of the geological faults, which controls aspects of the migration of hydrocarbons and is the least understood factor in the petroleum system (Noufal A. & Obaid K., 2017). Geological faults have three hydraulic properties: 1) Faults as effective conduits for fluid migration. 2) Faults form barriers to fluid flow. 3) Some faults are barriers and others are conduits (Sorkhabi R. et al., 2015). Detection and monitoring of leaks through geological faults near carbon dioxide (CO_2) storage formations is through pressure transient analysis (PTA) (Shchipanov A. et al., 2018).Therefore, for To facilitate the interpretation of PTA, the pressure derivative or Bourdet derivative was studied in geological reservoirs with finite conductivity faults.(Escobar F. et al., 2013).It is because of that Bourdet derivative analysis is a fundamental tool for optimal interpretation of reservoir parameters(Charles D. et al., 2005), what would allow the geological characterization of said reservoir (Bermudez G., 2012).that is more to be a fast and cheap method to estimate variables such as: Effect of well storage(Wellbore storage), formation damage (skin effect), reservoir permeability and boundary effects (sealing failure, partially sealing failure, non-sealing failure, constant pressure)(Escobar F, 2003). Among the transient pressure tests; the restoration test or Build up test is one of the most used and simplest since it does not require careful supervision in terms of flow (Flow = 0) like other pressure test methods (Valencia R., 2008), since the constant flow condition is more easily achieved in the Buildup test (Horne R., 1995). The objective of this research work is to carry out a geological evaluation of the reservoir parameters.

By means of a transient pressure analysis (PTA) in a shallow aquifer, such as: a) Wellbore storage effect; b) Reservoir parameters; c) Boundary conditions, since due to the characteristics of the fluid (water) as well as its high permeability of the sands that are consolidated or little consolidated, it enables the signal to propagate more quickly in addition to or having an invasive character with the environment.

2. METHODOLOGY

2.1 Fluid Level Gauge

The fluid level meter is a tool that allowed to measure the length of the water column in a well, it has a steel sensor incorporated in one end, which emits a sound when it comes into contact with the water, it is manually operated, which allowed the reading to be recorded directly on the equipment's tape. Figure 2 shows the representation of the fluid level meter "Model 101 P2".

Figure 2. Fluid level gauge "Model 101 P2", "Mark Solins"

2.2 Calculation of the length of the water column of a well

For the development of the transient pressure test, one of its many applications was used, eg the calculation of the hydrostatic pressure corresponding to the water column (shallow aquifer), through the use of a Fluid Level Meter, This equipment presents an important characteristic, the measurement from a "reference point", which is located in the lower part. The following steps were carried out to calculate the height of the water column:

i. The depth of the water well was measured, which has a measurement of X meters, where A meters represents the distance between the surface of the well and the contact with the water, in addition B meters is represented by the distance between the contact with the water and the bottom of the well, this measurement does not consider the distance from the bottom of the well to the "Reference Point" which measures 8.5 centimeters, so the depth of the well has a measurement of (A+B) meters. In Figure 3. (left) the scheme of the total length of the well can be seen.

ii. The fluid level meter was raised to a height where it does not emit sound (the reference point is not in contact with the water), which registers a height of Y meters. Figure 3. (right) shows the procedure in which the fluid level gauge stops beeping.

Figure 3. Representation of the steps to obtain the water column

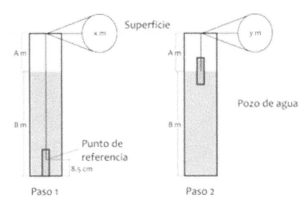

To determine the length of the water column "Y", equations (1), (2) and (3) were used. In addition, with this information the hydrostatic pressure was determined.

(A+B) meters = (X+0.085) meters (1)

A meters = Y meters (2)

Substituting these relationships, we get:

(B) meters = (X − A + 0.085) meters (3)

2.3 Field Procedure

1. The static level measurement (meters) was started while the water well was without any alteration and without turning on the motor pump, to then obtain the static pressure of the water column of the well.
2. The motorized pump was used, which sucked the water at a constant flow rate, this allowed knowing the flow rate that flows in the water well. Once the water level and the production flow of the well were restored, the flow meter was

installed. fluid level inside the water well, to obtain the measurement of the depth of the well and also the level of the water column.
3. The flow rate was kept constant, which made it possible to see the behavior of the well pressure as the test develops.
4. The static pressure values of the well were recorded at different time intervals, to then close the water pump and observe the pressure recovery.

2.4 Data Logging to the Walac 1 Well

The analysis unit (Walac 1 well) is located in the condominium association "Las Violetas" at kilometer 196 of the Panamericana Sur, geographic coordinates of latitude and longitude (-13.5029, -76.180402), in the district of Chincha Baja, Ica province and Ica region. For data collection, a Humboldt motor pump, model GP-80 of 7 hp, with a model UP-170 motor, suction and discharge diameter of 3 inches, was used, whose main function was to simulate well production. of water until a constant pressure exerted by the water column of the well and a constant pumping flow was achieved, which was monitored by the research team.

The initial value of the water column (meters) was recorded and consequently the hydrostatic pressure at the bottom of the well was obtained, in this way it was a reference value of the initial pressure at the bottom of the well since the water level of the well varies depending on the seasons of the year. Next, Fig. 4 can be seen, which is a representation of the height of the water column of the Walac 1 well, where the measurements of the depth of the water well are detailed, as well as the height of the water column. .

Figure 4. Scheme of the Walac 1 water well

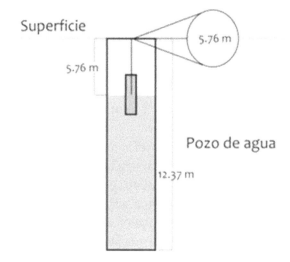

Next, the water motor pump was turned on with the aim of simulating the production of the well and starting the recording of the length measurements of the water column and, consequently, the pressure values exerted by said water column at the bottom of the well. For a period of time of 2 hours of pumping (well production) until a period of time in which the pressure and flow remained constant.

2.5 Simulation of Pressure Value Recording

The Bourdet derivative was simulated and obtained in the production stage (Drawdown) applying a smoothing of 0.8, which is an adequate value where the graph of the Bourdet derivative where imprecise values (noise) were eliminated without altering the representativeness of the data, consequently, a clearA transient pressure analysis technique was used, which consisted of analyzing the stages separately (production and restoration of the well) and interpreting it, then one over the other was superimposed considering the interpretation zones provided by the Bourdet graph.

3. ANALYSIS OF RESULTS

The Bourdet derivative graphs were analyzed in the production (Drawdown) and restoration (Build up) stages separately, considering the characteristics of the water well, ie the nature of the fluid (water); the restoration of the well (Build Up) is faster than in an oil well.

Figure 5 shows a constant pressure system based on the type curves, (Zhao Y & Zhang L.,2011) product of an active aquifer which exerts a direct influence on the response that the borders (faults) would exert. Specifically, this constant pressure effect acts quickly (45 minutes). This constant pressure system is the product of the influence of the internal currents of the aquifer that flows around the water well, consequently, it exerts a pressure accelerating the time of restoration of the water level in the well and therefore it does not allow to see the response of the geological fault (zone 3) that would exist a distance from the well (greater than the investigation radius and the respective investigation radius).

Figure 5. The representation of the Bourdet derivative in the water well, zone 1: the effect of the Storage of the well (early stage); zone 2: the effect of the reservoir (middle stage); zone 3: the border effect (late stage). Green crosses: pressure as a function of time recorded in the well, red circles: the derivative of pressure as a function of time, continuous black line represents the behavior of a derivative of pressure as a function of time and the continuous red line: represents the behavior of pressure as a function of time.

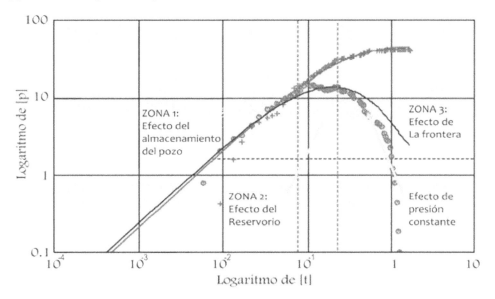

The analysis of the Bourdet graph or double logarithm shows the following:

Zone 1: Or also known as early time region (ETR), the transient pressure response moves through the proximity of the well or in the effect of the filling of the well: it does not allow to define this zone due to the internal currents of the aquifer that flow below the level of the water well this is determined by zone 1 (Storage effect of the well).

Zone 2: Or mean time region (MTR), gives us the response of the pressure from the well to the formation and characteristics of the reservoir. where the flow currents mask the response of the reservoir zone on the Bourdet plot. However, it can be seen in greater detail by superimposing the graphs of the analysis stage, Build Up and Drawdown, because as time passes the behavior of the bottom pressure stabilizes.

Zone 3: Or late time region (LTR), in which the investigation radius has reached the limits of the drainage area and is represented by the boundary effect, a strong effect of constant pressure is shown using the type curves of said characteristic caused by the active aquifer on the possible distant boundary effects of the well, such as: sealing or partially sealing fault, intersecting fault, etc.

4. CONCLUSION

The interpretation of the transient test in a shallow aquifer has the following conclusions:

1) **The well storage effect (wellbore storage):** The Buildup period prevails, masking the determination of the radial flow response, due to the active currents of the aquifer since at the beginning of the test the recorded pressure is that of the well and not that of the reservoir.

2) **Reservoir parameters (reservoir parameters):** The radial flow zone can be identified by interpreting the Drawdown period, and the characteristics pre-established in the simulation can be clearly appreciated by superimposing both periods.

3) **The effect of the border (boundary conditions):** A strong active aquifer effect is shown, which prevails over the sand response and other types of boundary effects due to the constant pressure system produced by the aquifer currents.

The results made it possible to determine the influence of underground water currents (aquifers) against the response provided by the well, in determining the nature of the boundaries or geological faults. This technique (approximation) is very useful for future projects of hydraulic fracturing, geothermal wells and CO_2

storage; because it is cheaper to carry out the study in a water well than in a well in oil production, if the test is carried out when the oil well is in production, it would imply stopping production for a period of time, which would represent operating costs additions in the production stage of the oil well. Performing transient pressure tests in a shallow aquifer is beneficial because the pressure response of the well,

SPECIAL THANKS

Special thanks to the "Applied Research and Technological Development Project 2018-I" E041-2018- 01-BM, to the "World Bank", to "CONCYTEC" and "FOND-ECYT" as financing entity of the Subproject: 61096 and contract 139.

REFERENCES

Bamberger, M., & Oswald, R. (2012, May). Impacts of Gas Drilling on Human and Animal Health. *New Solutions*, 22(1), 51–77. 10.2190/NS.22.1.e22446060

Bandach, A. (2018). Hydrocarbons Potential Of Peru. Academic Press.

Bartik, A., Currie, J., Greenstone, M., & Knittel, C. R. (2019, October). The Local Economic and Welfare Consequences of Hydraulic Fracturing. *American Economic Journal. Applied Economics*, 11(4), 105–155. 10.1257/app.20170487

Bermudez. (2012). Reservoir characterization through the interpretation of pressure tests, Capaya formation, Tácata and Tacat fields, Anzoátegui and Monagas states.

Buijze, L. (2017, December). Fault reactivation mechanisms and dynamic rupture modeling of depletion-induced seismic events in a Rotliegend gas reservoir. *Netherlands Journal of Geosciences*, 96(5), s131–s148. 10.1017/njg.2017.27

Charles, D., Rieke, H., & Pal, S. (2005). Pressure Transient Model Characterization of Sealing and Partially Communicating Strike-Slip Faults. *Proceedings - SPE Annual Technical Conference and Exhibition*. 10.2118/96774-MS

Colborn, T., Kwiatkowski, C., Schultz, K., & Bachran, M. (2011, September). Hazard Assessment Articles Natural Gas Operations from a Public Health Perspective. *Human and Ecological Risk Assessment*, 17(5), 1039–1056. 10.1080/10807039.2011.605662

Darrah, T. H., Vengosh, A., Jackson, R. B., Warner, N. R., & Poreda, R. J. (2014, September). Noble Gases Identify the Mechanisms of Fugitive Gas Contamination in Drinking-Water Wells Overlying the Marcellus and Barnett Shales. *Proceedings of the National Academy of Sciences of the United States of America*, 111(39), 14076–14081. 10.1073/pnas.1322107111 25225410

Environmental Protection Agency. (2016). Hydraulic Fracturing for Oil and Gas: Impacts from the Hydraulic Fracturing Water Cycle on Drinking Water Resources in the United States. Available: www.epa.gov/hfstudy

Escobar, F. (2003). Modern Well Pressure Analysis. Academic Press.

Escobar, F., Martinez, J.-A., & Montealegre-Madero, M. (2013). Pressure Transient Analysis for a Reservoir with a Finite-Conductivity Fault. *CT&F Ciencia, Tecnología y Futuro*, 5(June), 15. 10.29047/01225383.53

Hong, Z., Moreno, H. A., & Hong, Y. (2018). Spatiotemporal Assessment of Induced Seismicity in Oklahoma: Foreseeable Fewer Earthquakes for Sustainable Oil and Gas Extraction? Geosciences, 15. 10.3390/geosciences8120436

Hornbach, M. J., DeShon, H. R., Ellsworth, W. L., Stump, B. W., Hayward, C., Frohlich, C., Oldham, H. R., Olson, J. E., Magnani, M. B., Brokaw, C., & Luetgert, J. H. (2015, April). Causal factors for seismicity near Azle, Texas. *Nature Communications*, 6(1), 6728. Advance online publication. 10.1038/ncomms772825898170

Horne, R. (1995). Modern Well Test Analysis A Computer-Aided Approach (2nd ed.). Academic Press.

Jackson, R., Vengosh, A., Darrah, T. H., Warner, N. R., Down, A., & Poreda, R. J. (2013). Increased Stray Gas Abundance in a Subset of Drinking Water Wells near Marcellus Shale Gas Extraction. *Proceedings of the National Academy of Sciences*. 10.1073/pnas.1221635110

Kendall, Butcher, Stork, Verdon, Luckett, & Baptiste. (2019). How Big is a Small Earthquake? Challenges in Determining Microseismic Magnitudes. .10.3997/1365-2397.n0015

Ministry of Energy and Mines. (2001). Cuencas sedimentarias. *Atlas Mining and Energy in Peru*, 1, 1.

Morales, W., Porlles, J., Rodriguez, J., Taipe, H., & Arguedas, A. (2018). First Unconventional Play from Peruvian Northwest: Muerto Formation. *SPE/AAPG/SEG Unconventional Resources Technology Conference*, 14,, 10.15530/urtec-2018-2903064

Morelo, C. (2015). Shale Gas Development Perspective in Peru. National University of Piura. Available: http://repositorio.unp.edu.pe/handle/UNP/866

Noufal, A., & Obaid, K. (2017). *Sealing Faults: A Bamboozling Problem in Abu Dhabi Fields. Abu Dhabi International Petroleum Exhibition & Conference. Society of Petroleum Engineers*. SPE. 10.2118/188775-MS

Osborn, Vengosh, Warner, & Jackson. (n.d.). Methane Contamination of Drinking Water Accompanying Gas-Well Drilling and Hydraulic Fracturing. *Proc Natl Acad Sci*.

Raimi, D. (2017). *The Fracking Debate: The Risks*. Benefits, and Uncertainties of the Shale Revolution. 10.7312/raim18486

Rasmussen, S. G., Ogburn, E. L., McCormack, M., Casey, J. A., Bandeen-Roche, K., Mercer, D. G., & Schwartz, B. S. (2016, September). Association Between Unconventional Natural Gas Development in the Marcellus Shale and Asthma Exacerbations. *JAMA Internal Medicine*, 176(9), 1334–1343. 10.1001/jamainternmed.2016.243627428612

Reagan, M. T., Moridis, G. J., Keen, N. D., & Johnson, J. N. (2015). Numerical Simulation of the Environmental Impact of Hydraulic Fracturing of Tight/Shale Gas Reservoirs on near- Surface Groundwater: Background, Sase Cases, Shallow Reservoirs, Short-Term Gas, and Water Transport. *Water Resources Research*, 51(4), 2543–2573. Advance online publication. 10.1002/2014WR01608626726274

Shchipanov, Kollbotn, & Berenblyum. (2018). Fault Leakage Detection from Pressure Transient Analysis. .10.3997/2214-4609.201802990

Sorkhabi, R., Suzuki, U., & Sato, D. (2000). Structural Evaluation of Petroleum Sealing Capacity of Faults. *SPE Asia Pacific Conference on Integrated Modeling for Asset Management*, 9. 10.2118/59405-MS

Valencia. (2008). *Conventional Analysis And Interpretation Of Pressure Tests*. Academic Press.

Vera, R. (2016). *Eagle Ford Shale Play: Oil-Mining Industrial Geography in Southernn Texas, 2008-2015.SciELO Analytics*, 34.

Yesquen, S. (2020). Oil and gas after COVID 19: An Opportunity to Reconfigure the Sector. Academic Press.

Zhao, Y., & Zhang, L. (2011). Solution and Type Curve Analysis of Fluid Flow Model for Fractal Reservoir. *World Journal of Mechanics*, 1(5), 209–216. 10.4236/wjm.2011.15027

Section 3
Operation Module

Chapter 9
Cost-Effective Advanced Remote Diagnostics of Sucker Rod Pumping Wells From Dynamometric Charts:
A Deep Learning Approach

Joel Hancco Paccori
https://orcid.org/0009-0003-4219-5432
National University of Engineering of Peru, Peru

Manuel Castillo Cara
https://orcid.org/0000-0002-2990-7090
Polytechnic University of Madrid, Spain

Jesus Samuel Armacanqui-Tipacti
National University of Engineering of Peru, Peru

ABSTRACT

There is a growing number of oil production wells in the world that use rod pump units as an extraction system. In fact, this lifting method is become the preferred one for unconventional wells that are producing at the late-stage period, yet with still attractive rates of a few hundred bopd. Dynamometry basically consists of the visual interpretation of the shape of the load graph based on the position of the piston of the subsoil pump. This task is carried out by an operator and his experience is used for the correct interpretation that can be contrasted. With additional tests, diagnosis becomes very important because it allows optimizing production by adjusting rest

DOI: 10.4018/979-8-3693-0740-3.ch009

and production times, reducing operation and maintenance costs, avoiding failures and unscheduled stops, but many times there are not enough trained personnel for interpretation. In this context, dynamometry has been refined both in the acquisition and in the interpretation of dynamometric records.

1. INTRODUCTION

Mechanical pumping is the most widespread artificial lift system in oil fields and contributes a significant volume to global oil production. For this reason, different techniques have been developed for the diagnosis and optimization of its operation, from the simplest ones such as the manometric test to the most sophisticated ones such as dynamometry. Dy- namometry is the most widely used and accepted technique for the diagnosis of mechanical pumping units, its bases have been developed by Gilbert (1936) and Fagg (1950) more than 50 years ago. This technique is based on the interpretation of the shape of the dynamometric chart, which is the graph of the measured or estimated load versus position on the piston in the subsurface pump. The traditional method of interpretation consists of visual inspection performed by trained and experienced personnel, but many times due to the number of records generated primarily by continuous

With the development of image processing techniques and neural networks, automatic diagnosis methods based on one-to-one architecture and multiclass classification have been implemented. But extracting just one operating condition from a dynamometric chart wastes all the information that the dynamometric chart can provide. In general, most research related to the automatic diagnosis of dynamometric charts report metrics greater than 99% under the one-to-one architecture for operating conditions ranging from 6 to 30 classes. Works such as those carried out by Boguslawski, Boujonnier, Bissuel-Beauvais, Saghir and Sharma (2018) and Nazi, Ashenayi, Lea and Kemp (1994) allow reporting percentage values that indicate the probability of belonging to one or more operating conditions, even though these models have not been trained for the simultaneous diagnosis of several operating conditions. However, they open the possibility of diagnosing conditions simultaneously from a dynamometric chart. The contributions of this research to the task of automatic computer diagnosis are listed below:

- In this work, a data set is assembled that is made up of a dynamometric chart as an instance (x) and an objective (y) which is a label that contains all the operating conditions identified in the chart.

- Regarding training, the proposed models seek to simultaneously identify all operating conditions in a dynamometric chart, going from one to one to one to one to one to many architecture. In this way, more information is extracted from the dynamometric chart and consequently the operation of the unit.
- To evaluate the performance of these multiple classification models, it is proposed to make a modification to the traditional confusion matrix, with the objective of extracting metrics that allow measuring the performance of deep learning architectures.

2. PREVIOUS CONCEPTS

Concepts such as the basics of the mechanical pumping system, dynamometry and dynamometric charts are exposed.

2.1. Rod pump system

Mechanical pumping is the simplest, most efficient, and adaptable artificial lift system for most types of fluids. For this reason, it is the most widespread oil extraction system in the world. Mechanical pumping is a system that provides the energy necessary to lift the fluid from the bottom of the well to the surface. It consists of three parts: a surface unit, a system of rods that connect the surface installation with the bottom, and a pump. subsoil. Surface equipment has a drive unit which may be a gas or electric motor, a gearbox, counterweights, an outrigger, a "horse head", reins, a polished bar, etc. In the Figure 1 Conventional type pumping unit shown.

Figure 1. Conventional rod pump unit

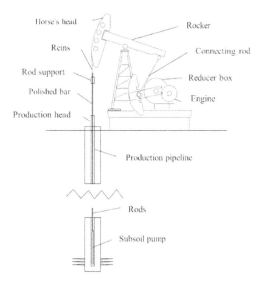

2.2. Dinamometría

The bases for dynamometric analysis for mechanical pumping units were developed by Gilbert (1936) and Fagg (1950). This technique consists of visual interpretation of the shape of the graph of the load as a function of the position measured on the piston of the subsurface pump. This graph represents the work done by the unit in each pumping cycle and allows inferring the operating condition of the mechanical pumping system, In Figure 2 the construction of a dynamometric chart for a condition of complete pump filling is shown.

Figure 2. Bottom dynamometric chart in a pumping cycle

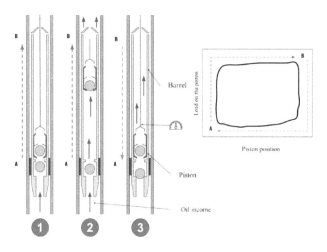

2.3. Dynamometric Charts

Dynamometric charts are capable of identifying a numerous number of operating conditions depending on their shape. In the Figure 3 some cards are shown theoretical dynamo- metric measurements of the most important operating conditions. However, in real conditions it is common to find a combination of several operating conditions present on a dynamometric chart.

Figure 3. Theoretical background dynamometric charts for each operating condition

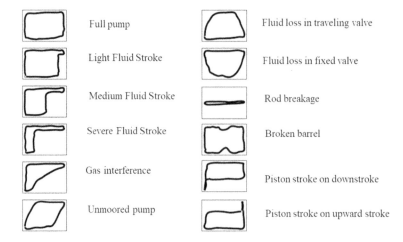

3. PREVIOUS WORK

Computer-assisted automatic diagnosis of dynamometric charts is a topic that has recently gained enormous importance due to the large number of records generated, this as a consequence of the cheaper acquisition systems, connectivity, the implementation of computer systems in the edge and the integration of these to the cloud Xia, Boyu, Lan and Lixia (2020), Peng (2019) and Carpenter (2020). This is reflected in the number of articles published in the last five years, the majority under the one-to-one architecture, that is, an operating state is identified from a dynamometric chart. Currently, a precision of over 99% has been achieved, making it an interesting alternative to the lack of production operators who can interpret dynamometric charts.

3.1. Dynamometric Diagnosis Using Classification Algorithms

The problem of interpreting dynamometric charts can be posed as a problem of image classification based on descritors Nascimento, Laurindo, Maitelli and Cavalcanti (2021), Ramez, Mahmoud and Ahmed (2020) and Tenorio-Trigoso, Castillo-Cara, Mondragón-Ruiz, Carrión and Caminero (2021).

Classification algorithms were explored such as *decision tree, random forest, Gradient eXtreme Gradient Boosted Trees, Regularized Logistic Regression, Random Forest Classifier, Light Gradient Boosted Trees con SoftMax, ExtraTrees Classifier y Light Gradient Boosted Trees* and *XGBoost* achieved metrics exceeding 99% under one-to-one architecture.

3.2. Dynamometric Diagnosis Using Deep Learning Algorithms

Most works related to deep learning focus on the diagnosis of mechanical pumping units by applying computer vision techniques such as convolutional neural networks: Wang, He, Li, Wang, Dou, Xu and Lipei (2021), Ramez et al. (2020), Cheng, Yu, Zeng, Osipov, Li and Vyatkin (2020), Peng (2019) and Ameni (2019).

In the same way, the indicators for these techniques exceed 99% accuracy and some models have been put into production in Chinese oil fields, Boguslawski et al. (2018).

4. METHODOLOGY

The methodology used in the research can be seen in Figure 4. First, the assembly of the data set is carried out from the dynamometric records. Then two diagnostic techniques are proposed under the one-to-many architecture, the first is a regression model and the second is a comparison model. Finally, the metrics are presented to evaluate the performance of the suggested models.

4.1. Data Mining in Dynamometric Records

The dynamometer records are the result of execution of diagnostic and optimization tests of mechanical pumping units, they are made up of a set of measurements such as: load, position, acceleration of the polished bar on the surface, pumping cycle time, load, position, acceleration on the piston of the subsoil pump and complementary data such as the dynamic level of fluid and the current in the supply line of the electric motor.

Figure 4. Flowchart of proposed methodology for the research

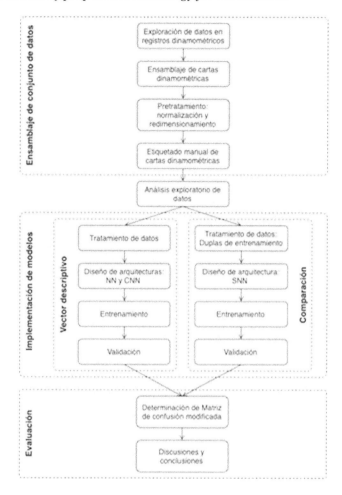

From this set of data, the variables necessary for the construction of the bottom dynamometric chart are: the position and load on the piston of the subsurface pump. These data are estimated from surface measurements.

4.2. Criteria for Identifying States of Operation

Inspired by the personnel training process for the interpretation of dynamometric charts, the operating conditions were divided into two groups: (i) operating conditions whose identification depends on a specific location on the dynamometric chart and (ii) operating conditions that For identification it is necessary to consider the entire chart .In the first group we can find conditions such as pump stroke on upstroke and downstroke, fluid loss in pump barrel, etc. In the second group we have complete pump filling, partial pump filling, gas interference, fluid loss in traveling or fixed valve, etc. See Figure 3.

4.3. Diagnostic Methods for Dynamometric Charts

Two techniques are proposed for the identification of operating conditions in a dynamometric chart: (i) the first is based on regression models to estimate a descriptive vector that contains the identified operating conditions, (ii) the second is based on the comparison models between theoretical conditions and the objective vector of the data set.

4.3.1. Diagnosis Using Descriptive Vector Estimation

This technique identifies the operating conditions on a dynamometric chart by estimating the descriptive vector. The descriptive vector contains in a binary way the presence or absence of one or more operating conditions.

The descriptive vector is obtained in two ways: (i) through dense neural networks and (ii) through convolutional neural networks, the architectures of these models can be seen in Figures 5 and 6 respectively.

Figure 5. Dense neural network architectures for descriptive vector estimation

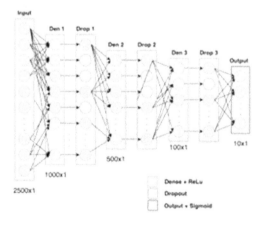

Figure 6. Convolutional neural network architectures for descriptive vector estimation

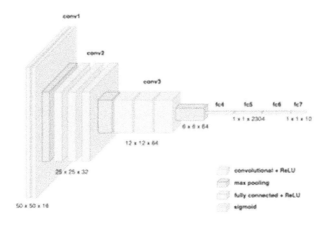

The objective of minimizing the error between the vector estimated by the model and the target vector.

$$MSE = 1 \sum_{i=1}^{n} (y_i - \hat{y}_i)^2 \qquad (1)$$

where:

n: Number of points
y_i: Observed vector
$\hat{y}_i$: Estimated vector

In Figure 7, you can see the training process of the diagnostic models by estimating the descriptive vector. The input data (X) is the cards Dynamometric and target (Y) labels are in a binary vector format. The dynamometric charts enter the neural network which estimates a model that has as its output the descriptive vector, the latter is compared with the objective vector through a loss function that is the Mean Square Error (MSE) according to the equation 1. The output of the loss function together with the training metrics modify the weights of the convolutional neural network.

Figure 7. Training the regression model for the estimation of the descriptive vector

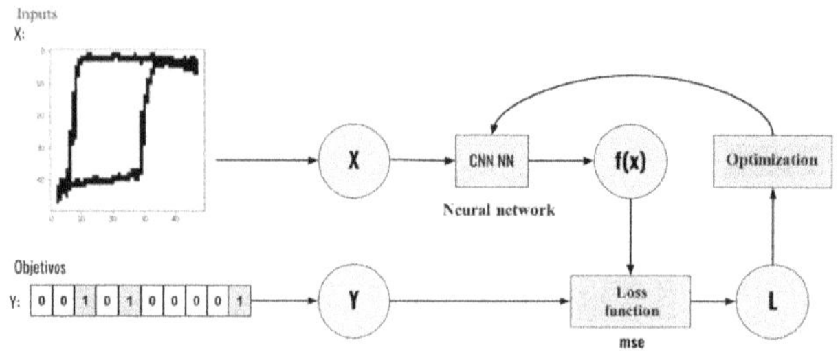

4.3.2. Diagnosis Using Comparison Techniques

This technique performs the diagnosis through the bi- nary comparison between the input data (dynamometric charts) and the theoretical operating conditions. For this purpose, Siamese neural networks are used where the paradigm of training changes, while in classification models training consists of determining the membership of an instance to a class, in comparison models training consists of determining whether the pair of instances belong to the same class or not. In the Figure 8 The architecture of a Siamese neural network is shown. The input data are pairs of images that may or may not belong to the same class; The images pass through convolutional layers that are identical and share the training weights. The output of the convolutional layers is transformed into vectors that contain the characteristics of the images. Then, both vectors are compared using the Euclidean distance, Equation 2. Finally, the value of the Euclidean distance goes through a binarization function whose output is binary, it is 1 if they are of the same class.

The comparison function is the Euclidean distance be- tween two points:

$$d(x_i, x_j) = \|F(x_i) - F(x_j)\| \qquad (2)$$

where:

$F(x_i)$: Feature vector of the first convolutional neural network.
$F(x_j)$: Feature vector of the second convolutional neural network.

Figure 8. Siamese convolutional neural network architecture

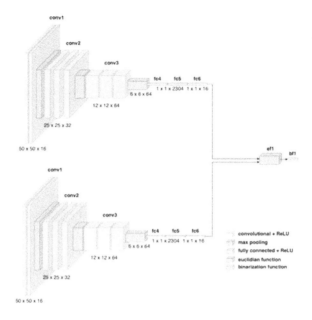

4.4. Metrics

The metrics to evaluate the performance of the models are obtained from a modified confusion matrix, see Figure 9. In this matrix, the double entry (actual and prediction values) is preserved and a row and a column are added. The row records the conditions that are present in the target and that the model has not been able to identify, while the column shows conditions that the model has identified but are not present in the actual record.

Figure 9. Modified confusion matrix for simultaneous multi- class classification

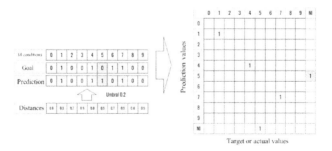

The suggested metrics are:

- Accuracy: It is the proportion of predictions that the model classified correctly with respect to the total.

$$Accuracy = \frac{TP + TN}{TP + TN + FP + FN} \qquad (3)$$

where:

TP: True positive, is the correct estimate of the positive class.
TN: True negative, it is the correct estimate of the negative class
FP: False positive, it is the incorrect estimation of the positive class.
FN: False negative, it is the incorrect estimation of the negative class.

- Precision: also called positive predictive value, is the proportion of positive identified conditions that are correct.

$$Precision = \frac{TP}{TP + FP} \qquad (4)$$

- Sensitivity: real success rate, It is the proportion of positive identifications that are correct.

$$Recall = \frac{TP}{TP + FN} \qquad (5)$$

- Specificity: rate of negative hits, it is the proportion of real negatives and is opposite to sensitivity.

$$Specificity = \frac{TN}{TN + FP} \qquad (6)$$

- F1-score: is the harmonic mean of the precision and sensitivity, estimates a measure of the precision and robust- ness of the model.

$$F1 = \frac{2(precision \cdot recall)}{precision + recall} = \frac{2TP}{2TP + FP + FN} \qquad (7)$$

- Macro F1-score: this metric considers all classes as basic elements of the calculation, each class has the same weight in the average, so there is no distinction between very sparsely populated classes. To obtain Macro

F1- Score, it is necessary to calculate the average macro precision and the average macro sensitivity according to Equations 8 and 9 respectively.

$$MAP = \frac{\sum_{k=1}^{K} Precision_k}{K} \qquad (8)$$

$$MAR = \frac{\sum_{k=1}^{K} Recall_k}{K} \qquad (9)$$

$$MacroF1 - Score = \frac{2 * (MAP * MAR)}{MAP + MAR} \qquad (10)$$

where:

MAP: Macro Average Precision
MAR: Macro Average Recall

5. EXPERIMENTAL RESULTS AND EVALUATION

The results are presented below.

Figure 10. Example of dynamometric charts processed from dynamometric records

5.1. Analysis of the Data Set

A data exploration was carried out in the dynamometric records, selecting the variables necessary for the construction of the background dynamometric chart. The dynamo- metric chart was graphed and resized to a size of 50x50. In the Figure 10 a sample of the data set is observed.

Subsequently, manual labeling was carried out by trained personnel following the process described in Figure 11. The operating conditions identified on the torque chart are recorded in a descriptive vector of length 10. Each position represents an operating condition where 1 means that the condition is present on the torque chart and 0 means that the condition is not present.

Figure 11. Manual labeling process of dynamometric charts according to the identified operating conditions

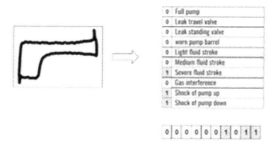

In the Figure 12 The number of operating conditions identified in the manual labeling process is shown. An imbalance is evident in the number of classes found, but as the conditions are dependent on each other, it is not possible to completely balance them due to their physical nature. For example, the pump hit condition on upstroke or downstroke does not appear independently, but is tied to a card with full bomb fill or fluid hit.

Table 1. Deep neural network architecture for the diagnosis of pumping units

Layer	Activation function	Input	Output
Input	ReLu	2500	2500
Dense 1	ReLu	2500	1000
Dense 2	ReLu	1000	500
Dropout 1		500	500
Dense 3	ReLu	500	100
Dropout 2		100	100
Output	Sigmoid	100	10

Figure 12. Number of operating conditions identified in the labeling process

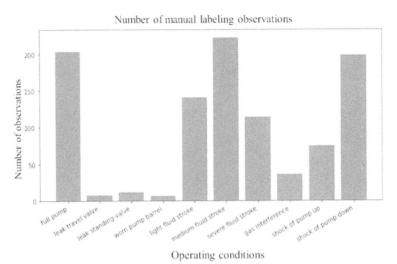

5.2. Implementation of Diagnostic Models

Two techniques were implemented for the identification of operating conditions in the dynamometric charts: (i) by estimating a descriptive vector using regression models and (ii) by comparison techniques using Siamese neural net- works.

5.2.1. Diagnosis by Estimating a Descriptive Vector

Scenario 1: Descriptive vector estimation using deep neural networks.

The deep neural network architecture shown in Table was implemented.1, the dynamometric chart of dimension 50x50 is resized into a vector of length 2500. This vector enters the neural network through the input layer whose activation function is the Rectified Linear Unit (ReLu) and an initialization of random weights, He normalization. Then, there are alternating layers with ReLu activation function and regularization layers. Finally, it goes through an output layer with a sigmoid activation function of dimension 10, which is the descriptive vector. A value close to 1 in the out- put vector indicates the presence of the observed condition.

For training, the data set was divided: 60% for training and 40% for validation. The Figure 13 shows the training result for 50 epochs where Figure 13 (a) shows the values of the loss function, Mean Squared Error (MSE) and Figure 13 (b) shows the

Mean Absolute Error (MAE) metric, with an Adam optimizer and a learning rate of 0.005. The convergence of both parameters is observed from epoch 30 onwards.

Figure 13. Deep neural network training and validation values for 50 epochs

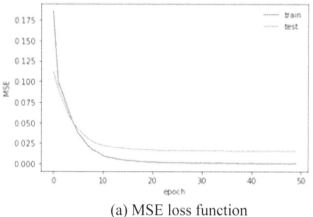

(a) MSE loss function

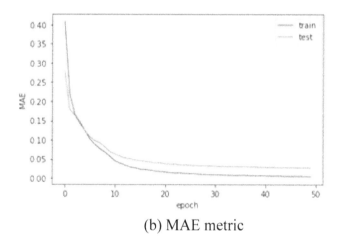

(b) MAE metric

In the Figure 14 Some examples of interpretation of dynamometric charts are observed using the architecture of deep neural networks to estimate the descriptive vector. It can be seen that the estimated values are not binary, but are between 0 and 1, so it will be necessary to determine a threshold to determine the existence or not of any condition. According to Figure 14 In the first dynamometric chart, two operating conditions can be visually identified: (i) complete pump filling (full pump) and (ii) pump shock on downward stroke (shock of pump down). The vector estimated by the model is [0.997, 0., 0.001, 0., 0.001, 0.001, 0.002, 0., 0.001,

0.999] where the values of positions 1 and 10 are close to 1, which indicates that these conditions have been identified. This estimated vector is plotted next to the dynamometric chart to visualize the diagnosis. In the second dynamometric chart the model identifies two operating conditions: (i) piston stroke on upward stroke (shock of pump up) and (ii) medium fluid stroke, both with values close to 1. Finally, the third Dynamometric chart shows a slightly more complex condition in which the model identifies up to three conditions:

(i) Severe fluid stroke, which is the highest value and is contrasted with the real diagnosis;
(ii) Medium fluid stroke (medium fluid stroke) with a lower magnitude compared to the severe fluid stroke; and
(iii) Pump shock on the downward stroke (shock of pump down) with a minimum value and which is far from a correct interpretation.

Figure 14. Examples of dynamometric chart diagnosis through descriptive vector estimation using deep neural networks

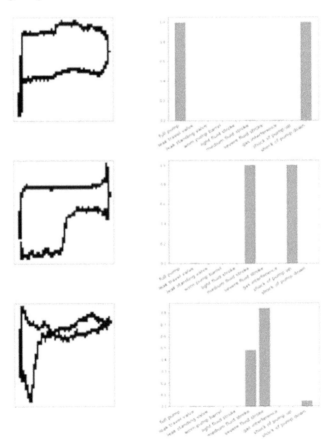

Because the performance of the model depends on the threshold selected to identify the operating conditions. The Macro F1-score metric was evaluated with different thresh- olds (0.5 to 0.9) and the one with the highest value (best overall performance) was chosen, which is 0.6 with a Macro F1-score indicator of 0.9291. These results are shown in Table 2.

The results of applying the 0.6 threshold can be seen in the modified confusion matrix shown in Figure 15. The unidentified row and column list conditions that the model has not been able to identify or has incorrectly identified.

For example, for the full pump operating condition, the model has found this condition in 69 dynamometric charts, but it has not been able to identify the condition in 3 dynamometric charts; on the contrary, it has found this condition in 8 dynamometric charts incorrectly.

Table 2. Evaluation of the deep neural network model using the Macro F1-score metric for different thresholds

Conditions	Metrics F1-score/umbral			
	0.5	0.6	0.7	0.9
full pump	0.9333	0.9262	0.9262	0.9252
leak travel valve	1.0000	1.0000	1.0000	0.6667
leak standing valve	1.0000	1.0000	1.0000	1.0000
worn pump barrel	1.0000	1.0000	1.0000	1.0000
light fluid stroke	0.9231	0.9231	0.9217	0.8411
medium fluid stroke	0.9535	0.9595	0.9412	0.9221
severe fluid stroke	0.9574	0.9565	0.9565	0.9318
gas interference	0.6667	0.7143	0.7143	0.6154
shock of pump up	0.8727	0.8929	0.8518	0.6808
shock of pump down	0.9125	0.9308	0.9114	0.8947
Macro F-1 score	0.9250	**0.9359**	0.9282	0.8627

Figure 15. Confusion matrix for diagnostic model based on deep neural networks with threshold of 0.6

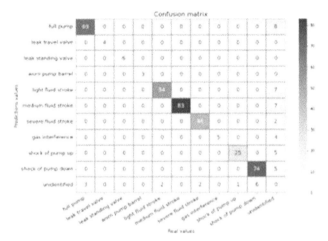

Scenario 2: Descriptive vector estimation using convolutional neural networks.

The proposed architecture is shown in Table 3 where there is an input layer of dimension 50x50x1. The Conv2D layers apply a convolution with a 3x3x1 kernel, a ReLu activation function, and padding: same. In the subsampling layers (Max-Pooling2D) a 2x2 kernel is applied and the dimensions are reduced. The flatten layer resizes the 12x12x16 matrix to a vector of length 2304. The dense layers apply a ReLu activation function and are alternated with Dropout regularization layers that have a parameter of 0.1. The output layer Output has a dimension of 10, which is the estimated descriptive vector, a sigmoid activation function is applied to have an output between 0 and 1.

For training, 60% of the data set is used and 40% is left for validation. The weights are initialized randomly, and the Adam optimizer is used with a learning rate of 0.001. Figure 16 (a) shows the loss function which is the mean squared error and Figure 16 (b) shows the metric which is the mean absolute error. In the Figure 16 The training process is shown for 150 epochs where the metric and the loss function reach stable values from epoch 80 onwards. Because the estimate of the descriptive vector varies in the range from 0 to 1, it is necessary to apply a threshold to binarize this vector. But the model is sensitive to the choice of this threshold, for this reason, the performance of the model is evaluated using different thresholds between 0.5. and 0.9. The results can be seen in the Table 4 where the threshold that has the best metric is 0.8 with a Macro F1- score of 0.9618. In the Figure 17 The confusion matrix generated using this threshold is shown.

Table 3. Convolutional neural network architecture for the diagnosis of pumping units

Layer	Activation F.	Input	Output	Kernel
Input		50x50x1	50x50x1	
Conv	ReLu	50x50x1	50x50x32	3x3x1
MaxPooling		50x50x32	25x25x32	2x2x1
Conv	ReLu	25x25x32	25x25x16	3x3x1
MaxPooling		25x25x16	12x12x16	2x2x1
Flatten		12x12x16	2304	
Dense	ReLu	2304	512	
Dropout		512	512	
Dense	ReLu	512	128	
Dropout		128	128	
Dense	ReLu	128	32	
Dropout		32	32	
Output	Sigmoid	32	10	

Figure 16. Convolutional neural network training and validation values for 150 epochs

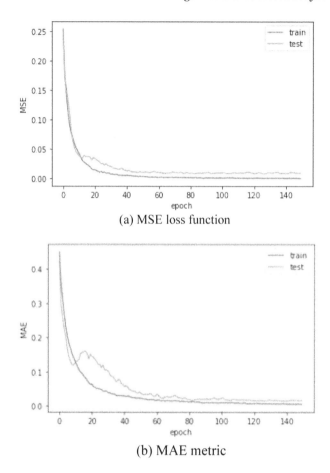

(a) MSE loss function

(b) MAE metric

Table 4. Evaluation of the convolutional neural network model using the Macro F1-score metric for different thresholds

Conditions	Metrics F1-score/umbral			
	0.6	0.7	0.8	0.9
full pump	0.9804	0.9804	0.9804	0.9804
leak travel valve	1.0000	1.0000	1.0000	1.0000
leak standing valve	1.0000	1.0000	1.0000	1.0000
worn pump barrel	1.0000	1.0000	1.0000	1.0000
light fluid stroke	0.9756	0.9756	0.9756	0.9756

continued on following page

Cost-Effective Advanced Remote Diagnostics of Sucker Rod Pumping Wells

Table 4. Continued

Conditions	Metrics F1-score/umbral			
	0.6	0.7	0.8	0.9
medium fluid stroke	0.9944	0.9944	0.9888	0.9888
severe fluid stroke	0.9574	0.9574	0.9677	0.9677
gas interference	0.8000	0.8000	0.8000	0.7143
shock of pump up	0.9153	0.9153	0.931	0.9310
shock of pump down	0.9434	0.9434	0.9494	0.9367
Macro F-1 score	0.9598	0.9598	**0.9619**	0.9553

Figure 17. Confusion matrix for diagnostic model based on convolutional neural networks with threshold of 0.8

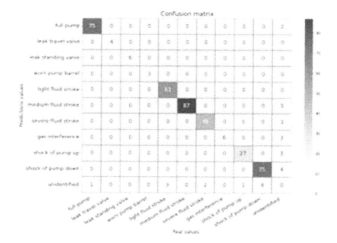

5.2.2. Diagnosis by Comparison Techniques

A mechanical pumping unit diagnostic model based on comparison algorithms was implemented and trained following the one-shot learning approach. This model is implemented under the Siamese convolutional neural network architecture where the reference image is a theoretical dynamometric chart. The Siamese neural network architecture can be visualized in Figure 18, this architecture consists of two stages: (i) The feature extraction stage and (ii) comparison stage using Euclidean distance.

One shot learning, in this technique the training of the Siamese neural network is developed by comparing two images (dynamometric charts). To do this, it is necessary to assemble the training pairs together with the target value. In the Figure 19 The assembly process of the training set is observed. Each of the dynamometric charts of the training set is compared with the 10 theoretical dynamometric charts of the operating conditions, if there is any coincidence. The value that accompanies this pair is 1, otherwise it is 0. For operating conditions where there is a spatial dependence, the comparison is made by focusing attention on the respective quadrant. For example, for the pump hit condition during the upstroke, the comparison is given in the upper right quadrant of the image and not across the entire dynamometric chart.

Figure 18. Architecture for the comparison of generated vectors generated by the Siamese neural network

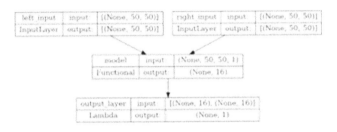

Figure 19. Assembly of training data set for the one-shot learning algorithm

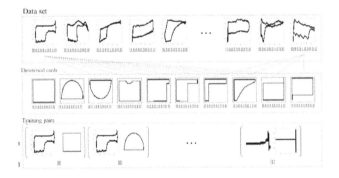

For training the Siamese neural network under the one- shot learning technique, the data set was divided into 60% for training and 40% for validation. The shared weights of the network are initialized with random values, the Adam optimizer was defined with a learning rate of 0.001, the Euclidean distance loss function was used and it was trained for 30 epochs. The training of the model considered the attention criteria for the conditions that are identified in specific places on the dynamometric chart, such as pump shock in upstroke and downstroke (shock of pump up and shock of pump down). The results are seen in Figure 20 where the values of the loss function for training and validation are shown, the loss function reaches convergence from epoch 25 onwards approximately where there is no longer a significant improvement.

Figure 20. Training Siamese neural network for Euclidean distance loss function for 30 epochs

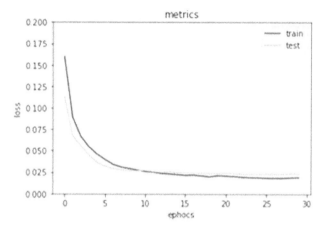

In the Figure 21 an example of diagnosis using the one- shot learning technique is shown. The test dynamometric chart is compared to the 10 theoretical images used in train- ing. The distances are located below each pair of images. The blue color indicates a distance below the threshold of 0.2 and red a distance above this threshold. A smaller distance indicates that the images are more similar or similar, therefore, choosing an appropriate threshold is crucial to properly identify the operating conditions present in the test dynamometric chart.

Figure 21. Euclidean distances of a dynamometric chart with respect to the theoretical ones, in blue distances below the threshold of 0.2

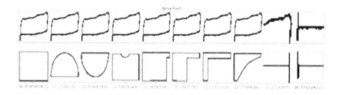

By plotting the distances estimated by the model with respect to each theoretical operating condition as shown in Figure 22 we can better visualize the function of the threshold. In this case, a threshold of 0.2 allows the two operating conditions present in the test dynamometric chart to be adequately identified: full pump and shock of pump down. If we move the threshold below 0.1, the model stops identifying the shock of pump down condition, so its performance decreases.

Figure 22. Plot of Euclidean distances and operating conditions estimated by the diagnostic model using Siamese neural networks and one-shot learning for a threshold of 0.2 (red line)

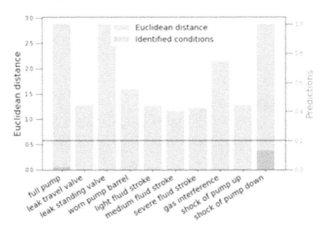

The performance of the one-shot learning model with different thresholds was evaluated using the Macro F1-score metric. The evaluation range is from 0.1 to 0.5 because Euclidean distance values close to 0 indicate greater similarity between the images that are compared. In Table 5 shows the evaluation result where the threshold of 0.2 is the one with the best Macro F1-score metric with a value of 0.9372. With this threshold, the confusion matrix shown in Figure 23 was determined.

Table 5. Evaluation of the convolutional neural network model using the Macro F1-score metric for different thresholds

Conditions	Metrics F1-score/umbral			
	0.1	0.2	0.3	0.4
full pump	0.9286	0.9116	0.8765	0.7553
leak travel valve	1.0000	1.000	1.0000	1.0000

continued on following page

Cost-Effective Advanced Remote Diagnostics of Sucker Rod Pumping Wells

Table 5. Continued

Conditions	Metrics F1-score/umbral			
	0.1	0.2	0.3	0.4
leak standing valve	0.7273	1.000	0.6154	0.5333
worn pump barrel	1.0000	1.000	1.0000	1.0000
light fluid stroke	0.9815	0.9402	0.8730	0.7801
medium fluid stroke	0.9730	0.9246	0.9200	0.8720
severe fluid stroke	0.9804	0.9524	0.9273	0.8095
gas interference	0.9412	0.9412	0.8649	0.7234
shock of pump up	0.6316	0.8627	0.8000	0.6944
shock of pump down	0.3929	0.825	0.8538	0.8085
Macro F1-score	0.8857	**0.9372**	0.8822	0.8126

Figure 23. Confusion matrix for diagnostic model based on Siamese neural networks and one-shot learning with threshold of 0.2

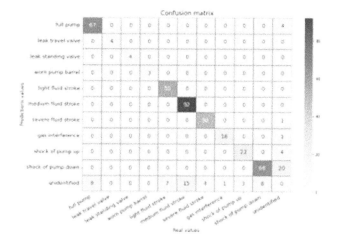

5.3. Discussions

From the diagnostic techniques through descriptive vector estimation (Scenario 1 and 2) and through comparison techniques, all models have managed to exceed the indicator of 0.9 for the Macro F1-score metric, which is an indicator of the robustness of these models. . But the model that has the best performance according to this metric is the Scenario 1 architecture with an indicator of 0.9619 above 0.9359 and 0.9372 for the Scenario 2 architecture and the Siamese neural networks architecture respectively. This shows the importance of spatial analysis through convolutional filters. On the other hand, the comparison model under the Siamese neural network architecture, despite using convolutional filters, is observed to have less performance than the descriptive vector estimation technique of Scenario 2. Specifically for the pump hit operating conditions in downward stroke (shock of pump down) and pump shock in upward stroke (shock of pump up), which are conditions that have a specific location on the dynamometric chart. In Table 4 it is observed that these operating conditions have the lowest indicators for all metrics. This is explained because the comparison model for conditions that have a specific location only analyzes the quadrant of interest and not the dynamometric chart as a whole, which leads to erroneous identifications, reducing its performance. Another aspect is shown by analyzing the Tables 2, 4 and 6 that contain the performance metrics: Ac- curacy, Precision, Recall, Specificity and F1- score and take into account counts the optimal thresholds for the models of Scenario 1, Scenario 2 and Siamese neural networks respectively. It is observed that the diagnostic techniques using the descriptive vector (Scenario 1 and 2) have good performance for most operating conditions, except for the gas interference condition; the opposite occurs for the comparison strategy. using Siamese neural networks that have better performance for this condition. This fact is because training using the comparison technique performs learning individually for each condition, which allows it to extract greater details from the conditions individually. Unlike the estimation technique using the descriptive vector, whose training is generalized, so it will be more difficult to extract features individually.

6. CONCLUSION AND FUTURE CHALLENGES

Of the points mentioned, according to the results of this research, we can mention:

- The set of data assembled for the interpretation - The development of the models allows training of multi-class and multi-objective classification algorithms. (one to many architecture).
- The modified confusion matrix proposed allows recording targets with multiple classes and makes it possible to extract metrics for evaluating the performance of classification models.
- The proposed techniques: Descriptive vector estimation and comparison, allow multiclass classification under the one-to-many architecture with a performance greater than 0.93 for the Macro F1-score metric, which indicates that they are robust models.

- The descriptive vector estimation technique measurement: When faced with convolutional neural networks, it has the best performance taking the Macro F1-score metric carefully and it has a greater tolerance to the threshold hyperparameter.

In addition, this work has contributed to the experimental knowledge of the oil industry and computational diagnosis techniques based on dynamometric charts, being tested in the automatic diagnostics system in oil fields of the United States, Mexico and Peru. Further work is proposed below:

- Increase in records of the data set of training: One of the main limitations to overcome is access to the data source to increase the number and diversity of records. After the gathering of records comes the manual labeling of the identified operating conditions, which must necessarily be done with the intervention of one or more production operators with the appropriate expertise.
- Comparison methods: As explained in the development of the present work, in the discussion section of comparison techniques, it is necessary to implement an architecture that considers focused interpretation and weights the dynamometric chart as a whole in the same way.
- Model verification in production: Based on the estimates generated in a production environment, it is necessary to evaluate the models, as well as expand the operating conditions identified with the data resulting from the well start-up.

REFERENCES

Ameni. (2019). Smart Sucker Pimp Failure Analysis with Machine Learning. Montan Universität Leoben - Thesis .

Boguslawski, Boujonnier, Bissuel-Beauvais, Saghir, & Sharma. (2018). IIoT Edge Analytics: Deploying Machine Learning at the Wellhead to Iden- tify Rod Pump Failure. SPE Middle East Artificial Lift Conference and Exhibition. .10.2118/192513-MS

Carpenter, C. (2020, March 1). Carpenter, 2020. Dynamometer-Card Classification Uses Machine Learn-ing. *Journal of Petroleum Technology*, 72(3), 52–53. Advance online publication. 10.2118/0320-0052-JPT

Cheng, Y., Zeng, O., & Li, V. (2020). Automatic Recognition of Sucker-Rod Pumping System Working Conditions Using Dynamometer Cards with Transfer Learning and SVM. *Sensors (Basel)*, 20(19), 5659. Advance online publication. 10.3390/s2019565933022966

Fagg. (1950). Dynamometer Charts and Well Weighing. Petroleum Transactions.

Gilbert. (1936). An Oil Well Pumping Dynagraph. API Drilling and Production Practice.

Nascimento, L., Maitelli, C., Maitelli, C., & Cavalcanti, A. (2021). Diagnostic of Operation Conditions and Sensor Faults Using Machine Learning in Sucker-Rod Pumping Wells. *Sensors (Basel)*, 21(13), 4546. Advance online publication. 10.3390/s2113454634283092

Nazi, A., & Lea, K. (1994). Application of Artificial Neural Network to Pump Card Diagnosis. *SPE Comp App*, 6(6), 9–14. Advance online publication. 10.2118/25420-PA

Peng. (2019). Artificial Intelligence Applied in Sucker Rod Pumping Wells: Intelligent Dynamometer Card Genertion, Diagnosis and Failure Detection Using Deep Neural Networks. SPE Annual Technical Conference and Exhibition. .10.2118/196159-MS

Ramez, M. (2020). Identification of Downhole Conditions in Sucker Rod Pumped Wells Using Deep Neural Networks and Genetic Algorithms. *SPE Production & Operations*, 35(2), 435–447. Advance online publication. 10.2118/200494-PA

Tenorio-Trigoso, A., Castillo-Cara, M., Mondragón-Ruiz, G., Carrión, C., & Caminero, B. (2021). An Analysis of Computational Resources of Event- Driven Streaming Data Flow for Internet of Things: A Case Study. *The Computer Journal*. Advance online publication. 10.1093/comjnl/bxab143

Wang, H., Li, W., & Dou, X. (2021). A Working Condition Diagnosis Model of Sucker Rod Pumping Wells Based on Deep Learning. *SPE Production & Operations*, 36(2), 317–326. Advance online publication. 10.2118/205015-PA

Xia, B., & Lan, L. (2020). Trend prediction of pumping well conditions using temporal dynamometer cards. *2nd International Conference on Machine Learning, Big Data and Business Intelligence (MLBDBI).* 10.1109/MLBDBI51377.2020.00039

Chapter 10
Use of Limonene as a Biodegradable Surfactant for the Inhibition and Removal of Paraffins in Oil Production Operations

Maria Rosario Viera-Palacios
National University of Engineering of Peru, Peru

Cesar Lujan Ruiz
National University of Engineering of Peru, Peru

Miguel Ángel Guzmán
https://orcid.org/0009-0000-4784-9181
Peruvian University of Applied Sciences, Peru

Guillermo Prudencio
https://orcid.org/0000-0002-8408-1797
National University of Engineering of Peru, Peru

Jesus Samuel Armacanqui-Tipacti
National University of Engineering of Peru, Peru

Luz Eyzaguirre-Gorvenia
National University of Engineering of Peru, Peru

ABSTRACT

Paraffin buildup is one of the most common problems in oil production operations. The agglomeration of these alkanes in the production tubing and the surface flow lines not only lowers the production rate but also results in an additional damage repair cost. It also represents a threat to both the operation of the equipment (rods and subsoil pumps) and to the environment, and it is a risk for the personnel who carry out the paraffin removal work, since the conventional chemicals used, such as xylene and toluene, are highly toxic and polluting. The presence of paraffins in

DOI: 10.4018/979-8-3693-0740-3.ch010

crude oil is frequent in unconventional fields or share oil fields, which have API gravity values greater than 35. In conventional fields, cases of paraffinic crude oil can be found; especially in the final part of the field's life cycle. In the present work, crude oil samples were used from a conventional field located in the Talara Basin, in Northern Peru, which had an API gravity (ASTM D1298) of 36.

1. INTRODUCTION

Paraffinic crude oils, despite having a high value for their derivatives, present themselves as a problem due to the complexity of their production (León, E., 2011). When these are at temperatures below the cloud point, paraffin crystallization and deposition occur in the pipes and flow lines, which cause partial or total blockage of these (Alnaimat, F. & Ziauddin, M., 2023).

In the Talara basin, Northwestern Peru, the crude oil is highly paraffinic, the fields are over 40 years old and have marginal production, so production is low; presenting the deposition of paraffins, increasing production costs (Viera, M., 2020). The increase in cost is due to the reduction in the volumetric efficiency of the well and the application of conventional removal treatments in both production tubing and flow lines.

Some conventional removal treatments, such as thermal and microbiological, present benefits such as increased hydrocarbon recovery and the decomposition of heavy fractions in the crude oil, respectively (González, D. et al., 2010). However, they are not widely used because the first is economically unfeasible for long sections of pipes due to the high energy consumption that it presents and the second, being susceptible to pressure and temperature, is limited by the operating conditions and the time in which production ceases.

On the other hand, the most used treatment to mitigate paraffin problems is chemical. There are 2 ways to address this drawback: prevention through the use of inhibitors and removal using solvents and surfactants (Khabullina, K., 2016). The most commonly used chemical agents for these purposes are aromatic organic compounds, such as xylene, toluene and benzene. However, these hydrocarbons are highly toxic and polluting (Tanirbergenova, S. et al., 2022), so they can affect the health of operating personnel and the environment. Therefore, the search for a replacement for these substances is a necessity.

The present research work proposes and evaluates limonene, a natural chemical agent extracted from lemon, as an alternative to conventional substances for the inhibition and removal of paraffins.

2. MATERIALS AND METHODS

2.1 Equipment and Materials for Paraffin Removal Testing

The oil samples used for the study were provided by the company Savia Perú from its LO16-28D well, which operates in Northwest Peru. This sample was used because it generated paraffin accumulation problems in the pipelines. It is important to mention that the sample did not present water content.

The experimentation was carried out in the laboratories of the Faculty of Petroleum and Natural Gas Engineering of the National University of Engineering of Peru.

2.2 Characterization of the Crude Oil Sample

For the characterization of the oil, the experimental procedures of the American Society for Testing and Materials (ASTM) were followed. The parameters analyzed were the water and sediment content present, API gravity, viscosity and pour point. The standards used for this purpose were ASTM D4007 (ASTM International, 2010), ASTM D1298 (ASTM International, 2010), ASTM D445 (ASTM International, 2010), and ASTM D97 (ASTM International, 2010), respectively. The determination of the chemical composition was achieved through Non-Standard SARA analysis.

For asphaltenes, the crude oil sample was weighed, and 100 ml of hexane was added. A millipore (filter paper) was weighed and placed in a filtration set to filter the mixture. Subsequently, the millipore was dried in an oven at 90 °C and cooled in a desiccator. The millipore was finally weighed and the percentage of asphaltenes present in the crude oil sample was determined (Formula 1):

$$\%Asf = \frac{((millipore\ weight + asphaltenes\ weight) - millipore\ weight)}{crude\ oil\ sample\ weight} \times 100 \qquad (1)$$

The filtered mixture passed through a chromatographic column; product of the hexane present, only the saturated ones will not remain trapped in it. The volume that passed through the column was collected in a previously weighed glass (Empty glass weight No. 1) and dried in an electric cooker until the sample was evaporated. The glass was weighed again (Final glass weight No. 1) and the percentage of saturates present in the crude oil sample was determined (Formula 2):

$$\%Sat = \frac{(glass\ weight\ N°1\ final - glass\ weight\ N°1\ empty)}{crude\ oil\ sample\ weight} \times 100 \qquad (2)$$

100 ml of toluene were measured and added to the mixture trapped in the chromatographic column, achieving the dragging of the aromatics to the outside. The volume that passed through the column was collected in a previously weighed glass (Empty glass weight No. 2) and dried in an electric cooker until the sample evaporated. The glass was weighed again (Final glass weight No. 2) and the percentage of aromatics present in the crude oil sample was determined (Formula 3):

$$\%Hoop = \frac{(glass\ weight\ N°2\ final - glass\ weight\ N°2\ empty)}{crude\ oil\ sample\ weight} \times 100 \qquad (3)$$

100 ml of a mixture of 60% toluene and 40% methanol were measured for washing the chromatographic column. The volume that passed through the column (representing the resin content) was collected in a previously weighed glass (weight empty glass No. 3) and dried in an electric cooker until the sample evaporated. The glass was weighed again (Final glass weight No. 3) and the percentage of resins present in the crude oil sample was determined (Formula 4):

$$\%Res = \frac{(glass\ weight\ N°3\ final - glass\ weight\ N°3\ empty)}{crude\ oil\ sample\ weight} \times 100 \qquad (4)$$

The Colloidal Instability Index (CII) was calculated with the percentage of organic fractions determined from the crude oil sample (Formula 5):

$$CII = \frac{(\%Sat + \%Asf)}{(\%Aro + \%Res)} \times 100 \qquad (5)$$

Finally, thanks to the SARA Analysis determined by liquid chromatography, the stability of the crude oil sample was evaluated using the Colloidal Stability Index (Asomaning, S. & Watkinson A.P., 2000):

- If CII > 0.9, asphaltenes are unstable in crude oil.
- If 0.7 < CII < 0.7, there is uncertainty about stability.
- If CII < 0.7, asphaltenes are stable in crude oil.

2.3 Selected Surfactant and Paraffin Removal Tests

2.3.1 Limonene-Based Chemical Agent

The experiments were carried out with the limonene-based surfactant (Melikyan et al., 1997), a chemical agent chosen for this research work due to its organic origin, biodegradability and non-toxic (EPA, 2011).

Table 1. Properties of the limonene-based chemical agent

Chemical Agent Properties	Values
flash point	> 95°F
pour point	-2°F
Kinematic viscosity (40°F)	1.137368 SUS
Specific gravity (60°F)	1.11
pH	13.4
Surfactant Type	Nonionic
Additives	None
Solubility	Miscible in oil, water, etc.

Note. Data extracted from the classification of the Environmental Protection Agency (EPA-USA)

2.3.2 Qualitative Paraffin Removal Tests: Presence of Crude Oil and Paraffins

2.3.2.1 Manual Paraffin Removal Tests: Presence of Crude Oil and Paraffins

The action of the chemical agent for paraffin removal was simulated through the oil circulation process in a well production pipe. Paraffin samples were cut into 1 cm side cubes and weighed before placing them in a glass beaker. The paraffin blocks were placed on the interior walls of the bottom of the glass, forming a circle (Figure 1). Then, the crude oil sample was poured, and a circular movement of the mixture was generated in the glass with the hand at a constant rate for 15 minutes. Finally, the reduction in the volume of paraffin in the glass was observed, obtaining qualitative results.

Figure 1. Circular placement of solid paraffins in a glass vessel (Own elaboration)

2.3.2.2 Experimental Dynamic Paraffin Removal Tests With a Testing Loop

A testing loop was designed to evaluate the performance of the chemical agent for paraffin removal considering a moving fluid under the operation of a low flow positive displacement pump. A constant flow pump, a manometer, a 40 cm long and 12.5 mm diameter tube, connections and a stopwatch were used. The connections were tested, and the flow rates calibrated (indicated by the pump manufacturer) depending on the operating speed using a blue fluid and measuring the time and length that the fluid traveled through the pipe until similar values were achieved in the repetitions.

Figure 2. Graphic representation of testing loop water footprint reduction (Own elaboration)

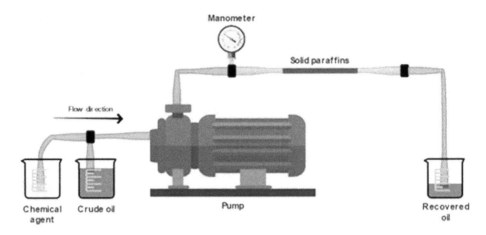

After calibration of the flow rates, a 25.5 mm glass tube, which has solid paraffins inside, was installed at the end of the testing loop (see Figure 2). Thus, paraffin removal tests were carried out with the chemical agent simulating the dynamic conditions of an oil production well, obtaining qualitative results.

2.3.3 Paraffin Removal Test With the Use of a Magnetic Stirrer

Paraffins were introduced in a circular manner at the bottom of two glass vessels, where the samples were weighed on an electronic balance. The two glasses were filled until they had a 150 ml sample of paraffin plus crude oil and chemical agent; but of the latter, 0.25 ml was used in one glass and 0.5 ml in the other. Dynamic conditions were initiated for both vessels using a magnetic stirrer; At the end, the paraffin samples introduced were weighed again before starting the removal test.

2.3.4 Quantitative Paraffin Removal Tests: Presence of Crude Oil and Paraffins

2.3.4.1 Determination of the Paraffin Removal Effect by Measuring the Weight of Paraffin Samples With and Without Chemical Agent

The effect of moving crude oil in contact with paraffin deposited on the walls of the pipes was simulated. Two experiments were carried out in two glass vessels: the first with 12 g of paraffin, 40 ml of crude oil and 0.75 ml of chemical agent inside, and the second with 12 g of paraffin and 40 ml of crude oil. Both samples were allowed to rest for 24 hours to be shaken at 300 RPM for 15 minutes with a magnetic stirrer (Figure 3). The contents were poured out and the samples were weighed and compared; Then, the experiment was repeated with the same conditions and the samples were weighed and compared again.

Figure 3. Paraffin and crude oil samples, with and without chemical agent, in magnetic stirrers at 300 RPM (Own elaboration)

2.3.4.2 Paraffin Inhibition Tests With the Use of a Magnetic Stirrer

Two 150 ml test glasses were prepared with 40 ml of crude oil sample in each one and the chemical agent was added at a concentration of 2% to one of them. A magnetic pill was dipped into both beakers, placing the samples on top of two magnetic stirrers. Two experiments were carried out: (i) with the crude oil sample subjected to a rotary movement of 400 RPM for 15 minutes and (ii) in the same way, but at 500 RPM. Subsequently, viscosity bath equipment (Figure 4) was used to determine this property in both vessels. For both situations, a comparison of the amount of paraffins precipitated after the rotary movement was carried out.

Figure 4. Viscosity bath equipment (Own elaboration)

3. RESULTS AND DISCUSSION

3.1 Characterization of the Crude Oil Sample

The characterization tests provided information on the properties and composition of the crude oil sample (see Table 2).

The results of the ASTM D4007 test indicate the absence of sediments in the crude oil sample used. This can be attributed to a low content of inorganic components and insoluble contaminants in its composition (Gómez, M. et al., 2003).

The API gravity (ASTM D1298), with a value of 36, indicates that the sample is a light crude oil. This result is optimal for the purpose of the research, because generally paraffinic crude oils have API gravity values greater than 35 (León, E., 2011).

The results of the ASTM D445 test show a low kinematic viscosity value, within the normal range. Relating this result to the ASTM D4007 test, it is understood that, since there are no sediments, the viscosity of the sample has no reason to increase (Gómez, M. et al., 2003).

The pour point (ASTM D97) shows a low temperature, but relatively close to 0°C. This result favors the purpose of the research, since when the temperature drop occurs in the crude oils at reservoir conditions, it is likely that they will cool below the pour point and the paraffinic precipitates will proceed to accumulate (Newberry, M. & Jennings, D., 2022).

The SARA method indicates that the sample has more than 50% of its content in saturates. This result is directly related to those of the ASTM tests, given that the higher the content of saturates in the crude oil, the higher the API gravity and crystallization temperature (León, E., 2011).

On the other hand, the colloidal stability index is greater than 0.9, so the asphaltenes in the crude oil are unstable. However, since these compounds are in the crude oil in such a small quantity, this value can be neglected.

The properties and composition shown in the results of the characterization tests are in accordance with the present investigation: a sample with a paraffinic nature.

Table 2. Results of the characterization of the crude oil sample

Properties and Composition	Results
Water and sediment in crude oil (ASTM D4007)	0.7%
API Gravity (ASTM D1298)	36
Kinematic viscosity (ASTM D445)	4,883 cSt
Pour Point (ASTM D97)	-12°C
Non-standardized SARA method	**Results**
Saturated	58.22%
Aromatics	31.22%
Resins	10.09%
Asphaltenes	0.47%
Colloidal stability index	1.4

Note. Own elaboration

3.2 Qualitative Paraffin Removal Tests: Presence of Crude Oil and Paraffins

3.2.1 Manual Paraffin Removal Tests: Crude Oil and Paraffin

The result of the manual test shows the potential that the chemical agent has when coming into contact with already precipitated paraffins.

The circular movements made with the hand simulated an almost non-existent flow, very close to a state of rest in the fluid (Al-Yaari, M., 2011). Under these conditions, the volume of solid paraffins was reduced by approximately 10%.

The percentage reduction indicates that it is prudent to continue experimentation under conditions similar to reality.

3.2.2 Experimental Dynamic Paraffin Removal Tests With a Testing Loop

The result of the experimentation with the testing loop shows a limitation that limonene has as an alternative for the treatment of paraffin precipitation.

By applying the chemical agent and activating the flow of crude oil through the test pipes at the same time, the paraffins were minimally removed. Given this result, it was considered necessary to repeat the experiment, but with different conditions.

3.2.3 Paraffin Removal Test With the Use of a Magnetic Stirrer

The results of the experimentation under dynamic conditions with the magnetic stirrer indicate how the maximum performance can be given to limonene.

Given the inconvenience that arose with the testing loop, the chemical agent was used in this experiment with two different concentrations and with a retention time of 24 hours of said compound with the mixture of crude oil and solid paraffins. The response of both scenarios can be seen in Table 3.

Table 3. Results of paraffin removal tests under dynamic conditions with a magnetic stirrer

Dynamic test	**Chemical agent concentration (%)**	
	0.17	0.33
Magnetic Stirrer	Release of solid paraffins. 70% effectiveness	Easily releases solid paraffins. 80% effectiveness

Note. Own elaboration

Like some chemical methods for paraffin removal, such as the application of xylene and toluene (aromatic compounds), a retention time is necessary to maximize the effectiveness of the chemical agent's action on solid paraffins (Lijian, D. et al., 2001).

On the other hand, the concentration of the chemical agent for optimal performance of its removal capacity is directly proportional to the amount of precipitated paraffins that needs to be removed. This proportion does not apply to all chemical treatments available for the inhibition and/or removal of paraffin waxes (Palermo, L. et al., 2013).

Based on the conclusions obtained from the analysis of this experiment, the parameters for adequate limonene performance are the retention time and concentration of the chemical with the mixture of crude oil and solid paraffins.

3.3 Quantitative Paraffin Removal Tests: Presence of Crude Oil and Paraffins

3.3.1 Determination of the Paraffin Removal Effect by Measuring the Weight of Paraffin Samples With and Without Chemical Agent

The results of the paraffin removal test with weighing before and after the experimentation are shown in Table 4.

In glass without chemical agent, some disparity is observed with the inhibition test. Since paraffins exist in a solid state, the rotary movements had a small removal effect, which occurred with a 5% paraffin dissolution. Although this result suggests an increase in the flow rate of the operating pipes as a treatment for clogging them, this proposal is ruled out in its entirety due to the increase in precipitation of new paraffin waxes that this action would cause.

In the glass with chemical agent, a reduction in the volume of solid paraffins was observed. With a dissolution percentage of more than 90%, this demonstrates the potential that limonene has as a treatment for removing solid paraffins precipitated in oil extraction operations.

Additionally, upon noticing a reduction in the solid paraffinic volume when the crude oil-paraffin mixture is subjected to rotary movements without any additional chemical, it follows that, in combination with limonene, a higher RPM can appreciate the usefulness of said substance.

Table 4. Results of paraffin removal tests with weighing comparison

Initial conditions		
Description	Sample	
	Just raw	Crude oil with chemical agent
Paraffin	12g	
Raw	40mL	
chemical agent	-	0.75mL
Final Conditions (24 hours of rest, at 300 RPM for 15 minutes)		
solid paraffin	11.40g	1.08g
Percentage of dissolved paraffin	5%	91%
Final observation	There was minimal paraffin removal	The paraffins were almost completely dissolved

Note. Own elaboration

Finally, with this experiment it is possible to reproduce the precipitation and inhibition conditions at the laboratory level and we have the complete parameters for optimizing the performance of this removal treatment: retention time and concentration of the chemical agent with the mixture of crude oil and paraffins. solid. The present work represents the continuation of the work carried out in (Armacanqui, J. S., 2016), where similar materials were used.

3.3.2 Paraffin Inhibition Tests With the Use of a Magnetic Stirrer

The results of the paraffin inhibition test are shown in Table 5.

These results indicate that, as long as the crude oil is in constant rotational motion, the paraffins will tend to precipitate and, therefore, deposit in the test glass between rotation speeds of 300 to 400 rpm (revolutions per minute), due to the forces that act on the liquid. In the well pipe, precipitation can be more aggressive with greater speed and turbulence of the crude oil flow (Armacanqui, J. S., 2016).

On the other hand, when the sample has been accompanied with the chemical agent, there was no precipitation; be it with 400 or even 500 rpm. This is attributed to the fact that said surfactant decreases the viscosity of the crude oil and alters the forces that act molecularly (Table 5), resulting in the centrifugal force not being a cause for the appearance of paraffins.

Table 5. Results of paraffin inhibition tests

Sample in glass	Viscosity (cSt 40°)	RPM (15 min)	
		400	500
Sample without chemical	4,883	Moderate paraffin precipitation	Excessive paraffin precipitation
Sample + chemical (2%)	4,215	No paraffin precipitation	No paraffin precipitation

Note. Own elaboration

Finally, it is to note that the paraffin deposition process is dependent on the temperature, pressure, and composition of the crude oil (Viera, M., 2020) and (Armacanqui, J. S., 2016). Therefore, it is suggested that the experiments be repeated, varying the operating conditions.

4. CONCLUSION

Limonene is presented as an effective paraffin inhibitor agent thanks to its action of reducing the intermolecular forces that govern these processes and the decrease in viscosity of crude oil. This capacity is not affected by the speed or turbulence of the oil flow; enabling a good performance without the need to modify the operating conditions.

Likewise, this biodegradable agent is shown to be very effective for removing solid paraffins. However, it has the removal time as a limitation that could be decisive when defining the operational treatment procedure to be used; since conventional chemicals (xylene, toluene, etc.) can act more quickly.

In this sense, the paraffin removal capacity is reduced by approximately 86% if it is applied with crude oil in circulation. However, if limonene has an adequate retention time (soaking time) with the fluid, it can considerably increase the percentage of paraffins removed.

Finally, the inclusion of a biodegradable, non-toxic and highly effective chemical treatment (such as limonene) against the accumulation and precipitation of paraffin waxes will not only ensure the health of workers and the environment; but also support the reduction of operational costs in oil extraction activities. This is specifically of importance for the high well count of unconventional oil wells that get sooner or later affected by the presence of paraffin. Further work is suggested to be carried out under field conditions. Pursuing these field tests would effectively contribute to a more sustainable operation of unconventional oil wells.

REFERENCES

Al-Yaari, M. (2011). Paraffin Wax Deposition: Mitigation & Removal Techniques. Paper presented at the SPE Saudi Arabia section Young Professionals Technical Symposium, Dhahran, Saudi Arabia. 10.2118/155412-MS

Alnaimat, F., & Ziauddin, M. (2020, January). Wax deposition and prediction in petroleum pipelines", J. Petroleum Sci. Eng., vol. 184, p. 106385. *Journal of Petroleum Science Engineering*, 184, 106385. Advance online publication. Retrieved August 16, 2023, from. 10.1016/j.petrol.2019.106385

Ariza León. (2011). On the characterization of crude oils what is key to diagnosing paraffin precipitation. *Revfue*.

Asomaning, S., & Watkinson, A. P. (2000). Petroleum Stability and Heteroatom Species Effects in Fouling of Heat Exchangers by Asphaltenes. *Heat Transfer Engineering*, 21(3), 3, 10–16. 10.1080/014576300270852

ASTM Standard D1298-12b. Standard Test Method for Density, Relative Density, or API Gravity of Crude Petroleum and Liquid Petroleum Products by Hydrometer Method. Annual Book of ASTM Standards, Vol. 05.01, ASTM International, West Conshohocken, PA, 2010. DOI: 10.1520/D1298-12BR17E01

ASTM Standard D4007-22. Standard Test Method for Water and Sediment in Crude Oil by the Centrifuge Method (Laboratory Procedure). Annual Book of ASTM Standards, Vol. 05.01, ASTM International, West Conshohocken, PA, 2010. DOI: 10.1520/D4007-22

ASTM Standard D445-21e2. Standard Test Method for Kinematic Viscosity of Transparent and Opaque Liquids (and Calculation of Dynamic Viscosity). Annual Book of ASTM Standards, Vol. 05.01, ASTM International, West Conshohocken, PA, 2010. DOI: 10.1520/D0445-21E02

ASTM Standard D97-17b. Standard Test Method for Pour Point of Petroleum Products. Annual Book of ASTM Standards, Vol. 05.01, ASTM International, West Conshohocken, PA, 2010. DOI: 10.1520/D0097-17BR22

Dong, L., Xie, H., & Zhang, F. (2001). Chemical Control Techniques for the Paraffin and Asphaltene Deposition. Paper presented at the SPE International Symposium on Oilfield Chemistry, Houston, Texas. 10.2118/65380-MS

Gómez. (2003). Formation of sediments during the hydrodisintegration of petroleum residues. Magazine of the Chemical Society of Mexico, 47(3), 260-266. https://www.scielo.org.mx/scielo.php?script=sci_arttext&pid=S0583-76932003000300010&lng=es&tlng=es

González García, D., Villabona Carvajal, C., Vargas Torres, H., Ariza León, E., Roa Duarte, C., & Barajas Ferreira, C. (2010). Methods for the Control and Inhibition of the Accumulation of Paraffinic Deposits. *UIS Engineering Magazine*, 9(2), 193–206.

Khaibullina, K. (2016). Technology to Remove Asphaltene, Resin and Paraffin Deposits in Wells Using Organic Solvents. Paper presented at the SPE Annual Technical Conference and Exhibition, Dubai, UAE. 10.2118/184502-STU

Melikyan. (1997). Surfactants based aqueous compositions with D-limonene and hydrogen peroxide and methods using the same, Patent Number: 5,602,090, Date of Patent: Feb. 11, 1997, United States Patent. Technical Product Bulletin #sw-61, US-EPA, OEM Regulations – Implementation Division https://www.epa.gov/emergency-response/epa-oil-field-solution

Newberry, M., & Jennings, D. W. (2022). Chapter 2 - Paraffin management, Editor(s): Qiwei Wang, In Oil and Gas Chemistry Management Series, Flow Assurance, Gulf Professional Publishing. 10.1016/B978-0-12-822010-8.00003-9

Palermo, L. C. M., Souza, N. F. Jr, Louzada, H. F., Bezerra, M. C. M., Ferreira, L. S., & Lucas, E. F. (2013). Development of multifunctional formulations for inhibition of waxes and asphaltenes deposition. *Brazilian Journal of Petroleum and Gas*, 7(4), 181–192. Advance online publication. 10.5419/bjpg2013-0015

Samuel Armacanqui, J. (2016). Testing of Environmental Friendly Paraffin Removal Products. Paper SPE 181161, presented at Latin America Heavy and Extra Heavy Oil Conference, Lima, Peru. 10.2118/181162-MS

Tanirbergenova, S., Ongarbayev, Y., Tileuberdi, Y., Zhambolova, A., Kanzharkan, E., & Mansurov, Z. (2022, June). Selection of Solvents for the Removal of Asphaltene–Resin–Paraffin Deposits. *Processes (Basel, Switzerland)*, 10(7), 1262. 10.3390/pr10071262

Viera, M. (2020). Removal of paraffins and improvement of the recovery factor through the use of multifunctional and biodegradable chemicals to increase productivity in oil wells. Master's Thesis, National University of Engineering.

Chapter 11
Using H2 to Increase the Energy Efficiency and Reduce the Containment Emissions in the Operation of Generators

Franco A. Cassinelli-Cisneros
National University of Engineering of Peru, Peru

Jesus Samuel Armacanqui-Tipacti
National University of Engineering of Peru, Peru

Ciro Ormeño-Aquino
National University of Engineering of Peru, Peru

Jose Carlos Rodriguez
San Luis Gonzaga University of Ica, Peru

José Rosendo Campos Barrientos
San Luis Gonzaga University of Ica, Peru

Mohamed Yehia
Suez Oil Company, Cairo, Egypt

Nelson Michael Villegas-Juro
National University of Engineering of Peru, Peru

Alfredo Vazquez-Barrios
National University of Engineering of Peru, Peru

Ricardo Hector Rodriguez-Robles
National University of Engineering of Peru, Peru

ABSTRACT

The application of gaseous hydrogen in combustion internal (CI) diesel dual-fuel (DDF) engine is not new, and it does not yet have broad commercial applications that

DOI: 10.4018/979-8-3693-0740-3.ch011

ensure the increase in performance and consequent reduction in costs to which it is normally associated. The effect of hydrogen on diesel internal combustion engines can be analyzed through the different parameters that an CI engine has grouped into four main aspects: performance, combustion characteristics, polluting emissions, and future challenges. The effect of hydrogen application is consequently a multivariate analysis that this study addresses individually on each parameter, explaining its probable causes and a necessary theoretical framework for it. The effect on each parameter is finally quantified in the best approximate framework for application in the petroleum industry, which is defined with a hydrogen contribution of between 10-20% and load conditions of at least 50%. The importance of this study lies in the empirical analysis and verification of the supply of hydrogen to DDF engines.

1. INTRODUCTION

Diesel engines designed for compression ignition (CI), are used in a wide range of power capacities and are known for their efficiency, reliability and durability, making them suitable for various applications such as transportation, agriculture, civil construction, electrical power generation, among others. These benefits are consequence of the following characteristics:

- High thermal efficiency, which usually leads to better fuel economy.
- Robustness and durability, which makes them suitable for heavy-duty applications, continuous operation for extended periods of time and high torque applications.
- Longevity, they often have long lifespans compared to gasoline engines, which contributes to their cost-effectiveness over time.
- Reliability, they are characterized by a high availability index and quick start that makes them suitable for electrical power backup supply.
- Modularity, they are available in modular configurations, allowing easy scalability according to application by combining multiple engines.
- Fuel flexibility, which allows them to operate with biodiesel, traditional and synthetic diesel, and dual-fuel mode, also known as DDF mode (Diesel dual-fuel), which this study will focus on based on hydrogen.
- Low carbon-based emissions.

Despite of these benefits, these engines have high level of smoke/particle matter (PM) and oxides of nitrogen (NO_x) emissions, which are critical to achieving good air quality and other environmental goals. European Union sets NO_x emission regulation, which is the combined emissions of NO and NO_2, in 80mg/km for new diesel passenger cars (European Commission, 2015) while the Environmental Protection Agency (EPA) sets the limit to 0.15lbs/mmBtu in the US for utilities (US EPA, 2024). Although several technologies have been tested to improve CI engine environmental sustainability, such as exhaust gas recirculation (EGR), retardation of injection timing, among others, neither of them is capable of provide efficiency improvements and emissions reduction simultaneously (Chintala & Subramanian, 2017). However, use of gaseous hydrogen (H_2) in CI engines is a viable option that can improve efficiency and reduce pollution gas emissions.

1.1. Hydrogen Application in Internal Combustion Engines

Hydrogen has the highest energy content per unit mass among all fuels, approximately 120-142 MJ/kg, depending on its specific form and store conditions, in comparison with diesel and gasoline, 35.8MJ/kg and 44.0MJ/kg respectively. As fuel, hydrogen alone, works better in spark ignition (SI) engines due to its high-octane number of about 120 (Salvi & Subramanian, 2015), which represents its resistance to ignition. However, it has some disadvantages such as power de-rating due to throttling loses, low thermal efficiency, low volumetric efficiency and high level of NO_x emissions (Tahir et al., 2015). In this application case these problems can be solved with improving engine control through better monitoring hardware and software or new technology need to be developed (Chintala & Subramanian, 2017).

In contrast, under DDF mode hydrogen is suitable for successful adoption as carbon free fuel for CI engines and higher combustion efficiency (Karim, 2015). Here are some of its properties in CI applications:

- High flame temperature, which improve efficient energy conversion but may increase NO_x emissions.
- High flame speed compared to hydrocarbon fuels, which means that it burns faster and lead to a rapid and efficient combustion because it approximates to constant volume combustion.
- Wide flammability range, it means that it can be ignited and burned in a broad range of air-fuel ratios and combustion conditions.
- Low ignition energy, meaning it requires low energy to ignite.
- High diffusivity, which allows hydrogen to mix rapidly with air resulting in a well air-fuel mixture and improving combustion efficiency.
- Short quenching distance, meaning in case of hydrogen that the flame is extinguished relatively quickly. It is an important parameter in design of combustion chambers to ensure that the flame remains stable and does not extinguish during combustion.

Using H2 to Increase the Energy Efficiency

Despite of these advantages, using hydrogen has some limitations on CI applications because of its *high autoignition temperature*, which means it does not autoignite easily under compression alone, this is important in terms of engine design and combustion timing control and is the main reason why it is used in DDF mode. This characteristic is represented in its high-octane number and low-cetane number, in comparison to conventional gasoline and diesel respectively. In this condition hydrogen can be used as the main fuel (major energy contributor) with a small amount of diesel (pilot fuel) that work as an ignition source (Chintala & Subramanian, 2017).

1.2. Generalities About DDF Technology

DDF engines are fuel versatile since they can work using hydrogen as main fuel and in case not it can work based just on diesel. Figure 1 shows valve timing and process diagram of DDF engine that was optimized based on better performance and lower emissions (Cameretti et al., 2022). Injection time of hydrogen is set to 43° to avoid scavenging losses since the outlet valve is already closed when injection starts (Chintala & Subramanian, 2015b).

Figure 1. Valve timing and process diagram depicting hydrogen fuel injection timings (Chintala & Subramanian, 2015b)

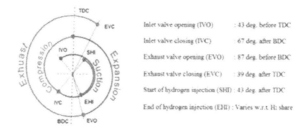

Usually, diesel fuel is injected directly into the combustion chamber at the end of the compression stroke to ignite hydrogen as main fuel, which is supply during expansion stroke (Chintala & Subramanian, 2015b). Hydrogen can be supplied using three different technics that play an important role in engine performance (Rahman et al., 2017):

- Manifold injection, this technique has several advantages that allows to optimize combustion efficiency, control power output, regulate emissions and improve overall engine performance among others:

- ○ Better control of gaseous fuel injection timing and duration.
- ○ Better air-fuel mixture.
- ○ Low temperature exposure of hydrogen injector.
- ○ Minor engine hardware modification or software improvements.
- ○ Allows to avoid backfiring problem because manifold can be design or modified in a way that its intake does not contain any combustible mixture.
- ○ Minor mechanical modifications if needed and low-cost implementation.

- Port in-cylinder injection and direct in-cylinder injection, which have important limitations:

- ○ Does not avoid backfiring problem.
- ○ Less durability and reliability of hydrogen injector because it is exposed to high temperature conditions and in consequence it must be fabricated with high temperature compatible materials.
- ○ Major mechanical modifications are usually needed which increment its implementation cost.

1.3. Oil Industry Application of Diesel Engines

Diesel engines play crucial role in oil industry in the different stages of extraction process. The following are some important applications:

- **Drilling rigs:** Diesel engines are used in exploration and production stages powering the necessary equipment such as drawworks, mud pumps and rotary TABLEs among others.
- **Power generation:** They are often used in remote sites to power essential equipment, lightning and accommodation facilities on site. In this context, they can be used as backup power source in case of main power source failure or electrical grid limitations ensuring continuous operation of essential equipment. They provide electrical energy for exploration camps, drilling sites and temporary facilities too.
- **Pumping systems:** There are several applications such as transferring crude oil, reservoir water injection for pressure maintenance, power compressors for gas and oil transport and processing, among others. They are also used in fluid and proppants injection pumps into reservoirs to produce hydraulic fracture (fracking) to enhance production.

- **Transportation and services:** They are used in heavy-duty vehicles such as trucks, tankers and trains for transport crude oil and refined products. They are also used in maintenance and service vehicles to provide equipment repair, well servicing and facility maintenance.

Drilling rigs are representative in the oil and gas industry in terms of electrical energy consumption because they are used in resource exploration, well construction, extraction, production enhancement, data collection, operation, decommissioning, among other important aspects of this industry. The amount of diesel fuel consumed by drilling rigs depends on several factors such as type of rig, depth and drilling operation specific characteristics, rig efficiency, operation stage and drilling techniques. There is an important differentiation in terms of drilling rigs power consumption in offshore and onshore applications, being 20-45m^3 and 7-7.5m^3 of diesel fuel the average consumption per day respectively (IPIECA, 2023), each one of those values represent approximately 300,000kWh and 120,000kWh of energy per day respectively (Beneš, 2021). Offshore applications have more power consumption variability because of the different types of rigs: jackup rig will use 20m^3 of diesel per day on average, anchored rig 32m^3, dynamic positioning rigs 45m^3 and light well interventions less than 20m^3 (IPIECA, 2023).

2. METHODOLOGY

Empirical information on the effect of adding gaseous hydrogen to DDF engines was collected considering the most possible diversity of engines characteristics. Along with this information, the reviewed works offer an analysis of the causes of these effects that also was included in the relevant sections, along with a theoretical framework for better understanding of hydrogen effect on each CI engine parameter. Each parameter theorical framework was written with a level of detail and simplicity sufficient for those who only have a superficial knowledge of the topic can also understand it.

The literature reviewed of DDF engines based on hydrogen analyze its application according to four main aspects performance, combustion characteristics, polluting emissions and future challenges that were grouped into sections 3.1, 3.2, 3.3 and 3.4 each of which end with a summary focus on medium and high load conditions which are the usual operating conditions of CI diesel engines applications in oil and gas industry. It is important to mention that from literature a differentiation is made in the effect of the use of hydrogen according to the load level, classifying it into 3 levels: *low* (from 0% to 30%), *medium* and *high* (from 70% to 100%). Based on certain restrictions that will be explain in section 3.4, the mentioned summary subsection fill focus on average lineal behavioral of the parameters in the range of 10-20% of hydrogen energy share to evaluate the impact of using hydrogen in conventional diesel engines in oil and gas industry and will lead to this study conclusions associated to the oil industry. In the following sections the term "air-fuel mixture" refers to air, diesel and hydrogen mixture considering this last two as pilot and main fuel respectively.

3. RESULTS AND DISCUSSION

3.1. Hydrogen Effect on DDF Engines Performance

3.1.1. Thermal Efficiency

Thermal efficiency is a measure of how well a system can convert thermal energy into useful energy defined according to the equation (1), where $\dot{m}$ represents the mass flow rate in $\frac{kg}{s}$, CV is the calorific value of fuel in $\frac{kJ}{kg \cdot K}$ and brake power is in kW:

$$Thermal\ eff. = \frac{Brake\ Power}{(\dot{m}_{H_2} * CV_{H_2}) + (\dot{m}_{diesel} * CV_{diesel})} * 100 \quad (1)$$

Most studies report an increase in thermal efficiency in medium and high load conditions, while it decreases in low load conditions (Chintala & Subramanian, 2014a; Deb et al., 2015; Masood et al., 2007). Decreasing thermal efficiency during low load conditions is due to (Lata & Misra, 2010):

- Low quantity of diesel because in low load conditions the hydrogen supplied represents most of the fuel in the air-fuel mixture.
- At the end of the compression stroke there is lower charge temperature because of the low flame velocity of air-fuel mixture.
- There is enough time to transfer heat to cylinder walls, delaying the ignition.

Table 1 and Figure 2 show a brief compilation of the effect of hydrogen use on DDF engines at low load conditions where hydrogen supply concentrations and motor characteristics are different.

Using H2 to Increase the Energy Efficiency

Table 1. Different thermal efficiency effects of hydrogen on DDF engines at low loads (1) from (Senthil Kumar et al., 2003), (2) from (Lata & Misra, 2010) and (3) from (Santoso et al., 2013)

Load		Thermal Efficiency	H_2 concentration	Concentration type
20%	LOW	13.40%	0.0%	Mass share
		8.00%	29.0%	
9%	LOW	19.57%	0.0%	Mass share
		17.37%	???	
10Nm BMEP 1.9bar	LOW	57.90%	50.0%	Energy share
		54.30%	90.0%	
		49.00%	97.0%	

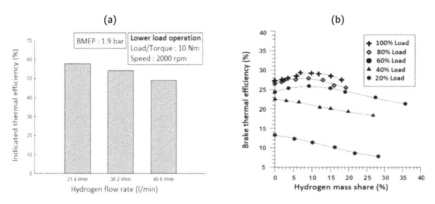

Figure 2. Brake thermal efficiency variation with respect to hydrogen flow rate (a) from (Santoso et al., 2013) and (b) from (Senthil Kumar et al., 2003)

In medium and high load conditions the increment in thermal efficiency is due to (Chintala & Subramanian, 2014b):

- Improvement in air-fuel mixture because a better distribution of hydrogen in combustion chamber due to its higher diffusivity (0.63 cm^2/s) than diesel (0.038 cm^2/s) and more time for air-fuel mixture because the hydrogen is inducted into chamber during suction stroke.
- Improvement in constant volume combustion because of hydrogen high flame speed and temperature.
- Reduction in combustion irreversibility and consequent energy losses.
- Reduction of unburned fuel quantity also due to high flame speed.

Table 2 and Figure 3 show a brief compilation of the effect of hydrogen use on DDF engines at high load conditions where hydrogen supply concentrations and motor characteristics are different.

Table 2. Different thermal efficiency effects of hydrogen on DDF engines (1) and (2) from (Geo et al., 2008) and (3) from (Yadav et al., 2014)

Load		Thermal Efficiency	H_2 energy share
100% BMEP 5.3bar	HIGH	29.90%	0.0%
		31.60%	10.1%
75% BMEP 3.99bar	HIGH	27.3%	0.0%
		28.70%	13.4%
75% BMEP 4.00bar	HIGH	23.35%	0.0%
		26.07%	16.4%

Figure 3. Brake thermal efficiency variation with respect to hydrogen flow rate (a) from (Geo et al., 2008) and (b) from (Varde & Frame, 1983)

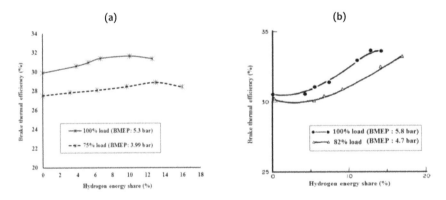

It is noticeable that despite of the change of thermal efficiency in relation to the hydrogen supply concentration is proportionally different for each study (slope), thermal efficiency reduction happens in any low load case and the opposite in medium and high load cases.

Using H2 to Increase the Energy Efficiency

3.1.2. Exergy Efficiency

Exergy of fuel input is the maximum useful work that can be done through combustion process, in consequence, exergy efficiency is the effectiveness of converting fuel exergy into actual useful work. Studies show that there is an increment in exergy efficiency at least in medium and high load conditions as shown in Table 3 and Figure 4 for a 7.4kW rated power and 1500RPM DDF engine (Chintala & Subramanian, 2014a):

Table 3. Exergy efficiency effects of hydrogen on DDF engines (Chintala & Subramanian, 2014a)

Load		Exergy Efficiency	H_2 energy share
100%	HIGH	29.10%	0.0%
		31.70%	31.7%

Figure 4. Exergy efficiency variation with respect to hydrogen energy share (Chintala & Subramanian, 2014a)

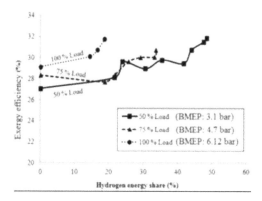

This significant increase in exergy efficiency can be attributed to (Chintala & Subramanian, 2014a):

- Increase in combustion efficiency by decrease in total irreversibility of the system as compared to conventional mode.
- The combustion irreversibility of DDF engine decreases proportionally to the increase of hydrogen amount substitution because this generates high temperature which reduces entropy.

It is important to mention in this context that the irreversibility in the DDF engine decrease from 3.14kW in dieselbase mode to 1.73kW with hydrogen dual mode, which means that there are less losses which can be attributed to an improvement in flame speed and temperature.

3.1.3. Volumetric Efficiency

This parameter measures how well a cylinder is filled with air-fuel mixture during intake or induction stroke and is defined as the relation between actual intake compared to the maximum amount of gas that could theoretically fill the cylinder. Experimental studies have shown that the volumetric efficiency is reduced with increasing the amount of hydrogen in dual-mode (Nag et al., 2019; Varde & Frame, 1983), mainly because a good part of the intake flow is replaced with hydrogen, which has less density. Studies show that there is a reduction in volumetric efficiency in medium and high load conditions as shown in Table 5 and Figure 5.

Table 5. Volumetric efficiency effects of hydrogen on DDF engines from (Geo et al., 2008)

Load		Volumetric Efficiency	H_2 energy share
100%	HIGH	91.0%	0.0%
		85.0%	13.4%

Figure 5. Volumetric efficiency variation with respect to hydrogen energy share (a) from (Geo et al., 2008) and (b) from (De Morais et al., 2013)

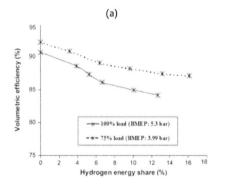
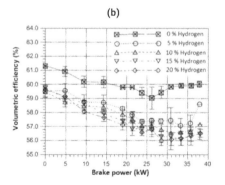

Using H2 to Increase the Energy Efficiency

This reduction happens also when increment load on the engine, probably because in these conditions the intake temperature also increases, in consequence the air-fuel mixture density goes down. Other parameters may influence decreasing volumetric efficiency such engine configuration and operation conditions.

3.1.4. Exhaust Gas Temperature

This parameter is used to indirectly measure combustion efficiency, a higher value may suggest a more complete combustion, also reflecting how good was the air-fuel mixture. Exhaust gas temperature can be indicative of general engine health and performance, allowing to detect underlying problems if it has abnormal values. Hydrogen has a general effect of increasing the exhaust gas temperature, as shown in Table 6 and Figure 6.

Table 6. Exhaust gas temperature effects of hydrogen on DDF engines from (Geo et al., 2008)

Load		Exhaust gas temperature	H_2 energy share
100%	HIGH	364°C	0.0%
		427°C	10.1%

Figure 6. Exhaust gas temperature variation with respect to hydrogen energy share (a) from (Geo et al., 2008) and (b) from (Varde & Frame, 1983)

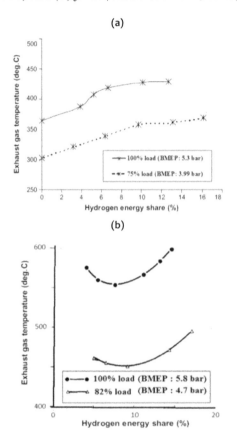

The increment in exhaust gas temperature is due to:

- The higher heat content of hydrogen respect to diesel.
- Hydrogen releases a large amount of heat at the end of the expansion stroke (Geo et al., 2008) that can be seen in Table 7.
- Rapid combustion of hydrogen because of its high flame

Table 7. Exhaust gas energy effects of hydrogen on DDF engines from (Chintala & Subramanian, 2014a)

Load		Exhaust gas energy	H_2 energy share
75%	HIGH	24.0%	0.0%
		25.7%	33.6%

There can be cases like the ones shown in figure 6.b where the exhaust gas temperature slightly decreases before increasing. This happens because of the late combustion hydrogen during expansion stroke producing limitations during heat release in combustion process.

3.1.5. Summary of Hydrogen Effects on DDF Engines Performance

Briefly the effect of hydrogen substitution in DDF engines performance is shown in Table 8.

Table 8. Effect of hydrogen substitution in DDF engines performance parameters

Parameter	Load conditions		
	Low	Medium	High
Thermal efficiency	Decrease	Increase	Increase
Exergy efficiency	Decrease	Increase	Increase
Volumetric efficiency	Decrease	Decrease	Decrease
Exhaust gas temperature	Increase	Increase	Increase

Based on the data collected, an average of the improvement values has been calculated for each parameter of this section, which are shown in the Table 8. These parameters have been calculated for less than 20% hydrogen energy share and medium and high load conditions. If the improvement value is negative means that the parameter worsens with the increase of hydrogen energy share.

Table 9. Average percentage improvement per engine performance parameter for less than 20% hydrogen energy share in medium and high load conditions

Parameters	Average percentage improvement
Thermal efficiency	+0.15% per % of H_2 energy share
Exergy efficiency	+0.10% per % of H_2 energy share
Volumetric efficiency	-0.47% per % of H_2 energy share
Exhaust gas temperature	+6.24°C per % of H_2 energy share

3.2. Hydrogen Effect on DDF Engines Combustion Characteristics

3.2.1. Heat Release Rate

To better understand heat release rate is important to describe the combustion process in a DDF engine, in this case from three phases as shown in Figure 7 (Karagöz et al., 2016; Karim, 2003; Lata & Misra, 2010):

- The first phase involves a major combustion of diesel as pilot fuel and a minor one of hydrogen as main fuel because an ignition delay period among them. In this phase different combustion centers are generated as consequence of diesel diffusion.
- The second phase involves a major combustion of remaining hydrogen and minor one of diesel. The hydrogen ignites in the vicinity of diesel spray zone.
- The last phase involves the total combustion of both pilot and main fuels, from the spray zone to the full combustion chamber as flame propagates and sometimes autoignition happens.

Figure 7. Heat release rate diagram for conventional DDF combustion (Tahir et al., 2015)

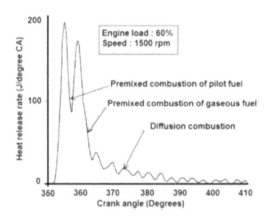

There are some authors that join the last two phases together just like diesel models because of the rapid transition between them both (Masood et al., 2007). In any case, under low load conditions there is less hydrogen in the air-fuel mixture which makes it very lean, in consequence most heat comes from the diesel and a

Using H2 to Increase the Energy Efficiency

successive rapid combustion occurs in the first and second phase as shown in figure 8.a, so that the hydrogen away from the pilot diesel combustion centers releases very low heat. At this condition appear some other problems such as increasing ignition delay and uncomplete fuel combustion. This engine behavior depends largely on pilot fuel quantity, type of gaseous fuel, operating conditions and type of engine. At high and medium load conditions the main heat is released from the hydrogen main fuel even away from the pilot diesel combustion centers as can be seen in Figure 8.b in the second phase, which favors the heat release when increasing hydrogen energy share (Liew et al., 2010).

Figure 8. Different components of energy release rate in a DDF engine under (a) light load and (b) heavy load conditions (Karim, 2003)

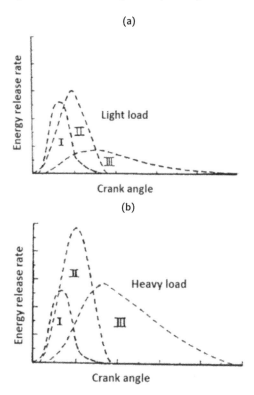

3.2.2. In-Cylinder Pressure and Combustion Reaction Rate

In-cylinder pressure reflects the dynamics of the combustion process because it is influenced by the rate at which the air-fuel mixture ignites, burns and expands producing useful work as heat is released during combustion reaction. In general, in DDF mode in-cylinder pressure increases proportionally to the hydrogen supply in high and medium load conditions while it decreases at low load conditions (Geo et al., 2008; Jamrozik et al., 2020; Karagöz et al., 2016; Masood et al., 2007; Santoso et al., 2013). Low load conditions can be seeing in Table 10 and Figure 9.

Table 10. Combustion reaction rate effects of hydrogen on DDF engines from (Santoso et al., 2013)

Load		Combustion reaction rate	In-cylinder peak pressure	H_2 energy share
BMEP 1.9bar	LOW	$17.8 s^{-1}$	73bar	50%
		$7.6 s^{-1}$	64bar	90%

Figure 9. Combustion reaction rate (a) and in-cylinder pressure (b) with respect to degree CA at lower load (Santoso et al., 2013)

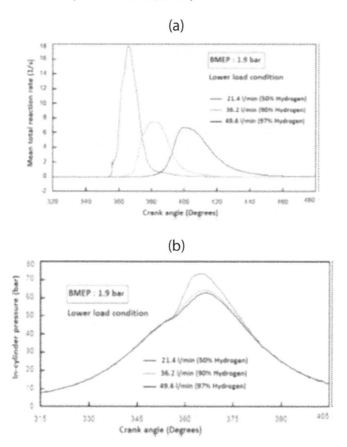

Under low load conditions, the proportion of diesel is reduced in relation to the injected hydrogen and the temperature inside the cylinder is low, which consequently produces a late start of combustion (Santoso et al., 2013) as shown in figure 9.a, poor spray characteristics and poor air-fuel mixing characteristics (Jamrozik et al., 2020), which lead to a significant reduction in combustion reaction rate and a smaller number of ignition chamber centers (Jamrozik et al., 2020).

Figure 10. In-cylinder pressure variation with respect to degree CA at high load (a) from reference (Miyamoto et al., 2011) (b) from (Chintala & Subramanian, 2015a)

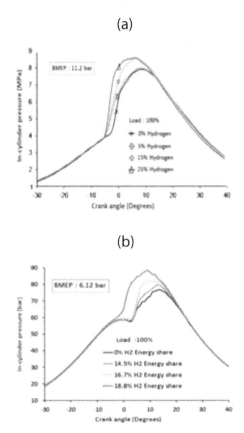

In contrast, at high and medium load conditions the in-cylinder pressure increases as shown in Figure 10, as consequence of higher temperature conditions and hydrogen high flame speed which leads to a rapid and complete oxidation process. Further evidence on this ambivalent behavior in dual motors is presented Table 11 and Figure 11 based on in-cylinder peak pressure (Liew et al., 2010), which is going to be analyzed in further sections.

Figure 11. Difference between in-cylinder pressure curves at low (a) and high (b) loads (Polk et al., 2013)

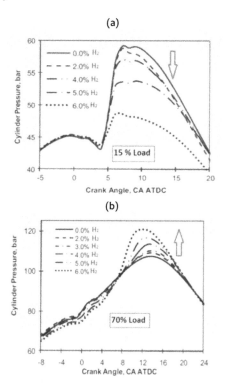

Table 11. In-cylinder peak pressure effects of hydrogen on DDF engines (1) and (2) from (Liew et al., 2010)

LOAD		In-cylinder peak pressure	H_2 energy share
70%	HIGH	108bar	0.0%
		122bar	6.0%
15%	LOW	59bar	0.0%
		48bar	6.0%

3.2.3. In-Cylinder Peak Pressure

In-cylinder peak pressure is a key parameter in engine development and optimization which involves a balance of high efficiency, power output increasing and decreasing emissions while ensuring reliability and durability of engine components. This parameter increases at high and medium loads with increasing hydrogen energy share while follows an inverse trend at low loads (Geo et al., 2008; Liew et al., 2010; Varde & Frame, 1983). Some important experimental results are shown in Table 12 and Figure 12.

Table 12. Different thermal efficiency effects of hydrogen on DDF engines (1) and (3) from (Geo et al., 2008) and (2) from (Liew et al., 2010)

LOAD		In-cylinder peak pressure	H_2 energy share
100%	HIGH	76bar	0.0%
		80bar	12.5%
???	HIGH	110bar	3.5%
		121bar	6.0%
20%	LOW	53.5bar	0.0%
		49.5bar	29.0%

Figure 12. In-cylinder peak pressure variation with respect to hydrogen substitution (a) from (Geo et al., 2008) (b) from (Senthil Kumar et al., 2003)

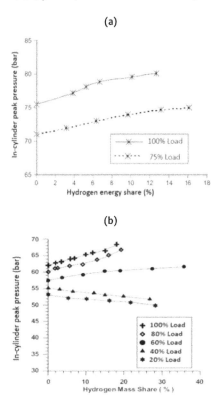

3.2.4. In-Cylinder Temperature

As previously analyzed, in-cylinder temperature increases during high and medium load conditions, as shown in Table 13 and Figure 13, along with increasing hydrogen energy share as consequence of increasing in-cylinder pressure and hydrogen high flame temperature and speed, which lead to more heat releasing and faster combustion.

Using H2 to Increase the Energy Efficiency

Table 13. In-cylinder peak temperature effects of hydrogen on DDF engines with biodiesel blend B20 from (Subramanian & Chintala, 2013)

LOAD		In-cylinder peak temperature	H_2 energy share
100%	HIGH	1774 K	0.0%
		1845 K	20.4%

Figure 13. Variation of in-cylinder temperature with respect to degree crank angle under DDF mode at 100% load (Subramanian & Chintala, 2013)

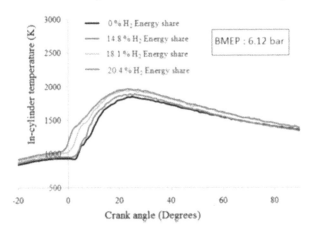

3.2.5. Ignition Delay

It refers to the time between ignition start and the moment when rapid combustion releases heat at high rate. In DDF engines this parameter depends directly on the autoignition capacity of hydrogen and air-fuel mixture conditions, which affect the combustion stability leading to knocking or misfire issues, in case of too short or too long ignition delay respectively, and temperature and pressure operation abnormal conditions. Experimental studies have found that ignition delay decreases with increasing hydrogen energy share at high and medium load conditions (Miyamoto et al., 2011), while it increases at low loads (Karim, 2003).

Increase hydrogen energy share can produce an increment in free radical concentration which leads to improve pre-ignition chemical conditions reducing ignition delay if there is enough diesel fuel. In case of low load conditions there is a fluctuation of ignition delay when increasing hydrogen energy share which can be due to (Lata & Misra, 2011):

Using H2 to Increase the Energy Efficiency

- A reduction in oxygen partial pressure because of increment of hydrogen share, which lead to not achieve the pressure and temperature preconditions for diesel oxidation.
- A reduction of in-cylinder temperature because of increment of overall specific heat of total air-fuel charge, setting poor pre-ignition reactions.

As load goes higher the peak ignition delay goes down and needs less hydrogen share to achieve it before it begins to decrease. A certain convergence can be seen in the figure 14 when surpassing the 40% of hydrogen share which may be due to an increase in hydrogen partial pressure and consequent rapid combustion.

Figure 14. Ignition delay variation with respect to hydrogen fuel substitution at different loads (Lata & Misra, 2011)

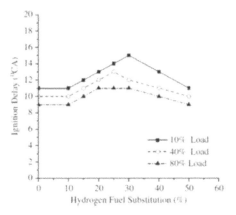

3.2.6. Combustion Duration

Many experimental studies show a tendency of combustion duration to decrease with increasing hydrogen share because of its high flame velocity (Lata et al., 2011; Liew et al., 2010; Santoso et al., 2013). Also, increasing hydrogen share lead to enhance reaction rate due to increment free radical concentration, specially at high pressure and temperature, which happens at high and medium load conditions, setting right the pre-ignition conditions. The experimental results show that:

- There is no significant impact on combustion duration at low load conditions based on 4-5% volume share of hydrogen. When surpasses this threshold the

combustion duration increases dramatically (Liew et al., 2010) as seen in Figure 15.a.
- Some studies reported fluctuations in combustion duration which does not lead to any special conclusion (Lata et al., 2011) as seen in Figure 15.b.
- At medium and high load conditions is plausible to say that combustion duration decreases (Liew et al., 2010). A drastic reduction of combustion durations occurs when increment just 3% of hydrogen volume share as seen in Figure 15.c.

Figure 15a. Combustion duration variation with respect to hydrogen fuel substitution: From (Liew et al., 2010)

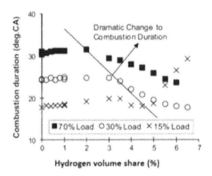

Figure 15b. Combustion duration variation with respect to hydrogen fuel substitution: From (Lata et al., 2011)

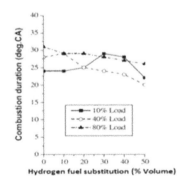

Figure 15c. Combustion duration variation with respect to hydrogen fuel substitution: From (Geo et al., 2008)

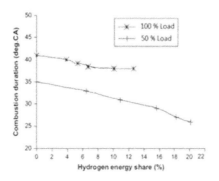

3.2.7. Combustion Efficiency

Most of previous parameters are closely and directly related to the combustion process and phases, its efficiency and engine power output. Combustion efficiency is defined as the ratio of total energy released during combustion, measured in terms of pressure, temperature and heat released, respect to the total energy content of the fuel, usually provided in terms of *lower heating value* (LHV) or *higher heating value* (HHV), both expressed in energy units per unit mass. Combustion efficiency depends directly on chemical oxidation rate, which are defined according to the Arrhenius equation (2), where *RR* is the reaction rate, *T* is the in-cylinder temperature in K, E_a is the activation energy in kJ, R_u is the universal gas constant in $\frac{kJ}{kg-mol*K}$, and k_1, k_2, k_3 are specific constants.

$$RR = k_1 * e^{-\frac{E_a}{R_u T}} * [fuel]^{k_2} * [oxidizer]^{k_3} \qquad (2)$$

Equation (2) shows that *RR* increases exponentially with increasing combustion temperature which at the same time is proportional to hydrogen share, especially in high and medium load conditions, as shown in the experimental results from Figure 16 and Table 14.

Figure 16. Combustion efficiency variation with respect to hydrogen substitution (a) from (Liew et al., 2010) (b) (a) from (Chintala & Subramanian, 2015b)

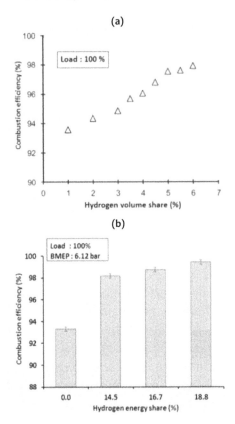

Table 14. Combustion efficiency effects of hydrogen on DDF engines from (Liew et al., 2010)

LOAD		Combustion efficiency	H_2 energy share
100%	HIGH	93.4%	1.0%
		98.0%	6.0%

3.2.8. Pressure Rise Rate

It is a measure of how rapid pressure increases in combustion chamber, mathematically defined as the derivative of pressure with respect to time. This parameter influences engine performance and combustion stability, consequently, is a key factor in combustion control and optimization, also used to analyze knocking tendency. Along with pressure and peak pressure, these are parameters that influence formation of oxide nitrogen because they can lead to higher temperatures as they increase. In general, increasing of hydrogen share increases pressure rise rate at high load conditions, while at medium and low loads it decreases, as shown in the experimental results from Table 15 and Figure 17.

Table 15. Pressure rise rate effects of hydrogen on DDF engines (1) and (2) from (Geo et al., 2008)

LOAD		Maximum rate of pressure rise	H_2 energy share
100%	HIGH	5.2 bar/°CA	0.0%
		6.1 bar/°CA	12.5%
25%	LOW	2.9 bar/°CA	0.0%
		2.3 bar/°CA	25.0%

Figure 17. Maximum rate of pressure rise with respect to hydrogen energy share at different loads (Geo et al., 2008)

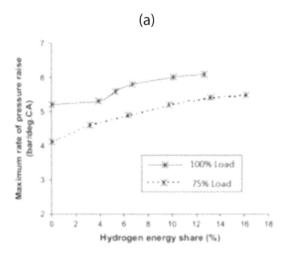

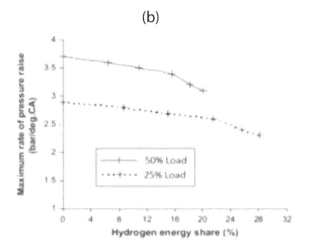

3.2.9. Summary of Hydrogen Effect on DDF Engines Combustion Characteristics

Briefly the effect of hydrogen substitution in DDF engines combustion characteristics is shown in Table 16. Some of the parameters have not been quantitatively evaluated in the previous subsections, so average of the improvement values has been calculated only for the ones quantified, which are shown in Table 17. These parameters have been calculated for less than 20% hydrogen energy share and medium and high load conditions.

Table 16. Effect of hydrogen substitution in DDF engines combustion characteristics

Parameter	Load conditions		
	Low	Medium	High
Heat release rate	Decrease	Increase	Increase
In-cylinder pressure	Decrease	Increase	Increase
Combustion rate	Decrease	Increase	Increase
In-cylinder peak pressure	Decrease	Increase	Increase
In-cylinder temperature	***	Increase	Increase
Ignition delay	***	Decrease	Decrease
Combustion duration	Depends on hydrogen energy share	Decrease	Decrease
Combustion efficiency	***	Increase	Increase
Pressure rise rate	Decrease	Decrease	Increase

Table 17. Average percentage improvement per engine combustion characteristics for less than 20% hydrogen energy share in medium and high load conditions

Parameters	Average percentage improvement
In-cylinder peak pressure	+2.33bar per % of H_2 energy share
In-cylinder peak temperature	+3.48K per % of H_2 energy share
Combustion efficiency	+0.92% per % of H_2 energy share
Maximum pressure rise rate	0.07bar per % of H_2 energy share

3.3. Hydrogen Effect on DDF Engines Emission Characteristics

3.3.1. Co Emission

CO is a residue of incomplete combustion, consequently, is primarily influenced by the air-fuel mixture and in-cylinder temperature (Heywood, 2018). Diesel-based mode has a high equivalence ratio with a fuel-rich air-fuel mixture, which tends to produce CO because there is not enough oxygen to completely burn the fuel. The equivalence ratio decreases significantly with hydrogen addition and consequently the CO emissions, usually at any load conditions as shown in Table 18.

Table 18. co emission effects of hydrogen on DDF engines from (Geo et al., 2008)

LOAD		CO emission	H_2 mass share
100%	HIGH	3.14g/kWh	0.0%
		2.31g/kWh	10.1%

Studies reported near zero CO emissions at 20%, 40% and 60% load conditions with 28% hydrogen energy share (Senthil Kumar et al., 2003) as shown in the Figure 18. Higher temperatures favor oxidation process, that is probably the main reason why CO emissions are reduced by oxidizing into CO_2.

Figure 18a. Variation of CO emissions with respect to hydrogen substitution: From (Senthil Kumar et al., 2003)

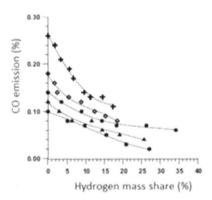

Figure 18b. Variation of CO emissions with respect to hydrogen substitution: From (Miyamoto et al., 2011)

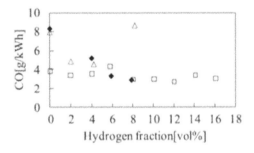

3.3.2. HC Emission

Same as CO, HC emissions are due to incomplete combustion of air-fuel mixture and many studies show that there is a significant reduction in hydrogen based DDF engines (Chintala & Subramanian, 2014c; Geo et al., 2008; Nag et al., 2019; Varde & Frame, 1983), as shown in the Table 19 and Figure 19. Emission reduction is due to increase of in-cylinder temperature and peak pressure which produce a rapid and more efficient combustion.

Table 19. HC emission effects of hydrogen on DDF engines (1) from (Geo et al., 2008), (2) and (3) from (Senthil Kumar et al., 2003)

LOAD		HC emission	H_2 mass share	Reference
100%	HIGH	0.55g/kWh	0.0%	(Geo et al., 2008)
		0.44g/kWh	10.1%	
100%	HIGH	130ppm	0.0%	(Senthil Kumar et al., 2003)
		70ppm	18.0%	
20%	LOW	40ppm	0.0%	
		20ppm	18.0%	

Figure 19a. Variation of HC emissions with respect to hydrogen substitution: From (Senthil Kumar et al., 2003)

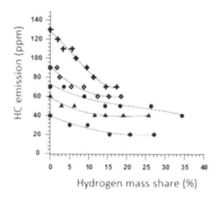

Figure 19b. Variation of HC emissions with respect to hydrogen substitution: From (Miyamoto et al., 2011)

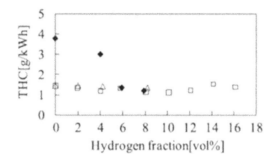

3.3.3. NO_x Emission

In general, increasing hydrogen energy share on DDF engines at high and medium load conditions increases NO_x emissions while at low load conditions NO_x decreases, as shown in figure 20.a. At medium and high load condition shown in figure 20.b, NO_x emissions decrease to a certain minimum before increase again, but the general tendency remains. This is consequence of the high flame speed, high flame temperature and high diffusivity of hydrogen which increases in-cylinder temperature and pressure (S. K. Sharma et al., 2015), oxidizing atmospheric nitrogen.

Figure 20a. Variation of NOx emissions with respect to hydrogen substitution: From (Senthil Kumar et al., 2003)

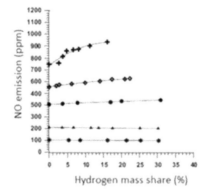

Figure 20b. Variation of NOx emissions with respect to hydrogen substitution: From (Miyamoto et al., 2011)

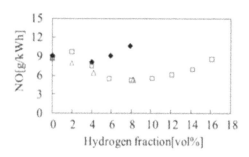

Table 20. NOx emission effects of hydrogen on DDF engines (1) from (Senthil Kumar et al., 2003) and (2) from (Dhole et al., 2014)

LOAD		NO_x emission	H_2 mass share
100%	HIGH	740ppm	0.0%
		930ppm	18.0%
13%	LOW	1.8g/kWh	0.0%
		0.67g/kWh	50.0%

On the other side, at low load conditions such as 20% and 40% it is noticeable in Figure 20.a NO_x emissions are almost invariable respect to hydrogen mass share. However, other study reported a general decrease of NO_x emissions as seen in Table 20 at low load conditions but with a high hydrogen mass share of at least 50% (Dhole et al., 2014). This is probably due to the presence of water when more than 1% of hydrogen is supplied which decreases in-cylinder peak temperature (Uludamar et al., 2016). This phenomenon does not happen at medium and high load conditions when the temperature is too high.

For a specific load, in variable speed engines, NO_x emissions are higher at low speed than at higher speed because of less residence time at high velocities (Pan et al., 2022), which does not allow temperature and chemical conditions to involve all chamber. This is partially explained because of the charge stratification which produces rich oxygen regions in the combustion chamber where temperature is enough to oxidate nitrogen, in consequence, air-fuel mixture homogeneity is a factor to be considered.

3.3.4. Smoke and PM Emission

The main cause of smoke and PM is incomplete combustion (Heywood, 2018), which can be due to (Chintala & Subramanian, 2017):

- Heterogenous air-fuel mixture, where air and fuel are not uniformly distributed, which cause insufficient oxygen or excessive fuel in some regions of combustion chamber leading to an incomplete combustion.
- A decrease in air-fuel ratio which cause a general oxygen deficit.
- The above conditions are aggravated at high and medium load conditions due to increasing of fuel supply to cover load demand.

Hydrogen supply to CI engines produces a general decrease in smoke and PM emissions at all load conditions (Chintala & Subramanian, 2015a) as shown in Table 21 and Figure 21.a, which can be due to (Miyamoto et al., 2011):

- Increasing of homogenous air-fuel mixture because of high diffusivity of hydrogen, creating a more uniform chamber fuel distribution. In case of excessive fuel regions, hydrogen does not react incompletely so there are no incomplete combustion residuals.
- Reduction of carbon content since there is less hydrocarbons in air-fuel mixture.
- In high and medium load conditions combustion temperature and pressure increases with hydrogen supply which means, as previously reviewed, an increase in combustion efficiency and enhancement of oxidation conditions.

Table 21. Smoke emission effects of hydrogen on DDF engines (Chintala & Subramanian, 2015a)

LOAD		Smoke opacity	H_2 mass share
100%	HIGH	15.4%	0.0%
		3.8%	18.8%

Figure 21a. Variation of smoke and PM emissions with respect to hydrogen substitution: From (Senthil Kumar et al., 2003)

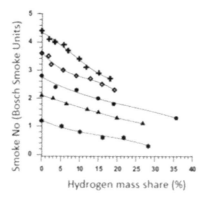

Figure 21b. Variation of smoke and PM emissions with respect to hydrogen substitution: From (Miyamoto et al., 2011)

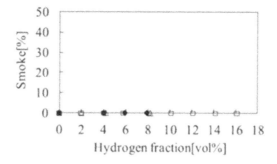

Other studies found almost zero smoke emissions with hydrogen substitution as seen in figure 21.b. Increasing hydrogen share decreases smoke emissions at high loads as seen in Figure 21.a, where the average slope is steeper at high load conditions compared to low load. One way to quantify this phenomenon is through carbon-hydrogen ratio, which when exceeding unity implies the formation of smoke. Since hydrogen fuel does not have carbon, the carbon-hydrogen ratio becomes almost zero and thus smoke and PM is not formed. Experimental results show this PM decreasing tendency in Table 22 and Figure 22. Figure 22 shows substantial reduction in PM emission in the first 2% of hydrogen supply, but beyond that reduction rate decrease significantly.

Using H2 to Increase the Energy Efficiency

Table 22. Pm emission effects of hydrogen on DDF engines (1) (Tsolakis et al., 2005) (2) (Liew et al., 2012)

LOAD		Particle matter	H_2 volume share
???	LOW	98.66mg/cm³	0.0%
		77.64mg/cm³	20.0%
10%	LOW	0.26g/kWh	0.0%
		0.06g/kWh	7.0%

Figure 22. Variation of PM emission with respect to hydrogen volume share (Liew et al., 2012)

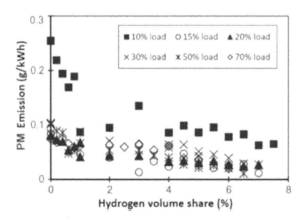

3.3.5. Greenhouse Gases Emission

DDF engines are a promising solution to GHG emissions along with others such as fuel economy improvement, alternative fuels, traffic management system improvements and system weight reduction (Subramanian & Chintala, 2013). In case of CO_2 studies showed significant reduction along with hydrogen supply at any load condition as shown in Table 23.

continued on following page

Using H2 to Increase the Energy Efficiency

Table 23. Continued

Table 23. co2 emission effects of hydrogen on DDF engines (1) (Liew et al., 2012) (2) (Miyamoto et al., 2011)

LOAD		CO_2 emission	H_2 mass share
10%	LOW	920g/kWh	0.0%
		240g/kWh	7.5%
???	???	688g/kWh	0.0%
		425g/kWh	10.0%

Other studies reported reduction from 190g/kWh in diesel-based mode to 104g/kWh in DDF mode with a hydrogen consumption of 0.05g/s (Korakianitis et al., 2010). As seen in Figure 23 as load goes down the effect of hydrogen increases making the slope steeper.

Figure 23. CO2 emission variation with respect to hydrogen volume share (Liew et al., 2012)

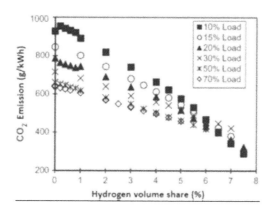

In case of methane there is not enough research to conclude what is the effect of hydrogen on this gas emissions. One study showed a 22% reduction of CH_4 emissions with 20% hydrogen energy share (Subramanian & Chintala, 2013).

3.3.6. Summary of Effect on DDF Engines Emission Characteristics

Briefly the effect of hydrogen substitution in DDF engines pollutant emissions is shown in Table 24. Most of the values in Table 25 are calculated based on slope of a lineal approximation of the empiric curves. These parameters have been selected and calculated for less than 20% hydrogen energy share and medium and high load conditions.

Using H2 to Increase the Energy Efficiency

Table 24. Effect of hydrogen substitution in DDF engines pollutant emissions

Parameter	Load conditions		
	Low	Medium	High
CO emission	Decrease	Decrease	Decrease
HC emission	Decrease	Decrease	Decrease
NOx emission	Do not change	Mainly increase	Increase
Smoke and PM emission	Decrease	Decrease	Decrease
Greenhouse gases	Decrease	Decrease	Decrease

Table 25. Average percentage improvement per engine pollutant emission for less than 20% hydrogen energy share in medium and high load conditions

Parameters	Average percentage improvement
CO emission	-0.43g/kWh per % of H_2 energy share
HC emission	-0.20g/kWh and -2.45ppm per % of H_2 energy share
NO_x emission	+7.51ppm per % of H_2 energy share
Smoke	-0.62% of opacity per % of H_2 energy share
CO_2 emission	-35g/kWh per % of H_2 energy share

3.4. Limitations and Solution Strategies for Hydrogen DDF Engines Applications

3.4.1. High Levels of No_x Emission

The more in-cylinder temperature increases, the greater nitrogen oxides formation will be because this condition is necessary for nitrogen oxidation. One way to minimize nitrogen oxide is using ultra-lean combustion which works identically to a low temperature combustion (S. K. Sharma et al., 2015). There are some strategies to solve this problem:

a) **Exhaust gas recirculation (EGR):** Consists of recirculate a portion of exhaust gases, previously cooled, back into the combustion chamber and mix it with inlet air-fuel mixture. Its main objective is to reduce oxygen concentration and consequently reduce in-cylinder temperature because the exhaust gases are inert in the combustion process.

b) **Injection time retarding:** Involves timing retardation of fuel injection and delay the autoignition occurring after the TDC. This method alters heat release pattern reducing peak pressure and temperature.

Using H2 to Increase the Energy Efficiency

c) **Compression ratio decrease:** Increase compression ratio also increases in-cylinder peak pressure and thermal efficiency which leads to a temperature increment and NO_x formation as shown in Table 26 Consequently, reducing the compression rate prevents the conditions necessary for the formation of NO_x.

d) **Homogenous charge compression ignition (HCCI):** Consists of thoroughly mix air-fuel until it becomes a homogeneous mixture before enters to the combustion chamber. Since CI engines usually take advantage of stratified charge to ignites, when air-fuel mixture is uniform is more difficult to set autoignition timing, making in-cylinder temperature and pressure crucial parameters. This technology is usually used to reduce NO_x, smoke and PM due to reduction of stratified in-cylinder temperature in diesel-base mode (P. Sharma & Dhar, 2018), which can be of 98% NO_x and 95% smoke reduction (Asad et al., 2015). In DDF mode there is evidence of increasing premixed charge quantity, by adding hydrogen or diesel-hydrogen, or increasing mixture preparation time would lead to PHCCI mode (Chintala & Subramanian, 2017).

e) **Water injection:** Involves injecting water or water-methanol mixture into the combustion chamber and is usually controlled by an electronic control unit. This method takes advantage of the heat needed to change water phase from liquid to vapor, which absorbs heat leading to cooling effect. Additionally, this method can improve combustion because water vapor increases the overall inlet air density but depending on input method and mixture conditions HC, CO and smoke can increase (Serrano et al., 2019). The following results were found for diesel-based mode with water injection application:

- Nitrogen oxide reduction from 1034ppm to 643ppm at 100% load (Subramanian, 2011)
- Using 60-65% water mass substitution of fuel 50% of nitrogen oxide reduction was achieved (Tauzia et al., 2010)
- Water injection of 3kg/h resulted in 50% nitrogen oxide reduction using manifold injection (Tesfa et al., 2012)
- Using 20% water mass substitution of fuel 34% of nitrogen oxide reduction was achieved (Gonca, 2014)

Table 26. Effects of increasing compression ratio in a DDF engine of 3.7kW – 1500RPM (Masood et al., 2007)

Parameters	Values
Compression ratio change	16.35:1 – 24.5:1

continued on following page

Table 26. Continued

Parameters	Values
In-cylinder peak pressure	+42%
Thermal efficiency	+27%
NO_x emissions	+38%
HC emissions	-17%
CO emissions	-21%
PM emissions	-16%

3.4.2. Knocking at High Amount of Hydrogen Substitution

Knocking is phenomenon characterized by the spontaneous ignition of air-fuel mixture before the intended timing that led to combustion efficiency reduction, increasing of wear and tear, and potential engine damage. Hydrogen autoignition depends on in-cylinder temperature, in-cylinder pressure and pressure rise rate, all of them increased by hydrogen combustion conditions. In term of engine design parameters hydrogen autoignition is also related to compression ratio, injection timing and engine load operating conditions (Dimitriou et al., 2018). It is also known that hydrogen has high polytropic index (Szwaja, 2019), which means the polytropic process of air-fuel combustion is closer to an isothermal process which at high in-cylinder temperature conditions can produce knocking. Much less hydrogen energy share is used in DDF engines to prevent knocking, so new technology needs to be identified and assessed to achieve high hydrogen energy share substitution.

This phenomenon also produces higher peak pressure, higher peak temperature and overheating of in-cylinder walls. Knocking produces a reduction of power output which decreases logarithmically with the inverse of intake absolute temperature in DDF engines with gaseous fuel and liquid pilot fuel (Karim, 2003). The hydrogen properties associated with knocking are lower ignition energy, wider flammability range and short quenching distance (Sivabalakrishnan & Jegadheesan, 2014). Several studies have analyzed the upper limit of hydrogen energy share before knocking problem:

- Chintala and Subramanian found 18.8% as the upper limit at 100% load and compression ratio of 19.5:1 based on autoignition timing delay (Chintala & Subramanian, 2015b). This is shown in Figure 24 where the red curve represents 18.8% hydrogen substitution, and which rises in the TDC limit from autoignition in around -4 crank angle.

Using H2 to Increase the Energy Efficiency

- Szwaja et al. found 17% as the upper limit at 100% load and compression ratio of 17:1 with HCCI DDF combustion based in high frequency component of in-cylinder pressure (HFCP) analysis (Szwaja & Grab-Rogalinski, 2009), as shown in Figure 25.
- Also, Chintala and Subramanian found 19% as the upper limit at same load and compression ratio conditions but based on ringing intensity analysis (Chintala & Subramanian, 2015b).
- Miyamoto reported autoignition of air-fuel mixture in 8% higher volume fraction of hydrogen (Miyamoto et al., 2011)

Figure 24. Autoignition of hydrogen-air mixture at higher hydrogen energy share (Chintala & Subramanian, 2015b)

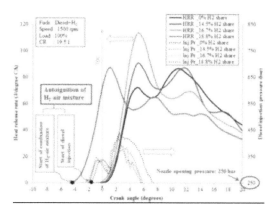

Figure 25. Maximum limit of hydrogen energy shares in DDF engines (Szwaja & Grab-Rogalinski, 2009)

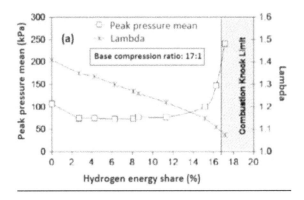

On the other hand, knocking can be prevented by the following strategies:
a) **Compression ratio reduction:** Allows to increase hydrogen energy share without this problem. The tendency is shown in Figure 26 and other studies have shown the same results, knocking problem intensifies with increasing compression ratio (Lim et al., 2013; Meng et al., 2022).
b) **Homogenous charge compression ignition (HCCI):** Due to the uniformity of the air-fuel mixture generated before entering the combustion chamber, this strategy can be an alternative to prevent mixture autoignition before reaching the TDC since there is no stratified charge.
c) **Water injection:** In-cylinder temperature reduction can prevent air-fuel mixture autoignition before TDC and enhance hydrogen energy share.

Figure 26. Maximum limit of hydrogen energy shares in DDF engines (Chintala & Subramanian, 2015b)

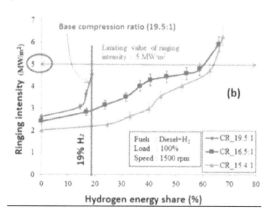

3.4.3. Hydrogen Energy Share Limitations

High amounts of hydrogen supply in diesel engine can cause abnormal rate of pressure rise, higher in-cylinder peak pressure and temperature, and loss of output or available work (Chintala & Subramanian, 2014b; Meng et al., 2022; Miyamoto et al., 2011; Polk et al., 2013). The reviewed studies show that the lower the charge level, the greater the hydrogen energy share (Chintala & Subramanian, 2014a), so in high load conditions the hydrogen contributes lower energy to the engine as shown in Table 27. Other predominant factor on upper limit hydrogen energy share is the injection method, being found that manifold injection is more susceptible to backfiring and knocking problems than direct port injection which allows better control of injection timing.

Table 27. Effect of load conditions on maximum hydrogen energy share (Chintala & Subramanian, 2014a)

Load	Maximum H_2 energy share
100%	18.8%
50%	48.4%

To increment the upper limit of hydrogen supply and energy share several strategies and methods have been tested. TABLE XXVIII shows comparative results of conventional hydrogen DDF engine mode with the method indicated in the first column using the evaluation parameter of the third column, which indicates the change in the method parameter. Table 28 shows that compression ratio (CR) needs to be reduced to cool in-cylinder temperature and prevent knocking. The last example is a combination of CR reduction and water injection which produces the maximum increase of energy share 60% from conventional DDF mode.

Table 28. Effect of different methods to maximize hydrogen energy share

Method	Method value	Evaluation Parameter (EP)	EP value	Reference
Partial HCCI	***	Hydrogen energy share	18.8%	(Chintala & Subramanian, 2014b)
			24.4%	
Water injection	***	Hydrogen volume substitution	38%	(Mathur et al., 1993)
			66%	
		Hydrogen energy share	20%	(Chintala & Subramanian, 2014c)
			39%	
Compression ratio (CR) reduction	From 24.5:1 to 16.35:1	Hydrogen substitution	0.096kg/h	(Masood et al., 2007)
			0.138kg/h	
CR reduction	From 19.5:1 to 16.5:1	Hydrogen energy share	19%	(Chintala & Subramanian, 2015b)
			59%	
CR reduction & Water injection	• CR from 19.5:1 to 16.5:1 • Water consumption 340g/kWh	Hydrogen energy share	19%	(Chintala & Subramanian, 2016)
			79%	

4. CONCLUSION

An average hydrogen energy share between 10-20% can be set based on previous described limitations and at least medium load operation conditions, as it is usual in oil and gas industry diesel engines. We can quantitatively summarize the effect of hydrogen in dual diesel engines based on the Tables 27, 28, 29 also calculate an approximation of the amount of improvement based on the Tables 27, 28, 29 which was made considering an interval average of 15% of hydrogen energy share.

Using H2 to Increase the Energy Efficiency

Table 29. Summary of qualitative and quantitative effect of hydrogen in average diesel engine oil and gas industry operation conditions

Parameter Type	Parameter	10-20% H_2 energy share and high load conditions	
Engine performance	Thermal efficiency	Increase	+2.25%
	Exergy efficiency	Increase	+1.50%
	Volumetric efficiency	Decrease	-7.05%
	Exhaust gas temperature	Increase	+93.60°C
Combustion characteristics	In-cylinder peak pressure	Increase	+34.95bar
	In-cylinder peak temperature	Increase	+52.20K
	Combustion efficiency	Increase	+13.80%
	Maximum pressure rise rate	Increase	+1.05bar
Pollutant emissions	CO emission	Decrease	-6.45g/kWh
	HC emission	Decrease	-3.00 g/kWh
	NO_x emission	Increase	+112.65ppm
	Smoke and PM emission	Decrease	-9.30% of opacity
	Greenhouse gases (CO_2)	Decrease	-525.00g/kWh

It is necessary to carry out more studies to quantify the cost reduction associated with the operational optimization of a diesel engine by supplying 15% hydrogen energy share, but it is possible to affirm based on all the information and data reviewed that this optimization exists and is a consequence of the improvement in thermal efficiency, exergy efficiency and combustion efficiency and their associated parameters, such as in-cylinder temperature and pressure, pressure rise rate, among others. In this context, the amount of diesel fuel will be reduced which has a positive consequence itself in the associated costs.

It is especially important to consider the reduction of all carbon-based pollutants such as carbon monoxide, unburned hydrocarbons, smoke and particulate matter, as well as greenhouse gases, in one of the most polluting industries in the world (Beneš, 2021). Regarding the worsened parameters, NO_x stands out, for which some application alternatives have been reviewed to control the temperature and pressure of the combustion chamber without reducing the positive effect of hydrogen on other parameters, but more empirical research in this type of applications is necessary and develop of new hydrogen specific technology.

REFERENCES

Asad, U., Zheng, M., Ting, D. S. K., & Tjong, J. (2015). Implementation challenges and solutions for homogeneous charge compression ignition combustion in diesel engines. *Journal of Engineering for Gas Turbines and Power*, 137(10), 101505. Advance online publication. 10.1115/1.4030091

Beneš, P. (2021). How much energy is needed to power a combustion car? LinkedIn. https://www.linkedin.com/pulse/how-much-energy-needed-power-combustion-car-petr-benes/

Cameretti, M. C., De Robbio, R., Mancaruso, E., & Palomba, M. (2022). CFD Study of Dual Fuel Combustion in a Research Diesel Engine Fueled by Hydrogen. *Energies*, 15(15), 5521. Advance online publication. 10.3390/en15155521

Chintala, V., & Subramanian, K. A. (2014a). Assessment of maximum available work of a hydrogen fueled compression ignition engine using exergy analysis. *Energy*, 67, 162–175. Advance online publication. 10.1016/j.energy.2014.01.094

Chintala, V., & Subramanian, K. A. (2014b). Experimental investigation on effect of enhanced premixed charge on combustion characteristics of a direct injection diesel engine. *International Journal of Advances in Engineering Sciences and Applied Mathematics*, 6(1–2), 3–16. Advance online publication. 10.1007/s12572-014-0109-7

Chintala, V., & Subramanian, K. A. (2014c). Hydrogen energy share improvement along with NOx (oxides of nitrogen) emission reduction in a hydrogen dual-fuel compression ignition engine using water injection. *Energy Conversion and Management*, 83, 249–259. Advance online publication. 10.1016/j.enconman.2014.03.075

Chintala, V., & Subramanian, K. A. (2015a). An effort to enhance hydrogen energy share in a compression ignition engine under dual-fuel mode using low temperature combustion strategies. *Applied Energy*, 146, 174–183. Advance online publication. 10.1016/j.apenergy.2015.01.110

Chintala, V., & Subramanian, K. A. (2015b). Experimental investigations on effect of different compression ratios on enhancement of maximum hydrogen energy share in a compression ignition engine under dual-fuel mode. *Energy*, 87, 448–462. Advance online publication. 10.1016/j.energy.2015.05.014

Chintala, V., & Subramanian, K. A. (2016). Experimental investigation of hydrogen energy share improvement in a compression ignition engine using water injection and compression ratio reduction. *Energy Conversion and Management*, 108, 106–119. Advance online publication. 10.1016/j.enconman.2015.10.069

Chintala, V., & Subramanian, K. A. (2017). A comprehensive review on utilization of hydrogen in a compression ignition engine under dual fuel mode. In Renewable and Sustainable Energy Reviews (Vol. 70). 10.1016/j.rser.2016.11.247

De Morais, A. M., Mendes Justino, M. A., Valente, O. S., Hanriot, S. D. M., & Sodré, J. R. (2013). Hydrogen impacts on performance and CO2 emissions from a diesel power generator. *International Journal of Hydrogen Energy*, 38(16), 6857–6864. Advance online publication. 10.1016/j.ijhydene.2013.03.119

Deb, M., Sastry, G. R. K., Bose, P. K., & Banerjee, R. (2015). An experimental study on combustion, performance and emission analysis of a single cylinder, 4-stroke DI-diesel engine using hydrogen in dual fuel mode of operation. *International Journal of Hydrogen Energy*, 40(27), 8586–8598. Advance online publication. 10.1016/j.ijhydene.2015.04.125

Dhole, A. E., Yarasu, R. B., Lata, D. B., & Priyam, A. (2014). Effect on performance and emissions of a dual fuel diesel engine using hydrogen and producer gas as secondary fuels. *International Journal of Hydrogen Energy*, 39(15), 8087–8097. Advance online publication. 10.1016/j.ijhydene.2014.03.085

Dimitriou, P., Kumar, M., Tsujimura, T., & Suzuki, Y. (2018). Combustion and emission characteristics of a hydrogen-diesel dual-fuel engine. *International Journal of Hydrogen Energy*, 43(29), 13605–13617. Advance online publication. 10.1016/j.ijhydene.2018.05.062

European Commission. (2015). Press corner. FAQ - Air pollutant emissions standards. European Commission. https://ec.europa.eu/commission/presscorner/detail/mt/MEMO_15_5705

Geo, V. E., Nagarajan, G., & Nagalingam, B. (2008). Studies on dual fuel operation of rubber seed oil and its bio-diesel with hydrogen as the inducted fuel. *International Journal of Hydrogen Energy*, 33(21), 6357–6367. Advance online publication. 10.1016/j.ijhydene.2008.06.021

Gonca, G. (2014). Investigation of the effects of steam injection on performance and NO emissions of a diesel engine running with ethanol-diesel blend. *Energy Conversion and Management*, 77, 450–457. Advance online publication. 10.1016/j.enconman.2013.09.031

Heywood, J. B. (2018). Internal Combustion Engine Fundamentals. In Internal Combustion Engine Fundamentals Second Edition.

IPIECA. (2023). Drilling rigs. Energy efficiency compendium.

Jamrozik, A., Grab-Rogaliński, K., & Tutak, W. (2020). Hydrogen effects on combustion stability, performance and emission of diesel engine. *International Journal of Hydrogen Energy*, 45(38), 19936–19947. Advance online publication. 10.1016/j.ijhydene.2020.05.049

Karagöz, Y., Sandalcl, T., Yüksek, L., Dalklllç, A. S., & Wongwises, S. (2016). Effect of hydrogen-diesel dual-fuel usage on performance, emissions and diesel combustion in diesel engines. *Advances in Mechanical Engineering*, 8(8). Advance online publication. 10.1177/1687814016664458

Karim, G. A. (2003). Combustion in gas fueled compression: Ignition engines of the dual fuel type. *Journal of Engineering for Gas Turbines and Power*, 125(3), 827–836. Advance online publication. 10.1115/1.1581894

Karim, G. A. (2015). Dual-fuel diesel engines. In Dual-Fuel Diesel Engines. 10.1201/b18163

Korakianitis, T., Namasivayam, A. M., & Crookes, R. J. (2010). Hydrogen dual-fuelling of compression ignition engines with emulsified biodiesel as pilot fuel. *International Journal of Hydrogen Energy*, 35(24), 13329–13344. Advance online publication. 10.1016/j.ijhydene.2010.08.007

Lata, D. B., & Misra, A. (2010). Theoretical and experimental investigations on the performance of dual fuel diesel engine with hydrogen and LPG as secondary fuels. *International Journal of Hydrogen Energy*, 35(21), 11918–11931. Advance online publication. 10.1016/j.ijhydene.2010.08.039

Lata, D. B., & Misra, A. (2011). Analysis of ignition delay period of a dual fuel diesel engine with hydrogen and LPG as secondary fuels. *International Journal of Hydrogen Energy*, 36(5), 3746–3756. Advance online publication. 10.1016/j.ijhydene.2010.12.075

Lata, D. B., Misra, A., & Medhekar, S. (2011). Investigations on the combustion parameters of a dual fuel diesel engine with hydrogen and LPG as secondary fuels. *International Journal of Hydrogen Energy*, 36(21), 13808–13819. Advance online publication. 10.1016/j.ijhydene.2011.07.142

Liew, C., Li, H., Liu, S., Besch, M. C., Ralston, B., Clark, N., & Huang, Y. (2012). Exhaust emissions of a H2-enriched heavy-duty diesel engine equipped with cooled EGR and variable geometry turbocharger. *Fuel*, 91(1), 155–163. Advance online publication. 10.1016/j.fuel.2011.08.002

Liew, C., Li, H., Nuszkowski, J., Liu, S., Gatts, T., Atkinson, R., & Clark, N. (2010). An experimental investigation of the combustion process of a heavy-duty diesel engine enriched with H2. *International Journal of Hydrogen Energy*, 35(20), 11357–11365. Advance online publication. 10.1016/j.ijhydene.2010.06.023

Lim, G., Lee, S., Park, C., Choi, Y., & Kim, C. (2013). Effects of compression ratio on performance and emission characteristics of heavy-duty SI engine fuelled with HCNG. *International Journal of Hydrogen Energy*, 38(11), 4831–4838. Advance online publication. 10.1016/j.ijhydene.2013.01.188

Masood, M., Mehdi, S. N., & Reddy, P. R. (2007). Experimental investigations on a hydrogen-diesel dual fuel engine at different compression ratios. *Journal of Engineering for Gas Turbines and Power*, 129(2), 572–578. Advance online publication. 10.1115/1.2227418

Mathur, H. B., Das, L. M., & Patro, T. N. (1993). Hydrogen-fuelled diesel engine: Performance improvement through charge dilution techniques. *International Journal of Hydrogen Energy*, 18(5), 421–431. Advance online publication. 10.1016/0360-3199(93)90221-U

Meng, H., Ji, C., Yang, J., Chang, K., Xin, G., & Wang, S. (2022). Experimental understanding of the relationship between combustion/flow/flame velocity and knock in a hydrogen-fueled Wankel rotary engine. *Energy*, 258, 124828. Advance online publication. 10.1016/j.energy.2022.124828

Miyamoto, T., Hasegawa, H., Mikami, M., Kojima, N., Kabashima, H., & Urata, Y. (2011). Effect of hydrogen addition to intake gas on combustion and exhaust emission characteristics of a diesel engine. *International Journal of Hydrogen Energy*, 36(20), 13138–13149. Advance online publication. 10.1016/j.ijhydene.2011.06.144

Nag, S., Sharma, P., Gupta, A., & Dhar, A. (2019). Experimental study of engine performance and emissions for hydrogen diesel dual fuel engine with exhaust gas recirculation. *International Journal of Hydrogen Energy*, 44(23), 12163–12175. Advance online publication. 10.1016/j.ijhydene.2019.03.120

Pan, S., Wang, J., Liang, B., Duan, H., & Huang, Z. (2022). Experimental Study on the Effects of Hydrogen Injection Strategy on the Combustion and Emissions of a Hydrogen/Gasoline Dual Fuel SI Engine under Lean Burn Condition. *Applied Sciences (Basel, Switzerland)*, 12(20), 10549. Advance online publication. 10.3390/app122010549

Polk, A. C., Gibson, C. M., Shoemaker, N. T., Srinivasan, K. K., & Krishnan, S. R. (2013). Analysis of Ignition Behavior in a Turbocharged Direct Injection Dual Fuel Engine Using Propane and Methane as Primary Fuels. *Journal of Energy Resources Technology*, 135(3), 032202. Advance online publication. 10.1115/1.4023482

Rahman, M. A., Ruhul, A. M., Aziz, M. A., & Ahmed, R. (2017). Experimental exploration of hydrogen enrichment in a dual fuel CI engine with exhaust gas recirculation. *International Journal of Hydrogen Energy*, 42(8), 5400–5409. Advance online publication. 10.1016/j.ijhydene.2016.11.109

Salvi, B. L., & Subramanian, K. A. (2015). Sustainable development of road transportation sector using hydrogen energy system. In Renewable and Sustainable Energy Reviews (Vol. 51). 10.1016/j.rser.2015.07.030

Santoso, W. B., Bakar, R. A., & Nur, A. (2013). Combustion characteristics of diesel-hydrogen dual fuel engine at low load. *Energy Procedia*, 32, 3–10. Advance online publication. 10.1016/j.egypro.2013.05.002

Senthil Kumar, M., Ramesh, A., & Nagalingam, B. (2003). Use of hydrogen to enhance the performance of a vegeTABLE oil fuelled compression ignition engine. *International Journal of Hydrogen Energy*, 28(10). Advance online publication. 10.1016/S0360-3199(02)00234-3

Serrano, J., Jiménez-Espadafor, F. J., & López, A. (2019). Analysis of the effect of different hydrogen/diesel ratios on the performance and emissions of a modified compression ignition engine under dual-fuel mode with water injection. Hydrogen-diesel dual-fuel mode. *Energy*, 172, 702–711. Advance online publication. 10.1016/j.energy.2019.02.027

Sharma, P., & Dhar, A. (2018). Effect of hydrogen supplementation on engine performance and emissions. *International Journal of Hydrogen Energy*, 43(15), 7570–7580. Advance online publication. 10.1016/j.ijhydene.2018.02.181

Sharma, S. K., Goyal, P., & Tyagi, R. K. (2015). Hydrogen-fueled internal combustion engine: A review of technical feasibility. In International Journal of Performability Engineering (Vol. 11, Issue 5). https://doi.org/10.23940/ijpe.15.5.p491.mag

Sivabalakrishnan, R., & Jegadheesan, C. (2014). Study of Knocking Effect in Compression Ignition Engine with Hydrogen as a Secondary Fuel. *Chinese Journal of Engineering*, 2014, 1–8. Advance online publication. 10.1155/2014/102390

Subramanian, K. A. (2011). A comparison of water-diesel emulsion and timed injection of water into the intake manifold of a diesel engine for simultaneous control of NO and smoke emissions. *Energy Conversion and Management*, 52(2), 849–857. Advance online publication. 10.1016/j.enconman.2010.08.010

Subramanian, K. A., & Chintala, V. (2013). Reduction of GHGs emissions in a biodiesel fueled diesel engine using hydrogen. ASME 2013 Internal Combustion Engine Division Fall Technical Conference, ICEF 2013, 2. 10.1115/ICEF2013-19133

Szwaja, S. (2019). Dilution of fresh charge for reducing combustion knock in the internal combustion engine fueled with hydrogen rich gases. *International Journal of Hydrogen Energy*, 44(34), 19017–19025. Advance online publication. 10.1016/j.ijhydene.2018.10.134

Szwaja, S., & Grab-Rogalinski, K. (2009). Hydrogen combustion in a compression ignition diesel engine. *International Journal of Hydrogen Energy*, 34(10), 4413–4421. Advance online publication. 10.1016/j.ijhydene.2009.03.020

Tahir, M. M., Ali, M. S., Salim, M. A., Bakar, R. A., Fudhail, A. M., Hassan, M. Z., & Abdul Muhaimin, M. S. (2015). Performance analysis of a spark ignition engine using compressed natural gas (CNG) as fuel. *Energy Procedia*, 68, 355–362. Advance online publication. 10.1016/j.egypro.2015.03.266

Tauzia, X., Maiboom, A., & Shah, S. R. (2010). Experimental study of inlet manifold water injection on combustion and emissions of an automotive direct injection Diesel engine. *Energy*, 35(9), 3628–3639. Advance online publication. 10.1016/j.energy.2010.05.007

Tesfa, B., Mishra, R., Gu, F., & Ball, A. D. (2012). Water injection effects on the performance and emission characteristics of a CI engine operating with biodiesel. *Renewable Energy*, 37(1), 333–344. Advance online publication. 10.1016/j.renene.2011.06.035

Tsolakis, A., Hernandez, J. J., Megaritis, A., & Crampton, M. (2005). Dual fuel diesel engine operation using H2. Effect on particulate emissions. *Energy & Fuels*, 19(2), 418–425. Advance online publication. 10.1021/ef0400520

Uludamar, E., Yildizhan, Ş., Aydin, K., & Özcanli, M. (2016). Vibration, noise and exhaust emissions analyses of an unmodified compression ignition engine fuelled with low sulphur diesel and biodiesel blends with hydrogen addition. *International Journal of Hydrogen Energy*, 41(26), 11481–11490. Advance online publication. 10.1016/j.ijhydene.2016.03.179

US EPA. (2024). Nitrogen Oxides Control Regulations. US EPA. https://www3.epa.gov/region1/airquality/nox.html

Varde, K. S., & Frame, G. A. (1983). Hydrogen aspiration in a direct injection type diesel engine-its effects on smoke and other engine performance parameters. *International Journal of Hydrogen Energy*, 8(7), 549–555. Advance online publication. 10.1016/0360-3199(83)90007-1

Yadav, V. S., Soni, S. L., & Sharma, D. (2014). Engine performance of optimized hydrogen-fueled direct injection engine. *Energy*, 65, 116–122. Advance online publication. 10.1016/j.energy.2013.12.007

Chapter 12
Water Footprint Reduction in Engineering, Laboratory, and Administrative Buildings:
A Field Case of the Scientific University of the South

Tiffany Krisel Billinghurst
https://orcid.org/0000-0002-2215-6044
Scientific University of the South, Peru

Alfredo David Lescano Lozada
Alwa, Peru

ABSTRACT

The water footprint is a tool that indicates the direct and indirect consumption of fresh water, whether in a production process, product, service, building, institution, geopolitical area, economic sector, or a person. In the headquarter of the oil and gas (O&G) corporations, also called exploration and production (E & P) companies, typically there are administration, engineering, and laboratory activities related to the business bottom line a close analogy to these types of buildings are educational buildings, such as universities. Therefore, the outcome of the presented work could be applied to the O&G buildings as well. The objective of this research was to calculate the water footprint of the Villa Campus of the Scientific University of the South for its activities in 2019, using the methodology of the Water Footprint Network, which is considered a world standard for the evaluation of the water footprint and based mostly on the work of Professor Dr. Arjen Hoekstra, who introduced the concept of

DOI: 10.4018/979-8-3693-0740-3.ch012

the water footprint in 2002.

INTRODUCTION

The limited availability of fresh water is a global risk for the satisfaction of people's basic needs and economic growth. In 2019, the World Economic Forum (WEF) included the water crisis in its report among the top ten risks in terms of impacts that companies face (World Economic Forum, 2019).

Increasingly, various companies have paid attention to the importance of measuring the Water Footprint (Arjen Y. Hoekstra, 2015), a tool that indicates direct and indirect water consumption, considering its three types, according to origin: Blue Footprint, Green Footprint and Gray Footprint (Arjen Y. Hoekstra, Chapagain, Aldaya, & Mekonnen, 2021). The blue footprint measures the water directed for supply minus the water used during the supply chain (Arjen Y. Hoekstra, Mekonnen, Chapagain, Mathews, & Richter, 2012); The green footprint refers to the precipitation water that is stored by the plant (Mekonnen & Hoekstra, 2011); and the gray footprint determines the water required to assimilate the contaminants present (Jeswani & Azapagic, 2011).

The Scientific University of the South (UCSUR), being located on the Pacific slope, suffers from poor water availability. Which, in addition, borders the Protected Natural Area and Ramsar site "Pantanos de Villa Wildlife Refuge", of national and international importance for its high biodiversity (Amaro & Goyoneche, 2017). Under this support, the calculation and analysis of sustainability of the water consumption of the Villa 1 and Villa 2 campuses of the higher educational institution was considered relevant, by measuring their Water Footprint; and subsequent proposal of measures for its reduction, focused on activities that demand greater consumption of the resource. In this way, the final calculation, the formulation of optimization proposals and their dissemination, aim to develop better awareness and subsequent action regarding the impact on the water consumption of the members of the UCSUR. It is expected that, in accordance with its humanistic vision and environmental approach, the institution will guide and allow these measures to be implemented for the development of a more sustainable campus.

MATERIALS AND METHODS

Study Area

The present research had as its study area the Villa Main Campus, made up of Villa 1 and Villa 2 of the Scientific University of the South (12°13'17.50" S 76°58'36.12"W) located at kilometer 19 of the Panamericana South in the Villa el Salvador district of the department of Lima (Figure 1.). It has a total area of 7.36 hectares, of which 2.84 hectares correspond to the occupied area and 4.52 hectares to green areas.

Figure 1. Satellite image of the area occupied by the UCSUR Villa Main Campus (delimited by the orange line polygon)

Sample

In 2019, there was a university population of 17,623 people made up of undergraduate students, teachers, administrative staff and workers from external companies (cafeteria, security, food sales stands, cleaning and gardening. Two samples were required to be determined, one for the population of 15,799 undergraduate students (population 1), and another for the population of teachers, administrators, and workers of external companies with a total of 1,824 people (population 2). Which were used to infer the consumption of water, paper, and fuel; through two different surveys for each population, because each population has different water and paper consumption habits.

To determine the population sample, being a quantitative variable and knowing the total number of people, the statistical formula (1) must be used (López-Roldán & Fachelli, 2017).

$$n = \frac{N \times z^2 \times \sigma^2}{e^2 \times (N-1) + z^2 \times \sigma^2} \quad (1)$$

where:

n: Sample size
N: Size of the university population
z^2: Value determined based on the confidence level (1.64) squared
σ^2: Variance of the quantitative variable
e^2: Sampling error squared (0.10)

A variance of 0.1 and a confidence level of 90% have been defined. Two samples were determined, one of 267 for the surveys directed at population 1 and 236 for the surveys directed at population 2, of which all were answered and applied in 2021.

METHODOLOGY ADAPTED FROM THE WATER FOOTPRINT NETWORK

This research was adapted and used the methodology proposed by the Water Footprint Network for the calculation and evaluation of the Water Footprint of the educational institution. This methodology consists of four phases (Figure 2) (Aldaya, Chapagain, Hoekstra, & Mekonnen, 2012).

Figure 2. Flowchart of the water footprint evaluation stages

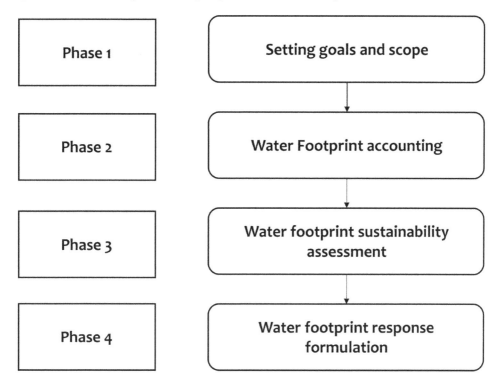

Definition of Objectives and Scope

The objective is to calculate the water footprint of the Villa Campus of the Scientific University of the South for its activities in 2019. In addition, it is expected to contribute to raising awareness of the impact of water consumption of the members of the university through the dissemination of the calculation results; the formulation of proposals to optimize water consumption and that these measures are implemented by the institution.

The scope of water footprint accounting includes the calculation of the direct and indirect footprint. In the calculation of the direct water footprint, its three components will be included: blue footprint, green footprint, and gray footprint. On the other hand, to calculate the indirect footprint, four elements will be considered: energy consumption, the fuel used in the university buses and also the transportation used by the members, the consumption of paper by the entire population. university, and the consumption of water containers; A monthly periodicity of the information collected for the calculation was considered; Furthermore, there are no import or export materials.

Water Footprint Accounting

The water footprint of a company, in this case UCSUR, corresponds to the total volume of water that is used directly or indirectly for its operation. Some of the data required for the quantification process had to be requested from the Faculty of Environmental Sciences of the Institution. The rest was obtained through the two surveys aimed at the university population.

Direct Water Footprint

For a company, this term is also known as "operational water footprint of general activities" and identifies the water consumption necessary for the continuous execution of a company (Arjen Y. Hoekstra et al., 2021). For the calculation, its three components were considered: blue footprint, green footprint and gray footprint.

Blue Water Footprint

For a better understanding of its calculation, the water balance of the system (UCSUR) was represented, considering its inputs and outputs (Figure 3).

Water Footprint Reduction

Figure 3. Water balance of the blue footprint

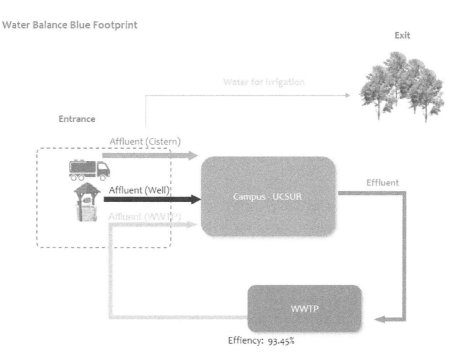

This footprint is divided into two (2), the blue footprint 1 (H1) (3), which is obtained from the subtraction of the water inputs (called tributaries (Afl)) from three sources, with the effluent, which is obtained through surveys. And, the blue footprint 2 (H2) (4), is the one coming from irrigation of green areas, which is equal to the minimum value between the effective irrigation (IE) and the irrigation requirement (RR), data obtained from the CROPWAT 8.0 software.

For the calculation in CROPWAT 8.0, the meteorology data required are: maximum and minimum temperature, precipitation, humidity percentage, wind speed and sunshine hours, these must be monthly and averaged and are measured by the UCSUR Meteorological Station, and delivered by the faculty of environmental sciences upon request.

$$HA = H1 + H2 \tag{2}$$

$$H1 = (AflCis + AflPozo + AflPTAR) - Ef \tag{3}$$

$$H2 = Min(IE, RR) \times 10 \tag{4}$$

The factor 10 fulfills the function of converting the depth of water (in millimeters) into volumes of water per land surface [m³/ha]. In turn, it is multiplied by the total area of green areas (AV) of the university to obtain the value in units of water flow m³/year.

Effective irrigation (IE) (5) is the difference between total crop evaporation (ETC), data obtained from CROPWAT 8.0, and green evapotranspiration (EV), data obtained in the calculation of the green footprint.

$$IE = ETC - EV \tag{5}$$

The first source of water is cistern water (AflCis) extracted from two fountains in the districts of Lurín and San Juan de Miraflores. The second and third comes from water extracted from two wells (AflPozo), and from water that is treated by the university's Wastewater Treatment Plant (WWTP). Regarding these last two sources, the institution does not monitor the volume entering the system, so it had to be calculated by the water balance.

The approximate efficiency percentage of 93.45% will be considered for the output flow of the WWTP, which was calculated based on the results of monitoring the BOD and COD parameters at the entrance and exit of the WWTP. From this percentage, the third entry will be estimated, based on the volume of effluent water (Efl) (calculated through surveys) that is treated, and enters the system again (AflPTAR). On the other hand, the well water (6) is calculated with the input volume data from the first two sources, the water directed for irrigation (HR), the effluent, and the percentage of 80% that is considered potable water. consumed, according to section 1.8. Sewer Contribution Flow of Technical Standard OS.100 of the National Building Regulations (Ministerio de Vivienda, 2006)The water directed for irrigation is equal to the irrigation requirement (RR) between the efficiency (%Ef) of the current (traditional) irrigation method (7), data provided by the Ministry of Agriculture (Ministerio de Agricultura y Riego, 2015).

$$Efl = (AflCis + AflPozo + AflPTAR - HR) \times 80\%$$

$$AflPozo = \frac{Efl}{80\%} + HR - AflPTAR - AflCis \tag{6}$$

$$HR = \left(\frac{RR}{\%E}\right)) \tag{7}$$

Green water footprint

This footprint refers to the amount of water that is stored by the plant from precipitation, which does not become runoff, but remains on the surface of the vegetation, and is evapotranspirated [5]. It was calculated from data obtained by the CROPWAT 8.0 software: green water evapotranspiration (EV) and humidity deficit (DH) (8).

$$HV = (EV + DH) \times 10 \times AV \qquad (8)$$

Gray Water Footprint

It is defined as the volume of water necessary to assimilate a load of contaminants according to natural concentrations and the parameters established in environmental water quality legislation.

In this case, the parameters will be considered: oils and fats, Biochemical Oxygen Demand (BOD) and Chemical Oxygen Demand (COD); analysis of three measurements granted by the Environmental Engineering Career, upon request. It is calculated by dividing the load of each of the parameters by the difference of the maximum accepted value of the parameter (cmax), established in subcategory A1: Water that can be made drinkable with disinfection of Supreme Decree No. 004-2017 MINAM Standard of Environmental Water Quality, and the natural concentration (cnat) in the receiving body (9). The contaminant load is defined by the volume of the effluent multiplied by the concentration of parameters in it (cs), minus the volume of water entering the system and the natural concentration of the parameter (ce) in this (Arjen Y. Hoekstra et al., 2021).

$$HG = \frac{Efl \times cs - Afl \times ce}{cmáx - cnat} \qquad (9)$$

In the case of the treated effluent that enters the system again, it has the same concentration at the inlet and outlet, which translates into a negative footprint, without any contribution, so it will not be considered in the calculation.

Indirect Water Footprint

Also known as "water footprint of the supply chain of general activities" resulting from the general goods and services consumed by the company. To calculate this footprint, it was not necessary to distinguish its components, since they are included in the goods and have equivalent water footprint values [m^3/unit].

Sustainability Analysis of the Water Footprint

This stage is based on the analysis of the influence of the use of the institution's water resources, considering the social, economic, and environmental approaches. The study area and the area belonging to the adjacent districts of this area are considered: Lurín and San Juan de Miraflores.

For the environmental approach, Supreme Decree No. 003-2010-MINAM will be considered, which approves the Maximum Permissible Limits (LMP) for effluents from Domestic or Municipal Wastewater Treatment Plants (PTAR) (Ministerio del Ambiente, 2010), to compare the BOD parameters. COD, and oils and greases from the institution's WWTP effluent. Because not all the volume at the outlet is treated, it will also be compared with the Environmental Quality Standards for water, using subcategory A2 Waters that can be made drinkable with conventional treatment (Ministerio del Ambiente, 2017).

Regarding the social approach, information was collected from SEDAPAL's 2015-2044 Master Plan (SEDAPAL, 2014b) on the demand for drinking water in 2020 in two districts adjacent to the study area, Lurín and San Juan de Miraflores. This is because the fountains or reservoirs from which the drinking water purchased by the institution comes are located there. These demand data, expressed in m^3 per person per year, will be compared with the average water footprint per member per year.

In the case of the economic dimension, the sources of greatest contribution to the direct and indirect footprint will be evaluated under an efficiency approach.

Formulating Responses to the Water Footprint

Finally, measures to optimize campus water consumption will be proposed, based on the results of the water footprint calculation. For each one, the Net Present Value will be calculated with an Annual Effective Rate of 4%, established in 2017 by the Ministry of Economy and Finance aimed at reduction or mitigation environmental services projects (Ministerio de Economía y Finanzas, 2017) and with a projection to the year 2030, in order to determine the profitability and viability of the proposal (Knoke, Gosling, & Paul, 2020). In turn, Marginal Cost Curves will be constructed, commonly used in mitigation or reduction proposals, since it allows measures to be easily classified in terms of cost-effectiveness based on the marginal cost of reduction [PEN/m 3] and the capacity. of water reduction [m^3] (Clemente et al., 2014).

ANALYSIS OF RESULTS

Water Footprint

Table 1 shows the results of the calculation of the total water footprint of UCSUR in 2019, which was 774,322.19 m^3. The results are also differentiated by source (Figure 4). Among these, the indirect footprint represents a little more than half [53.7%] of the total water footprint. The source of consumption is gasoline, which has the largest indirect footprint with 108,944.20 m^3/year (Figure 4). Paper consumption is the second source with the largest indirect footprint with 106,576.25 m^3/year. And, the third is energy consumption with 90,964.75m^3/year.

While the direct footprint represents 46.3% of the total. The gray footprint being the largest footprint with 249,431.19m^3/year, followed by the blue footprint with 107,105.40 m^3/year, and the green footprint with 2,356.60 m^3/year.

Table 1. Results of water footprint accounting

Sources	Water Footprint [m^3/year]	Blue Water Footprint [m^3/year]	Green Water Footprint [m^3/year]	Gray Water Footprint [m^3/year]
Direct				
Water tankers	56,739.31	56,739.31	-	-
Green areas	52,722.69	50,366.09	2,356.60	-
Effluents	249,431.19	-	-	249,431.19
Direct Water Footprint	**358,893.19**	107,105.40	2,356.60	249,431.19
Indirect				
Energy consumption	90,964.75	-	-	-
Fuel consumption-Diesel 2	70,606.57	-	-	-
Fuel consumption - Bio	22,075.19	-	-	-
Fuel consumption-Gasoline	108,944.20	-	-	-
Fuel consumption-Ethanol	15,997.18	-	-	-
water canisters	264.88	-	-	-
Paper	106,576.25	-	-	-
Indirect Water Footprint	**415,429.00**	-	-	-

continued on following page

Water Footprint Reduction

Table 1. Continued

Sources	Water Footprint [m³/year]	Blue Water Footprint [m³/year]	Green Water Footprint [m³/year]	Gray Water Footprint [m³/year]
Total Water Footprint [m³/year]	774,322.19	-	-	-

Figure 4. Water footprint types

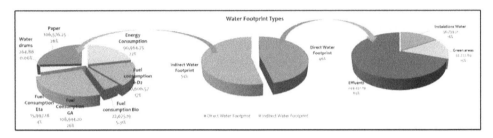

In addition, the result of the per capita water footprint of 43.75 m3 was evaluated with the results of two investigations also applied to the university sector (Figure 5). Castillo M. evaluated the water footprint of the Pontifical Catholic University of Peru and determined that the water footprint per member was 343.07 m³ (Castillo Valencia, 2014). On the other hand, Guamán D. and Illares R., in their study, determined that the water footprint per member of the Salesiana Polytechnic University located in Ecuador was 45.96 m³ (Guamán & Illares, 2019). The per capita water footprint of the PUCP is almost 8 times larger than that of the UCSUR, having only twice as many members as the UCSUR. On the other hand, the population of UCSUR doubles the number of members of the UPS, despite this, the per capita water footprint of UCSUR is smaller. With this we can infer that the treated water recirculation system of the UCSUR WWTP contributes to a lower water footprint.

Water Footprint Reduction

Figure 5. Water footprint per capita vs university population

WATER FOOTPRINT PER CAPITA VS UNIVERSITY POPULATION

[Bar and line chart showing Population and m³/year per capita for UCSUR, PUCP, UPS]

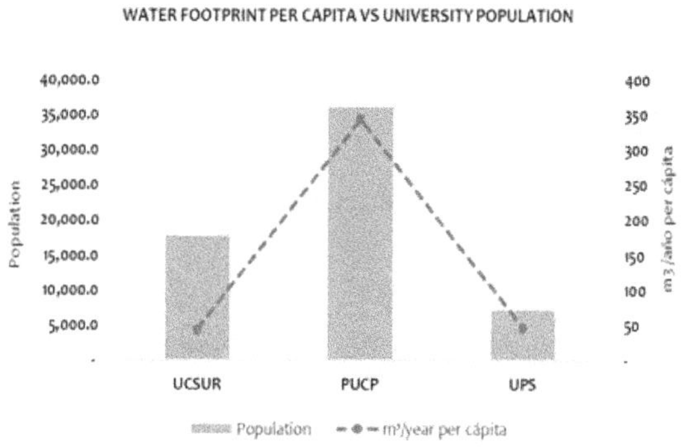

Sustainability of the Water Footprint

Environmental Focus

In Table 2, it is observed that the concentration of the parameters BOD, COD and oils and fats, Results of the laboratory analysis delivered by the Environmental Engineering Career are within the maximum permissible limit for effluents from a WWTP. In addition, they comply with the environmental quality standards of subcategory A2.

Table 2. Analysis of WWTP effluent

Parameter concentration [mg/L]	BOD	COD	Oils and fats
WWTP effluent	2	12.3	0.5
LMP Effluent WWTP	100	200	20
ECA Water (A2)	5	20	1.7

Although environmental regulations are complied with, the gray footprint is the largest of all sources (32.2% of the total water footprint). As mentioned above, the gray footprint of water treated by the WWTP is zero. Therefore, the impact on water pollution comes from supply through cisterns and wells. Therefore, it is recommended to increase the capacity and efficiency of the WWTP and reduce the volume of water acquired from other sources.

Social Focus

In the province of Lima, there is a high probability of a water crisis in the next 10 to 15 years, which would mean a reduction in the availability of drinking water by 30%, of which 21% would belong to the water supply service sector. drinking water (AQUAFONDO, 2020a).

Metropolitan Lima has approximately 10 million inhabitants (Instituto Nacional de Estadística e Informática, 2017), and is mainly supplied with water from the Chillón-Rímac-Lurín basin (SEDAPAL, 2014a). According to the United Nations, an area is under water stress when the per capita water availability is between 1,000 m^3 and 1,700 m^3; while when the availability of water is less than 1,000 m^3 per capita the population faces water scarcity (UNESCO, 2012). The per capita availability in the Chillón – Rímac – Lurín basins is 125 m^3 eight times lower than the water scarcity index (AQUAFONDO, 2016).

The demand for drinking water per person per year in the districts of Lurín and San Juan de Miraflores in 2020 was 61.88 m^3 and 88.79 m^3 respectively (SEDAPAL, 2014b). It was compared with the average water footprint per UCSUR member per year, which was 43.75 m^3.

Under this context, because the water footprint per member of the institution represents between 49% and 71% of the per capita water demand of the two districts, it should be considered that the purchase of water from these places exerts some pressure in their water supply. However, the drinking water service of the two mentioned districts would have to be evaluated in more detail, to determine if the institution could intervene negatively in this aspect, in the sense that these districts have periods of shortage of drinking water, intervening in the quality of service, and therefore in their quality of life. This, given that a water crisis is expected and currently, the capital is already in a situation of water shortage.

Economic Approach

The direct footprint contributes most to the institution's water footprint. Considering the drinking water rate of 6.71 PEN/m^3 for the non-residential class (SUNASS, 2021), for every 1000 m^3 of water that is reduced annually, 6,710.00 PEN would be saved.

Two of the sources with the highest percentage contribution to the indirect footprint are energy and paper consumption. Which represent 47.55% of the total indirect water footprint of the institution.

The energy rate, in this case, is 0.36 PEN/kWh, which is the average of the two rates that correspond to the institution (MT3 and BT3), which is 36.48 cmt. S/. /kWh (OSINERGMIN, 2022). Applying the electricity saving proposal, indicated in the

Water Footprint Reduction

Optimization Measures section, which is based on the replacement of luminaires with more efficient ones, 102,880.44 kWh and 37,530.78 PEN per year would be saved.

Finally, the administrative area is assigned 1,800,000 sheets of paper per year. The savings per thousand of paper would be 37.80 PEN, and 68,040.00 PEN for the total number of sheets assigned per year.

Optimization Measures

The measures were proposed based on the types of water footprint that represented greater water consumption. Such is the case of the blue footprint (107,105.40 m^3/year) and the indirect footprint (415,429.00 m^3/year).

Two of the sources with the largest footprint, the gray footprint, are being omitted due to lack of information necessary for the design and estimation of the investment of an expansion of the capacity and efficiency of the current Wastewater Treatment Plant. And, the indirect water footprint of the source of gasoline consumption, because the consumption of this fuel comes from the transportation used by the members of the institution, with the exception of the university buses that use diesel; Therefore, intervention in the choice of transportation by the university population is considered out of our reach.

Figure 6 shows the volume of water saved per year that the proposed measures aim to achieve.

Figure 6. Water savings by reduction measure [m3 /year]

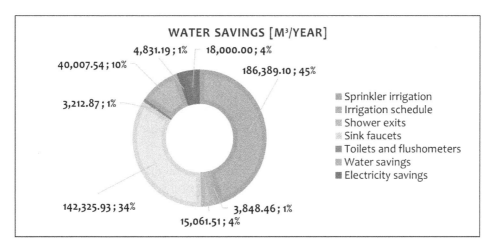

The eight proposed measures are:
a)   Sprinkler irrigation

It is based on changing the current irrigation method, which is conventional, to the sprinkler irrigation technique. Conventional irrigation has an application efficiency range of 10-30%, while sprinkler irrigation ranges from 50-90% (Ministerio de Agricultura y Riego, 2015). For the calculation, the average of both efficiency percentages by type of irrigation has been considered: 20% and 70%.

In the study by Ascencios & Peña, in 2014, they estimated an investment of 3,537,416.79 PEN and maintenance and operation cost of 434,722.39 PEN for the transition to a sprinkler irrigation system in the 10.7 hectares of the Olivar forest, this was calculated based to the irrigation requirement of the lawn, which is greater than that of the olive tree (Ascencios & Peña, 2014).

Dimensioned to the 4.41 hectares of green areas that the university has, an investment of 1,457,944.68 PEN and annual maintenance and operation cost of 179,170.63 PEN are estimated.

Water savings were calculated by dividing the irrigation requirement, data obtained from CROPWAT 8.0 software, and the efficiency percentages by type of irrigation. Then, the results are subtracted to obtain the volume of water in m^3 saved per year by the proposed method. Having 186,389.10 m^3 that would be saved per year, and a Net Present Value of 5,532,387.53PEN. In addition, a gain of 29.68 PEN per m^3 of water saved.

b) Irrigation schedule

The proposal to change the irrigation schedule is proposed based on schedules where solar radiation is lower, because it is considered one of the main factors for evaporation to occur (Encalada López, 2018). Therefore, less radiation means less evaporation of water from the soil surface.

The irrigation schedule is from Monday to Friday and differs by time of year, from September to April it is from 6:00 am to 8:00 am and from 3:00 pm to 5:00 pm. And, from May to August, it is watered three times a day from 6:00 am to 8:00 am, from 11:00 am to 1:00 pm and from 3:00 pm to 5:00 pm. On the other hand, the proposed schedule would be from 5:00 am to 7:00 am, and from 6:00 pm to 8:00 pm for the months between September and April. While for the months of May to August it has hours from 05:00 am to 07:00 am, from 09:00 am to 11:00 am, and from 06:00 pm to 08:00 pm.

The investment of this measure is calculated based on the extra hours per year that a gardening operator will have, which would be 12,300.00 PEN.

The water saved per year is calculated by comparing the evaporation rate of the water surface (Ep) of the current schedule and the proposed one for twelve months, which is calculated with the Penman-Monteith Formula (7) (Guevara, 2006). The meteorological data were taken from the UCSUR meteorological station. This measure represents a saving of 3,848.46 m^3/year.

Water Footprint Reduction

$$Ep = (\Delta/(\Delta + \gamma)) \times ((ho - G)/HV) + \{(\gamma/(\Delta + \gamma)) \times f(V) \times (e(a) - e(d))\} \quad (7)$$

where:

Δ: Slope of saturation vapor pressure curve [kPa/°C]
γ: Psychrometric constant [kPa/°C]
ho: Net radiation [MJ/m2]
G: Soil heat flux [MJ/m2]
HV: Latent heat of vaporization [MJ/kg]
f(V): Function of wind speed [mm/day kPa]
e(a): Saturation vapor pressure at mean air temperature [kPa]
e(d): Vapor pressure at mean air temperature [kPa]

The total water evaporated, for one year, with the current irrigation schedule was 303,288.41 m^3/year. While with the proposed schedule it was calculated that the total water evaporation for one year would be 299,439.95 m^3/year.

Although 1.27% annual water savings is not shown as a significant saving, the NPV of this measure is 75,669.55PEN, and has a gain of 19.66 PEN per m^3 saved, which makes the change in irrigation schedule, profitable.

c) Shower exits

It was proposed to replace conventional shower outlets with ones with eco-sprays that have a saving of 30%.

The investment for the purchase of 26 shower outlets is 1,375.40 PEN, and will be made only once, since they have a lifespan of more than 10 years.

The water footprint of the use of showers by the university population during 2019 was 50,205.02m^3/year. Applying this measure would mean a saving of 15,061.51m^3 annually, taking into account its savings percentage. In addition, this measure will generate a total profit of S/.652,741.60 PEN by 2030, and 43.34 PEN for every m^3 saved.

d) Sink faucets

Like the previous measure, it is based on the replacement of the 159 sink faucets in the toilets, and involves an investment of 45,079.68 PEN.

The flow rate of the current taps is 0.00024 m^3/s and the water footprint of the use of sinks was 165,816.62m^3/year. The replacement valves have a flow rate of 0.000034 m³/s and a consumption of 23,490.69 m3 per year is calculated. Saving 142,325.93m^3 of water annually.

This measure is considered profitable due to its net current value of S/.6,137,329.01 PEN, and negative marginal cost of 43.12 PEN, which translates into the gain per m^3 of water saved.

e) Toilets and flushometers

3 per siphonage, information provided by the general services area of the university, is proposed for ones with a discharge of 0.0035 m³ per siphonage. Whose investment would be 195,216.60 PEN.

The water footprint of toilet uses in 2019 was 11,862.92 m³/year, applying this measure the projected water savings would be 3,212.87 m³ per year. However, this measure is not viable, since implementing it would represent a loss of 48,185.37PEN according to the economic indicator of net present value. And, a marginal cost of 15 PEN for every m³ of water saved, without generating any profit. In addition, the current facilities comply with supreme decree N°015-2015-VIVIENDA that approves the Technical Code of Sustainable Construction, which establishes that toilets and flushometers must have a maximum consumption of 4.8 liters for each siphonage to be considered as a water saving technology (Ministerio de Vivienda, 2015).

f) Water savings

Being aware and acting in favor of water conservation can mean reducing water consumption from showers, sinks, irrigation, among other end uses, between 18.6% and 53% (Willis, Stewart, Panuwatwanich, Williams, & Hollingsworth, 2011). This proposal is based on promoting responsible water consumption through the integration of visual persuasion material in toilet facilities. Also, the implementation of a campaign that promotes saving drinking water in the institution, which will be based on one talk per cycle given by a technician in the field, and sending audiovisual material to the entire university population through institutional email.

Posters will be implemented in all toilets, with messages aimed at saving water from showers and sinks. Investing 5,000.00 PEN. The first year, due to the implementation of posters; and 4,000.00 PEN for the water saving campaign. As well as the poster and campaign measure, it will be implemented every year. In addition, a person in charge will be appointed to send material, monitor facilities, and analyze this measure based on the results of the measurement of the water footprint of the hygienic services that is expected to be carried out annually. For this reason, 9,000.00 PEN will be invested per year, starting the following year after the proposal is implemented.

This measure aims to save 40,007.54 m³/year and is considered viable due to its NPV of 1,608,476.04 PEN, and generate a profit in soles for each m³ saved of 40.20 PEN /m³.

g) Electricity savings

This measure aims to contribute to the reduction of the university's indirect water footprint. For every MWh, 48.02 m³ are consumed (COES, 2019). Therefore, a detailed comparison of the Net Present Value of the different replacement alternatives was carried out for each of the 6 types of luminaires. Finally, the alternatives with greater efficiency were chosen, maintaining and/or improving lighting needs.

Water Footprint Reduction

In total, it is proposed to replace 1,331 luminaires, which means a saving of 102.88 MWh/year and 4,831.19 m³/year. To determine the profitability of this measure, the NPV of the replacement of the 13 types of luminaires was added, resulting in 234,039.52PEN, which classifies this measure as viable.

h) Paper savings

Like the previous measure, this one also hopes to contribute to the reduction of the indirect footprint. The water footprint factor of a sheet of paper is 0.01 m³, a factor determined by the researchers who pioneered the concept of water footprint, Arjen Y. Hoekstra and Ashok K. Chapagain (A. Y. Hoekstra & Chapagain, 2007). The administrative area of the institution is assigned 1,800,000 sheets per year.

This measure proposes eliminating the use of paper through technological alternatives, such as sending payment receipts by mail and digitizing administrative documentation. A person in charge will be appointed to monitor the proposal, with an investment of 4,500.00 PEN per year.

The application of this measure would result in a saving of 18,000.00 m³/year. Finally, it is considered profitable for its current net value of 748,212.69 PEN.

Finally, the Marginal Cost Curves were constructed for each proposed measure (Figure 7). From this, it is obtained that the total reduction in water volume per year generated by these measures is 413,676.61m³. And it also classifies the proposed measures based on their economic viability, in order from highest to lowest: Sink faucets, Sprinkler irrigation, Water savings, Paper savings, Shower exits, Electricity savings, Irrigation schedule and Toilets and flushometers.

Figure 7. Marginal cost curves of water consumption optimization measures

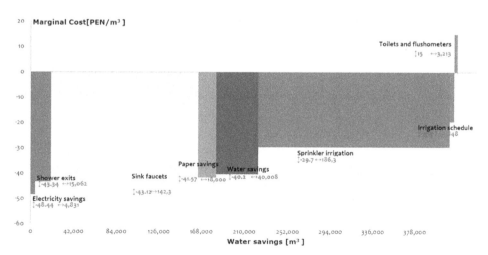

CONCLUSION AND RECOMMENDATIONS

- The water footprint in 2019 for the Villa Campus of the Scientific University of the South is 774,322.19m^3. The indirect footprint represents 53.7% with 415,429.00m^3 per year, and the direct footprint is 358,893.19m^3/year, being 46.3% of the total. Based on this result, the implementation of the proposed reduction measures is proposed, focused on the reduction of the sources that have the greatest contribution to the water footprint, which are the blue footprint, and the indirect footprint of paper and energy.
- Although the gray footprint of 249,431.19m^3/year is the largest of all, a proposal for a measure aimed at this has not been made, due to the lack of information on the capacity of the WWTP (influent and effluent flow)., data that would be needed for the proposal of an expansion in the capacity of the WWTP in order to reduce the entry of water from cisterns and wells, therefore reducing the gray footprint. Since the entry of water from the treated water of the WWTP does not contribute to the gray footprint. On the other hand, for future research, the study of the gray footprint of the university and the proposal of optimization measures for the WWTP are recommended.
- After the sustainability analysis, although the effluent does not exceed environmental regulations, the environmental approach is considered the most critical, because the largest water footprint is the gray one. Furthermore, if water quality is compromised, this situation can have social and economic repercussions. In the social sphere, it may cause disagreement with the supply of drinking water. Also, the university may be questioned by national authorities, generating a conflict, due to the proximity of the institution to the Pantanos de Villa Wildlife Refuge, in case there is any negative impact on the quality of the water and/or soil on this protected natural area. On the other hand, the economic aspect can also be affected, due to possible losses that may occur in the event of poor management of environmental impacts, due to the investment that mitigation of these impacts may entail.
- It was determined that the most cost-effective measure is the measure of replacing lavatory faucets, which represents a profit of 6,137,329.01PEN. Furthermore, the one that generates the greatest savings in water volume per year is the sprinkler irrigation measure with a saving of 186,389.10m^3/year. Finally, seven of the eight proposed measures qualify as viable. It is recommended to analyze the measure of changing toilets and flushometers, in the sense that co-benefits can be generated, not necessarily economic, since it does not generate significant economic gain.

Water Footprint Reduction

- Finally, the importance of proactively disseminating the results of the university's water footprint with the university population is highlighted in order to encourage the commitment to adopt conscious water consumption inside and outside the institution. As well as the implementation of water consumption optimization measures and monitoring, that is, the annual calculation of the water footprint. Likewise, certification of the university as a water-responsible company by the National Water Authority is expected.
- The presented results are applicable to similar buildings, such as Headquarters of Exploration and Production Companies, Field Camps, as well as to other large commercial and Governmental buildings, supporting a conscious use of water among the staff.

REFERENCES

Aldaya, M. M., Chapagain, A. K., Hoekstra, A. Y., & Mekonnen, M. M. (2012). *The Water Footprint Assessment Manual : Setting the Global Standard*. The Water Footprint Assessment Manual., 10.4324/9781849775526

Amaro, L., & Goyoneche, G. (2017). Anidación de aves en el refugio de vida silvestre los pantanos de villa 2007-2009, Lima-Perú. *The Biologist (Lima)*, 1(2). Advance online publication. 10.24039/rtb2017151151

AQUAFONDO. (2016). *Estudio de Riesgos Hídricos y Vulnerabilidad del Sector Privado en Lima Metropolitana y Callao en un contexto de Cambio Climático*.

AQUAFONDO. (2020a). *Estudio Crisis de agua: una amenaza silenciosa para el desarrollo económico*. Lima. Recuperado de https://www.cooperacionsuiza.pe/wp-content/uploads/2020/06/Estudio-Crisis-de-Agua-una-amenaza-silenciosa-para-eld-esarrollo-econ%C3%B3mico-1_compressed.pdf

Ascencios, D., & Peña, A. (2014). Diagnóstico, diseño y evaluación económica del sistema de riego por aspersión del bosque el Olivar. *Anales Científicos*, 75(1), 202–209. 10.21704/ac.v75i1.951

Castillo Valencia, M. (2014). *Huella hídrica del campus de la Pontificia Universidad Católica del Perú en el 2014* (Pontificia Universidad Católica del Perú). Pontificia Universidad Católica del Perú, Lima, Perú. Recuperado de https://tesis.pucp.edu.pe/repositorio/handle/20.500.12404/7633

Clemente, P., Fernández, M., Manuel, J., Rodríguez, G., Antonio, J., Vizán, A., & María, S. (2014). Análisis y comparación de medidas de ahorro energético mediante curvas de coste marginal. *Proceedings from the 18th International Congress on Project Management and Engineering*. Recuperado de https://dialnet.unirioja.es/servlet/articulo?codigo=8228220&info=resumen&idioma=SPA

COES. (2019). Estadísticas Anuales 2019. Recuperado 20 de abril de 2023, de https://www.coes.org.pe/Portal/publicaciones/estadisticas/estadistica?anio=2019

Direccion General de Salud Ambiental – Ministerio de Salud. *Reglamento de la Calidad del Agua para Consumo Humano. Decreto Supremo N.° 031-2010-SA.*, (2010).

Encalada López, D. A. (2018). *Análisis numérico y experimental de la evaporación en la interfase suelo - atmósfera*. Recuperado de https://upcommons.upc.edu/handle/2117/124922

Guamán, D., & Illares, F. (2019). *Análisis de la huella hídrica en el campus de la Universidad Politécnica Salesiana sede Cuenca mediante el uso de redes de telemetría. Universidad Politécnica Salesiana.* Cuenca, Ecuador. Recuperado de https://dspace.ups.edu.ec/bitstream/123456789/17729/1/UPS-CT008404.pdf

Guevara, J. M. (2006). La fórmula de Penman-Monteith FAO 1998 para determinar la evapotranspiración de referencia, ETo. *Terra. Nueva Etapa, 22*(31), 31–72. Recuperado de https://www.redalyc.org/articulo.oa?id=72103103

Hoekstra, A. Y. (2015). The water footprint of industry. *Assessing and Measuring Environmental Impact and Sustainability*, 221–254. 10.1016/B978-0-12-799968-5.00007-5

Hoekstra, A. Y., Chapagain, A. K., Aldaya, M. M., & Mekonnen, M. M. (2021). *Manual de evaluación de la huella hídrica. Establecimiento del estándar mundial.* Academic Press.

Hoekstra, A. Y., & Chapagain, A. K. (2007). Water footprints of nations: Water use by people as a function of their consumption pattern. *Water Resources Management*, 21(1), 35–48. 10.1007/s11269-006-9039-x

Hoekstra, A. Y., Mekonnen, M. M., Chapagain, A. K., Mathews, R. E., & Richter, B. D. (2012). Global Monthly Water Scarcity: Blue Water Footprints versus Blue Water Availability. *PLoS One*, 7(2), e32688. 10.1371/journal.pone.003268822393438

Instituto Nacional de Estadística e Informática. (2017). *Resultados Definitivos Censos Nacionales 2017.* Author.

Jeswani, H. K., & Azapagic, A. (2011). Water footprint: Methodologies and a case study for assessing the impacts of water use. *Journal of Cleaner Production*, 19(12), 1288–1299. 10.1016/j.jclepro.2011.04.003

Knoke, T., Gosling, E., & Paul, C. (2020). Use and misuse of the net present value in environmental studies. *Ecological Economics*, 174, 106664. 10.1016/j.ecolecon.2020.106664

López-Roldán, P., & Fachelli, S. (2017). El diseño de la muestra. *Metodología de la investigación social cuantitativa.* Recuperado de https://ddd.uab.cat/record/185163

Mekonnen, M. M., & Hoekstra, A. Y. (2011). The green, blue and grey water footprint of crops and derived crop products. *Hydrology and Earth System Sciences*, 15(5), 1577–1600. 10.5194/hess-15-1577-2011

Ministerio de Agricultura y Riego. (2015). *Manual para el cáculo de eficiencia para sistemas de riego.* Autor.

Ministerio de Economía y Finanzas. (2017). *Directiva N° 002-2017-EF/63.01 (Anexo N°3 Parámetros de evaluación social).* Author.

Ministerio de Vivienda, C. & Saneamiento. (2006). *Reglamento Nacional de Edificaciones. DS N° 011-2006-VIVIENDA.* Academic Press.

Ministerio de Vivienda, C. y S. (2015). *Decreto Supremo que aprueba el Código Técnico de Construcción Sostenible-Decreto Supremo N°015-2015-VIVIENDA.* Academic Press.

Ministerio del Ambiente. (2010). *Aprueba Límites Máximos Permisibles para los efluentes de Plantas de Tratamiento de Aguas Residuales Domésticas o Municipales. Decreto Supremo No 003-2010-MINAM.*

Ministerio del Ambiente. (2017). *Aprueban Estandares de Calidad Ambiental (ECA) para Agua y establecen disposiciones complementarias. Decreto Supremo N° 004-2017-MINAM..*

OSINERGMIN. (2022). Pliego Tarifario Máximo del Servicio Público de Electricidad. Recuperado 20 de abril de 2023, de https://www.osinergmin.gob.pe/Tarifas/Electricidad/PliegoTarifario?Id=150000

SEDAPAL. (2014a). *Plan maestro de los sistemas de agua potable y alcantarillado. Tomo I Diagnóstico.* Recuperado de https://cdn.www.gob.pe/uploads/document/file/4245313/Tomo%20I%20-%20Volumen%20I%20Diagnostico.pdf.pdf

SEDAPAL. (2014b). *Plan maestro de los sistemas de agua potable y alcantarillado. Tomo II Estimación Oferta Demanda de los Servicios.*

Superintendencia Nacional de Servicios de Saneamiento (SUNASS). (2021). *Resolución de Consejo Directivo No 079-2021-SUNASS-CD.*

UNESCO. (2012). *United Nations world water development report 4: managing water under uncertainty and risk* (Vol. 1). UNESCO. Recuperado de UNESCO website: https://unesdoc.unesco.org/ark:/48223/pf0000215644.locale=en

Willis, R. M., Stewart, R. A., Panuwatwanich, K., Williams, P. R., & Hollingsworth, A. L. (2011). Quantifying the influence of environmental and water conservation attitudes on household end use water consumption. *Journal of Environmental Management*, 92(8), 1996–2009. 10.1016/j.jenvman.2011.03.02321486685

World Economic Forum. (2019). *Global risks 2019 : insight report.* Recuperado de https://www3.weforum.org/docs/WEF_Global_Risks_Report_2019.pdf?_gl=1*ncy97p*_up*MQ.&gclid=CjwKCAjw3POhBhBQEiwAqTCuBso_PkjS56bwz-Qva6fl-Mb6cUJMDT5jrC1PJcvkSN8jQKW-ZogUyhoC8LoQAvD_BwE

Compilation of References

Ahmed, U., & Meehan, D. N. (2016). *Unconventional Oil and Gas Resources Explotation and Development*. T. & F. Group.

AIGLP. (2021). *Perú: el GLP es el segundo combustible de mayor consumo y éste año crecerá la importación*. Author.

Alameda García, D., Álvarez Álvarez, J. L., Hernández Zelaya, S. L., Fuertes Kronberg, C., García del Campo, L. A., Garcimartín Alférez, C., García Medina, J., Holguín Galarón, L. G., Llorente Valduvieco, J. I., Magdalena Miguel, L., Martín Martín, I., Matellán Pinilla, A., Reyes Reina, F. E., Rebollo Revesado, S., Rivas Herrero, L. A., & Sánchez Almaraz, J. (2021). Handbook of Sustainable Finance: A multidisciplinary approach. Academic Press.

Aldana, A. M. (1994). Estudio de la Macrofauna de los cuadrángulos de Piura, Sullana, Quebrada Seca, Paita, Talara, Negritos, Lobitos, Tumbes, Zorritos y Zarumilla. In *Geología de los Cuadrángulos de Paita, Piura, Talara, Sullana, Lobitos, Quebrada Seca, Zorritos, Tumbes y Zarumilla* (pp. 145–190). Academic Press.

Aldaya, M. M., Chapagain, A. K., Hoekstra, A. Y., & Mekonnen, M. M. (2012). *The Water Footprint Assessment Manual : Setting the Global Standard*. The Water Footprint Assessment Manual., 10.4324/9781849775526

Alizadeh, B., Sarafdokht, H., Rajabi, M., Opera, A., & Janbaz, M. (2012). Organic geochemistry and petrography of Kazhdumi (Albian-Cenomanian) and Pabdeh (Paleogene) potential source rocks in southern part of the Dezful Embayment, Iran. *Organic Geochemistry*, 49, 36–46. 10.1016/j.orggeochem.2012.05.004

Alnaimat, F., & Ziauddin, M. (2020, January). Wax deposition and prediction in petroleum pipelines", J. Petroleum Sci. Eng., vol. 184, p. 106385. *Journal of Petroleum Science Engineering*, 184, 106385. Advance online publication. Retrieved August 16, 2023, from. 10.1016/j.petrol.2019.106385

Alvarez, P., & Garrido, J. (1987). *Análisis de Roca Madre y Rocas Reservorio. Anexo 2*. Petroperu.

Alvarez, P., Garrido, J., & Aliaga, L. E. (1986). *Evaluación Geológica de la Cuenca Lancones. Perú Noroccidental. Estudios Geoquímicos del Cretáceo. Secciones Pan de Azucar, Cabrerías, Jaguay Negro, Huasimal, Corcobado, Gritón, Chilco*. Petroperu.

Al-Yaari, M. (2011). Paraffin Wax Deposition: Mitigation & Removal Techniques. Paper presented at the SPE Saudi Arabia section Young Professionals Technical Symposium, Dhahran, Saudi Arabia. 10.2118/155412-MS

Amaro, L., & Goyoneche, G. (2017). Anidación de aves en el refugio de vida silvestre los pantanos de villa 2007-2009, Lima-Perú. *The Biologist (Lima)*, 1(2). Advance online publication. 10.24039/rtb2017151151

Ambrose, R. J., Hartman, R. C., Diaz-Campos, M., Akkutlu, I. Y., & Sondergeld, C. H. (2012). Shale gas-in-place calculations Part I: New pore-scale considerations. *SPE Journal*, 17(1), 219–229. 10.2118/131772-PA

Ameni. (2019). Smart Sucker Pimp Failure Analysis with Machine Learning. Montan Universität Leoben - Thesis .

Andamayo, K. (2008). Nuevo estilo estructural y probables sistemas petroleros de la cuenca Lancones [Thesis: Ingeniero Geólogo]. Universidad Nacional Mayor de San Marcos.

Andamayo, K. (2008). *Nuevo estilo estructural y probables sistemas petroleros de la cuenca Lancones*. The National University of San Marcos.

AQUAFONDO. (2016). *Estudio de Riesgos Hídricos y Vulnerabilidad del Sector Privado en Lima Metropolitana y Callao en un contexto de Cambio Climático.*

AQUAFONDO. (2020a). *Estudio Crisis de agua: una amenaza silenciosa para el desarrollo económico*. Lima. Recuperado de https://www.cooperacionsuiza.pe/wp-content/uploads/2020/06/Estudio-Crisis-de-Agua-una-amenaza-silenciosa-para-eld-esarrollo-econ%C3%B3mico-1_compressed.pdf

Ariza León. (2011). On the characterization of crude oils what is key to diagnosing paraffin precipitation. *Revfue*.

Armacanqui, J. S. (2016). The Concept of the Sustainable Oil Field Development Applied to Heavy and Extra Heavy Oil Fields. *Society of Petroleum Engineers - SPE Latin America and Caribbean Heavy and Extra Heavy Oil Conference 2016*, 510–516. 10.2118/181169-MS

Armacanqui, J. S., De Fátima Eyzaguirre, L., Flores, M. G., Zavaleta, D. E., Camacho, F. E., Grajeda, A. W., Alfaro, A. D., & Viera, M. R. (2016). Testing of Environmental Friendly Paraffin Removal Products. *Society of Petroleum Engineers - SPE Latin America and Caribbean Heavy and Extra Heavy Oil Conference 2016*, 171–179. 10.2118/181162-MS

Armacanqui, J. S. (2015). The Innovations Friendly Organization: Effective Introduction of New Technologies and Innovations in Oil and Gas Companies. *Society of Petroleum Engineers - SPE North Africa Technical Conference and Exhibition 2015. NATC*, 2015, 1575–1587. 10.2118/175876-MS

Compilation of References

Arthur, M. A., Jenkyns, H. C., Brumsack, H.-J., & Schlanger, S. O. (1990). Stratigraphy, geochemistry, and paleoceanography of organic-carbon-rich Cretaceous sequences. In Ginsburg, R. N., & Beaudoin, B. (Eds.), *Cretaceous Resources* (Vol. 304). Events and Rhythms. 10.1007/978-94-015-6861-6_6

Asad, U., Zheng, M., Ting, D. S. K., & Tjong, J. (2015). Implementation challenges and solutions for homogeneous charge compression ignition combustion in diesel engines. *Journal of Engineering for Gas Turbines and Power*, 137(10), 101505. Advance online publication. 10.1115/1.4030091

Ascencios, D., & Peña, A. (2014). Diagnóstico, diseño y evaluación económica del sistema de riego por aspersión del bosque el Olivar. *Anales Científicos*, 75(1), 202–209. 10.21704/ac.v75i1.951

Asomaning, S., & Watkinson, A. P. (2000). Petroleum Stability and Heteroatom Species Effects in Fouling of Heat Exchangers by Asphaltenes. *Heat Transfer Engineering*, 21(3), 3, 10–16. 10.1080/014576300270852

ASTM Standard D1298-12b. Standard Test Method for Density, Relative Density, or API Gravity of Crude Petroleum and Liquid Petroleum Products by Hydrometer Method. Annual Book of ASTM Standards, Vol. 05.01, ASTM International, West Conshohocken, PA, 2010. DOI: 10.1520/D1298-12BR17E01

ASTM Standard D4007-22. Standard Test Method for Water and Sediment in Crude Oil by the Centrifuge Method (Laboratory Procedure). Annual Book of ASTM Standards, Vol. 05.01, ASTM International, West Conshohocken, PA, 2010. DOI: 10.1520/D4007-22

ASTM Standard D445-21e2. Standard Test Method for Kinematic Viscosity of Transparent and Opaque Liquids (and Calculation of Dynamic Viscosity). Annual Book of ASTM Standards, Vol. 05.01, ASTM International, West Conshohocken, PA, 2010. DOI: 10.1520/D0445-21E02

ASTM Standard D97-17b. Standard Test Method for Pour Point of Petroleum Products. Annual Book of ASTM Standards, Vol. 05.01, ASTM International, West Conshohocken, PA, 2010. DOI: 10.1520/D0097-17BR22

Atkinson, G. M., Eaton, D. W., & Igonin, N. (2020). Developments in understanding seismicity triggered by hydraulic fracturing. *Nature Reviews. Earth & Environment*, 1(5), 264–277. 10.1038/s43017-020-0049-7

B, D. (1986). *ORSTOM-INGEMMET Estudios Especiales*. Academic Press.

Bamberger, M., & Oswald, R. (2012, May). Impacts of Gas Drilling on Human and Animal Health. *New Solutions*, 22(1), 51–77. 10.2190/NS.22.1.e22446060

Bandach, A. (2018). Hydrocarbons Potential Of Peru. Academic Press.

Banerjee, A., Sinha, A. K., Jain, A. K., Thomas, N. J., Misra, K. N., & Chandra, K. (1998). A mathematical representation of Rock-Eval hydrogen index vs Tmax profiles. *Organic Geochemistry*, 28(1–2), 43–55. 10.1016/S0146-6380(97)00119-8

Barreriro, E., & Masarik, G. (2011). Los reservorios no convencionales, un "fenómeno global." Petrotecnia, 10–18.

Bartik, A., Currie, J., Greenstone, M., & Knittel, C. R. (2019, October). The Local Economic and Welfare Consequences of Hydraulic Fracturing. *American Economic Journal. Applied Economics*, 11(4), 105–155. 10.1257/app.20170487

Belyadi, H., Fathi, E., & Belyadi, F. (2017). *Hydraulic Fracturing in Unconventional Reservoirs Theories, Operations, and Economic Analysis*. Elsevier Inc.

Beneš, P. (2021). How much energy is needed to power a combustion car? LinkedIn. https://www.linkedin.com/pulse/how-much-energy-needed-power-combustion-car-petr-benes/

Bermudez. (2012). Reservoir characterization through the interpretation of pressure tests, Capaya formation, Tácata and Tacat fields, Anzoátegui and Monagas states.

Bianchi, C., & Jacay, J. (2011). *Distribución de Facies Anóxicas a través de la Margen Andina durante el Albiano*. VII Ingepet.

Boguslawski, Boujonnier, Bissuel-Beauvais, Saghir, & Sharma. (2018). IIoT Edge Analytics: Deploying Machine Learning at the Wellhead to Iden- tify Rod Pump Failure. SPE Middle East Artificial Lift Conference and Exhibition. .10.2118/192513-MS

Bolaños, Z. R. (2017). Reseña histórica de la exploración por petróleo en las cuencas costeras del Perú. *Boletín de la Sociedad Geológica del Perú*, 112, 1–13.

Bosworth, T. O., Woods, H., Vaughan, W. T., Cushman, J. A., Brock, T. A., & Hawkins, H. L. (1922). Geology of the Tertiary and Quaternary Periods in the North-West Part of Peru. *Geological Magazine*, 60(1), 43–45. Advance online publication. 10.1017/S0016756800002272

Bravo, J. J. (1921). *Reconocimiento de la región costanera de los departamentos de Tumbes y Piura*. Asociación Peruana para el Progreso de la Ciencia.

Brunauer, S., Emmett, P. H., & Teller, E. (1938). Adsorption of gases in multimolecular layers. *Journal of the American Chemical Society*, 60(2), 309–319. 10.1021/ja01269a023

Buijze, L. (2017, December). Fault reactivation mechanisms and dynamic rupture modeling of depletion-induced seismic events in a Rotliegend gas reservoir. *Netherlands Journal of Geosciences*, 96(5), s131–s148. 10.1017/njg.2017.27

Bustin, R. M., Bustin, A. M. M., Cui, A., Ross, D., & Pathi, V. M. (2008). *Impact of Shale Properties on Pore Structure and Storage Characteristics*. All Days. 10.2118/119892-MS

Cameretti, M. C., De Robbio, R., Mancaruso, E., & Palomba, M. (2022). CFD Study of Dual Fuel Combustion in a Research Diesel Engine Fueled by Hydrogen. *Energies*, 15(15), 5521. Advance online publication. 10.3390/en15155521

Campbell, C. J., & Laherrère, J. H. (1998). The End of Cheap Oil. *Scientific American*, 278(3), 78–83. 10.1038/scientificamerican0398-78

Compilation of References

Cao, T., Song, Z., Wang, S., Cao, X., Li, Y., & Xia, J. (2015). Characterizing the pore structure in the Silurian and Permian shales of the Sichuan Basin, China. *Marine and Petroleum Geology*, 61, 140–150. 10.1016/j.marpetgeo.2014.12.007

Carpenter, C. (2020, March 1). Carpenter, 2020. Dynamometer-Card Classification Uses Machine Learn-ing. *Journal of Petroleum Technology*, 72(3), 52–53. Advance online publication. 10.2118/0320-0052-JPT

Castillo Valencia, M. (2014). *Huella hídrica del campus de la Pontificia Universidad Católica del Perú en el 2014* (Pontificia Universidad Católica del Perú). Pontificia Universidad Católica del Perú, Lima, Perú. Recuperado de https://tesis.pucp.edu.pe/repositorio/handle/20.500.12404/7633

Castro-Ocampo, R. (1991). *El Cretaceo en la Cuenca Talara del Noroeste del Perú*. Universidad Nacional de Ingeniería - Peru.

Chalco, A. (1954). *Informe Geológico Preliminar de la Region Sullana - Lancones*. Empresa Petrolera Fiscal.

Chalmers, G. R. L., & Bustin, R. M. (2008, March). Lower Cretaceous gas shales in northeastern British Columbia, Part I: Geological controls on methane sorption capacity. *Bulletin of Canadian Petroleum Geology*, 56(1), 1–21. 10.2113/gscpgbull.56.1.1

Chalmers, G. R., Bustin, R. M., & Power, I. M. (2012). Characterization of gas shale pore systems by porosimetry, pycnometry, surface area, and field emission scanning electron microscopy/transmission electron microscopy image analyses: Examples from the Barnett, Woodford, Haynesville, Marcellus, and Doig uni. *AAPG Bulletin*, 96(6), 1099–1119. 10.1306/10171111052

Charles, D., Rieke, H., & Pal, S. (2005). Pressure Transient Model Characterization of Sealing and Partially Communicating Strike-Slip Faults. *Proceedings - SPE Annual Technical Conference and Exhibition*. 10.2118/96774-MS

Cheng, Y., Zeng, O., & Li, V. (2020). Automatic Recognition of Sucker-Rod Pumping System Working Conditions Using Dynamometer Cards with Transfer Learning and SVM. *Sensors (Basel)*, 20(19), 5659. Advance online publication. 10.3390/s2019565933022966

Chen, L., Dolado, J. J., Gonzalo, J., & Ramos, A. (2023). Heterogeneous predictive association of CO2 with global warming. *Economica*, 90(360), 1397–1421. 10.1111/ecca.12491

Chintala, V., & Subramanian, K. A. (2017). A comprehensive review on utilization of hydrogen in a compression ignition engine under dual fuel mode. In Renewable and Sustainable Energy Reviews (Vol. 70). 10.1016/j.rser.2016.11.247

Chintala, V., & Subramanian, K. A. (2014a). Assessment of maximum available work of a hydrogen fueled compression ignition engine using exergy analysis. *Energy*, 67, 162–175. Advance online publication. 10.1016/j.energy.2014.01.094

Chintala, V., & Subramanian, K. A. (2014b). Experimental investigation on effect of enhanced premixed charge on combustion characteristics of a direct injection diesel engine. *International Journal of Advances in Engineering Sciences and Applied Mathematics*, 6(1–2), 3–16. Advance online publication. 10.1007/s12572-014-0109-7

Chintala, V., & Subramanian, K. A. (2014c). Hydrogen energy share improvement along with NOx (oxides of nitrogen) emission reduction in a hydrogen dual-fuel compression ignition engine using water injection. *Energy Conversion and Management*, 83, 249–259. Advance online publication. 10.1016/j.enconman.2014.03.075

Chintala, V., & Subramanian, K. A. (2015a). An effort to enhance hydrogen energy share in a compression ignition engine under dual-fuel mode using low temperature combustion strategies. *Applied Energy*, 146, 174–183. Advance online publication. 10.1016/j.apenergy.2015.01.110

Chintala, V., & Subramanian, K. A. (2015b). Experimental investigations on effect of different compression ratios on enhancement of maximum hydrogen energy share in a compression ignition engine under dual-fuel mode. *Energy*, 87, 448–462. Advance online publication. 10.1016/j.energy.2015.05.014

Chintala, V., & Subramanian, K. A. (2016). Experimental investigation of hydrogen energy share improvement in a compression ignition engine using water injection and compression ratio reduction. *Energy Conversion and Management*, 108, 106–119. Advance online publication. 10.1016/j.enconman.2015.10.069

Clausthal, & Campbell, C. J. (2003, April 14). *Peak oil - A turning point for mankind*. Academic Press.

Clayton, J., Ryder, R. T., & Meissner, F. (1984). Geochemistry of Black Shales in Minnelusa Formation and Desmoinesian Age Rocks (Permian-Pennsylvanian), and Associated Oils, Powder River Basin and Northern DJ Basin, Wyoming and Colorado: ABSTRACT. *AAPG Bulletin*, 68(7), 935–935. 10.1306/AD4614C1-16F7-11D7-8645000102C1865D

Clemente, P., Fernández, M., Manuel, J., Rodríguez, G., Antonio, J., Vizán, A., & María, S. (2014). Análisis y comparación de medidas de ahorro energético mediante curvas de coste marginal. *Proceedings from the 18th International Congress on Project Management and Engineering*. Recuperado de https://dialnet.unirioja.es/servlet/articulo?codigo=8228220&info=resumen&idioma=SPA

COES. (2019). Estadísticas Anuales 2019. Recuperado 20 de abril de 2023, de https://www.coes.org.pe/Portal/publicaciones/estadisticas/estadistica?anio=2019

Colborn, T., Kwiatkowski, C., Schultz, K., & Bachran, M. (2011). Natural Gas Operations from a Public Health Perspective. *Human and Ecological Risk Assessment*, 17(5), 1039–1056. 10.1080/10807039.2011.605662

Columbia Industries. (2024). *Rig Walking Systems*. https://www.columbiacorp.com/oilfield-gas-industry-solutions/rig-walking-systems/

Compilation of References

Cooles, G. P., Mackenzie, A. S., & Quigley, T. M. (1986). Calculation of petroleum masses generated and expelled from source rocks. *Organic Geochemistry*, 10(1–3), 235–245. 10.1016/0146-6380(86)90026-4

Cooper, B. S., & Barnard, P. C. (1984). Source Rocks and Oils of the Central and Northern North Sea. In Demaison, G. J., & Murris, R. J. (Eds.), *Petroleum Geochemistry and Basin Evaluation* (Vol. 35). American Association of Petroleum Geologists. 10.1306/M35439C17

Corporation & Company. (1975)., *Well Completion Report 14-36-2X Tapiche*. Academic Press.

Cruzado, C. J., & Aliaga, L. E. (1985). *Micropaleontología del Cretáceo del Área Pazul*. Petroperu.

Curtis, M. E., Cardott, B. J., Sondergeld, C. H., & Rai, C. S. (2012, December). Development of organic porosity in the Woodford Shale with increasing thermal maturity. *International Journal of Coal Geology*, 103, 26–31. 10.1016/j.coal.2012.08.004

Dahl, B., & Augustson, J. H. (1993). The influence of Tertiary and Quaternary sedimentation and erosion on the hydrocarbon generation in Norwegian offshore basins. In A. G. Dore (Ed.), *Basin Modelling: Advances and Applications* (pp. 419–431). Elsevier.

Dahl, B., Bojesen-Koefoed, J., Holm, A., Justwan, H., Rasmussen, E., & Thomsen, E. (2004). A new approach to interpreting Rock-Eval S2 and TOC data for kerogen quality assessment. *Organic Geochemistry*, 35(11–12), 1461–1477. 10.1016/j.orggeochem.2004.07.003

Dahl, B., & Meisingset, I. (1996). Prospect resource assessment using an integrated system of basin simulation and geological mapping software: examples from the North Sea. In Doré, A. G., & Sinding-Larsen, R. (Eds.), *Quantification and Prediction of Hydrocarbon Resources* (pp. 237–251). Elsevier. 10.1016/S0928-8937(07)80021-X

Dahl, B., & Yükler, A. (1991). The Role of Petroleum Geochemistry in Basin Modeling of the Oseberg Area, North Sea. In Merrill, R. K. (Ed.), *Source and migration processes and evaluation techniques*. American Association of Petroleum Geologists. 10.1306/TrHbk543C6

Darrah, T. H., Vengosh, A., Jackson, R. B., Warner, N. R., & Poreda, R. J. (2014, September). Noble Gases Identify the Mechanisms of Fugitive Gas Contamination in Drinking-Water Wells Overlying the Marcellus and Barnett Shales. *Proceedings of the National Academy of Sciences of the United States of America*, 111(39), 14076–14081. 10.1073/pnas.132210711125225410

De Morais, A. M., Mendes Justino, M. A., Valente, O. S., Hanriot, S. D. M., & Sodré, J. R. (2013). Hydrogen impacts on performance and CO2 emissions from a diesel power generator. *International Journal of Hydrogen Energy*, 38(16), 6857–6864. Advance online publication. 10.1016/j.ijhydene.2013.03.119

Dean, W. E., Claypool, G. E., & Thide, J. (1984). Accumulation of organic matter in Cretaceous oxygen-deficient depositional environments in the central Pacific Ocean. *Organic Geochemistry*, 7(1), 39–51. 10.1016/0146-6380(84)90135-9

Deb, M., Sastry, G. R. K., Bose, P. K., & Banerjee, R. (2015). An experimental study on combustion, performance and emission analysis of a single cylinder, 4-stroke DI-diesel engine using hydrogen in dual fuel mode of operation. *International Journal of Hydrogen Energy*, 40(27), 8586–8598. Advance online publication. 10.1016/j.ijhydene.2015.04.125

Demaison, G. J., & Moore, G. T. (1980). Anoxic environments and oil source bed genesis. *Organic Geochemistry*, 2(1), 9–31. 10.1016/0146-6380(80)90017-0

Deslances, N. (2022, September 27). *Almost 400 new mines needed to meet future EV battery demand, data finds*. https://techinformed.com/almost-400-new-mines-needed-to-meet-future-ev-battery-demand-data-finds/

DGSI. (1999). VOL 3. Datos analíticos de rocas de afloramiento. In W. G. Dow (Ed.), *Estudios de Investigación Geoquímica del Potencial de Hidrocarburos. Lotes del Zócalo Continental y de Tierra*. PERUPETRO S.A.

Dhole, A. E., Yarasu, R. B., Lata, D. B., & Priyam, A. (2014). Effect on performance and emissions of a dual fuel diesel engine using hydrogen and producer gas as secondary fuels. *International Journal of Hydrogen Energy*, 39(15), 8087–8097. Advance online publication. 10.1016/j.ijhydene.2014.03.085

Dimitriou, P., Kumar, M., Tsujimura, T., & Suzuki, Y. (2018). Combustion and emission characteristics of a hydrogen-diesel dual-fuel engine. *International Journal of Hydrogen Energy*, 43(29), 13605–13617. Advance online publication. 10.1016/j.ijhydene.2018.05.062

Direccion General de Salud Ambiental – Ministerio de Salud. *Reglamento de la Calidad del Agua para Consumo Humano. Decreto Supremo N.° 031-2010-SA.*, (2010).

Dong, L., Xie, H., & Zhang, F. (2001). Chemical Control Techniques for the Paraffin and Asphaltene Deposition. Paper presented at the SPE International Symposium on Oilfield Chemistry, Houston, Texas. 10.2118/65380-MS

Dunham, R. J. (1962). Classification of Carbonate Rocks According to Depositional Textures. *AAPG Bulletin*, 108–121.

Du, X., Carlson, K. H., & Tong, T. (2022). The water footprint of hydraulic fracturing under different hydroclimate conditions in the Central and Western United States. *The Science of the Total Environment*, 840, 156651. Advance online publication. 10.1016/j.scitotenv.2022.15665135700779

E, A., & N., V. I.-P. (1974). *The volcanism of the Northern part of Peruvian Altiplano and of the Oriental Cordillera*. Academic Press.

Eberli, G. P., Weger, R. J., Tenaglia, M., Rueda, L., Rodriguez, L., Zeller, M., McNeill, D. F., Murray, S., & Swart, P. K. (2017). The unconventional play in the Neuquén Basin, Argentina - Insights from the outcrop for the subsurface. *SPE/AAPG/SEG Unconventional Resources Technology Conference 2017*, 1–12. 10.15530/urtec-2017-2687581

Compilation of References

EIA. (2013). *Technically Recoverable Shale Oil and Shale Gas Resources: An Assessment of 137 Shale Formations in 41 Countries Outside the United States*. U.S. Energy Information Administration. Obtenido de https://www.eia.gov/analysis/studies/worldshalegas/archive/2013/pdf/fullreport_2013.pdf

EIA. (2015). *World Shale Resource Assessments*. U.S. Energy Information Administration.

EIA. (2018). *Peru*. https://www.eia.gov/international/overview/country/PER

Encalada López, D. A. (2018). *Análisis numérico y experimental de la evaporación en la interfase suelo - atmósfera*. Recuperado de https://upcommons.upc.edu/handle/2117/124922

Energy Information Administration. (2009, June 29). *Annual U.S. Field Production of Crude Oil*. U.S. Field Production of Crude Oil(Thousand Barrels). https://web.archive.org/web/20091107181532/http://tonto.eia.doe.gov:80/dnav/pet/hist/LeafHandler.ashx?n=pet&s=mcrfpus1&f=a

Energy Information Administration. (2015). *World Shale Resource Assessments*. https://www.eia.gov/analysis/studies/worldshalegas/

Energy Information Administration. (2022, March 18). *EIA projects that renewable generation will supply 44% of U.S. electricity by 2050*. Author.

Energy Information Administration. (2023). *Annual Energy Outlook 2023*. https://www.eia.gov/outlooks/aeo/pdf/AEO2023_Narrative.pdf

Environmental Protection Agency. (2016). *Hydraulic Fracturing for Oil and Gas: Impacts from the Hydraulic Fracturing Water Cycle on Drinking Water Resources in the United States*. Author.

Environmental Protection Agency. (2016). Hydraulic Fracturing for Oil and Gas: Impacts from the Hydraulic Fracturing Water Cycle on Drinking Water Resources in the United States. Available: www.epa.gov/hfstudy

EPA. (2023). *Inventory of U.S. Greenhouse Gas Emissions and Sinks: 1990-2021 – Main Report*. https://www.epa.gov/system/files/documents/2023-04/US-GHG-Inventory-2023-Main-Text.pdf

Escobar, F. (2003). Modern Well Pressure Analysis. Academic Press.

Escobar, F., Martinez, J.-A., & Montealegre-Madero, M. (2013). Pressure Transient Analysis for a Reservoir with a Finite- Conductivity Fault. *CT&F Ciencia, Tecnología y Futuro*, 5(June), 15. 10.29047/01225383.53

Espitalie, J. (1986). Use of Tmax as a Maturation Index for Different Types of Organic Matter. Comparison with Vitrinite Reflectance. *Thermal Modeling in Sedimentary Basins,* 475-496.

Espitalie, J., Deroo, G., & Marquis, F. (1985). La pyrolyse Rock-Eval et ses applications. Deuxième partie. *Revue de l'Institut Français du Pétrole*, 40(6), 755–784. 10.2516/ogst:1985045

Espitalié, J., Madec, M., & Tissot, B. P. (1980). Role of Mineral Matrix in Kerogen Pyrolysis: Influence on Petroleum Generation and Migration. *AAPG Bulletin*, 64(1), 59–66. 10.1306/2F918928-16CE-11D7-8645000102C1865D

Espitalié, J., Madec, M., Tissot, B. P., Mennig, J. J., & Leplat, P. (1977). Source rock characterization method for petroleum exploration. *Proceedings of the Annual Offshore Technology Conference,* 439–444. 10.4043/2935-MS

Espitalié, J., Marquis, F., & Barsony, I. (1984). Geochemical Logging. In Voorhees, K. J. (Ed.), *Analytical Pyrolysis—Techniques and Applications* (pp. 276–304)., 10.1016/B978-0-408-01417-5.50013-5

Espitalié, J., Marquis, F., & Sage, L. (1987). Organic geochemistry of the Paris Basin. In Brooks, J., & Glennie, K. (Eds.), *Petroleum geology of north West Europe* (pp. 71–86). Geological Society of London Graham & Trotman.

European Commission. (2015). Press corner. FAQ - Air pollutant emissions standards. European Commission. https://ec.europa.eu/commission/presscorner/detail/mt/MEMO_15_5705

Evaluación de la Materia Orgánica, Potencial de Hidrocarburos, & Madurez Térmica en el Noreste Peruano: Formación Muerto. (2018). LACCEI International Multi-Conference for Engineering, Education and Techonoly: "Innovatios in Education and Inclusion," 1–4.

Fagg. (1950). Dynamometer Charts and Well Weighing. Petroleum Transactions.

Fernández, J., Martínez, E., Calderón, Y., Hermoza, W., & Galdos, C. (2005). *Tumbes and Talara Basin; Hydrocarbon Evaluation PERUPETRO S.A.* Academic Press.

Fernández, J., Martínez, E., Calderón, Y., Hermoza, W., & Galdos, C. (2005). Tumbes and Talara basins hydrocarbon evaluation. Basin Evaluations Group Exploration Department, internal r(PERUPETRO S.A.), 130.

Fildani, A., Hanson, A. D., Chen, Z., Moldowan, J. M., Graham, S. A., & Arriola, P. R. (2005). Geochemical characteristics of oil and source rocks and implications for petroleum systems, Talara basin, northwest Peru. *AAPG Bulletin*, 89(11), 1519–1545. 10.1306/06300504094

Fischer, A. G. (1956). *Cretaceous of Northwest Peru. IPC Report WP-13.*

Flexer, A., & Rosenfeld, A. (1986). Relative Sea Level Changes During the Cretaceous in Israel. *AAPG Bulletin*, 70(11), 1685–1699. 10.1306/94886C9A-1704-11D7-8645000102C1865D

Florides, G. A., & Christodoulides, P. (2009). Global warming and carbon dioxide through sciences. *Environment International*, 35(2), 390–401. 10.1016/j.envint.2008.07.00718760479

Frizzell, D. L. (1944). *Summary Report on the Stratigraphy and Paleontology of Northwestern Peru.*

Gagnon, G. A., Krkosek, W., Anderson, L., McBean, E., Mohseni, M., Bazri, M., & Mauro, I. (2016). Impacts of hydraulic fracturing on water quality: A review of literature, regulatory frameworks and an analysis of information gaps. *Environmental Reviews*, 24(2), 122–131. 10.1139/er-2015-0043

Gamarra, S. (1987). *Evaluación Geológica Cuenca Lancones. Estudio de Materia Orgánica Formaciones Huasimal y Muerto - Cretaceo*. Petroperu.

Gao, S., Dong, D., Tao, K., Guo, W., Li, X., & Zhang, S. (2021). Experiences and lessons learned from China's shale gas development: 2005–2019. *Journal of Natural Gas Science and Engineering*, 85, 103648. 10.1016/j.jngse.2020.103648

García, B., Jaimes, F., Concha, R., Astete, I., & Chapilliquen, J. (2015). Evolución tectono-sedimentaria del dominio occidental de la Cuenca de Lancones. INGEMMET Revista Institucional, 12–21.

Garrido, J., & Aliaga, L. E. (1986). *Madurez de Roca Madre por Reflecancia de Vitrinita. Formaciones Muerto y Huasimal. Cretáceo Cuenca Lancones (Secciones: Corcobado, El Cortado, Angelitos, etc)*. Petroperu.

Geo, V. E., Nagarajan, G., & Nagalingam, B. (2008). Studies on dual fuel operation of rubber seed oil and its bio-diesel with hydrogen as the inducted fuel. *International Journal of Hydrogen Energy*, 33(21), 6357–6367. Advance online publication. 10.1016/j.ijhydene.2008.06.021

Gilbert. (1936). An Oil Well Pumping Dynagraph. API Drilling and Production Practice.

Glorioso, J. C., & Rattia, A. J. (2012). Unconventional Reservoirs: Basic Petrophysical Concepts for Shale Gas. In *SPE/EAGE European Unconventional Resources Conference and Exhibition* (p. 38). Society of Petroleum Engineers. 10.2118/153004-MS

Gómez. (2003). Formation of sediments during the hydrodisintegration of petroleum residues. Magazine of the Chemical Society of Mexico, 47(3), 260-266. https://www.scielo.org.mx/scielo.php?script=sci_arttext&pid=S0583-76932003000300010&lng=es&tlng=es

Gonca, G. (2014). Investigation of the effects of steam injection on performance and NO emissions of a diesel engine running with ethanol-diesel blend. *Energy Conversion and Management*, 77, 450–457. Advance online publication. 10.1016/j.enconman.2013.09.031

González García, D., Villabona Carvajal, C., Vargas Torres, H., Ariza León, E., Roa Duarte, C., & Barajas Ferreira, C. (2010). Methods for the Control and Inhibition of the Accumulation of Paraffinic Deposits. *UIS Engineering Magazine*, 9(2), 193–206.

Groen, J. C., Peffer, L. A. A., & Pérez-Ramírez, J. (2003). Pore size determination in modified micro- and mesoporous materials. Pitfalls and limitations in gas adsorption data analysis. *Microporous and Mesoporous Materials*, 60(1-3), 1–17. 10.1016/S1387-1811(03)00339-1

Guamán, D., & Illares, F. (2019). *Análisis de la huella hídrica en el campus de la Universidad Politécnica Salesiana sede Cuenca mediante el uso de redes de telemetría. Universidad Politécnica Salesiana*. Cuenca, Ecuador. Recuperado de https://dspace.ups.edu.ec/bitstream/123456789/17729/1/UPS-CT008404.pdf

Guevara, J. M. (2006). La fórmula de Penman-Monteith FAO 1998 para determinar la evapotranspiración de referencia, ETo. *Terra. Nueva Etapa, 22*(31), 31–72. Recuperado de https://www.redalyc.org/articulo.oa?id=72103103

Guo, H., He, R., Jia, W., Peng, P., Lei, Y., Luo, X., Wang, X., Zhang, L., & Jiang, C. (2018, March). Pore characteristics of lacustrine shale within the oil window in the Upper Triassic Yanchang Formation, southeastern Ordos Basin, China. *Marine and Petroleum Geology*, 91, 279–296. 10.1016/j.marpetgeo.2018.01.013

Harde, H. (2023). *Science of Climate Change How Much CO2 and the Sun Contribute to Global Warming: Comparison of Simulated Temperature Trends with Last Century Observations*. 10.53234/scc202206/10

Harde, H., & Salby, M. L. (2021). What Controls the Atmospheric CO_2 Level? *Science of Climate Change*. 10.53234/scc202106/22

Harde, H. (2019). What Humans Contribute to Atmospheric CO_2 Comparison of Carbon Cycle Models with Observations. *Earth Sciences (Paris)*, 8(3), 139. 10.11648/j.earth.20190803.13

Hendricks, W. (2020). *CNX's E-Frac deal with evolution*. Academic Press.

Heywood, J. B. (2018). Internal Combustion Engine Fundamentals. In Internal Combustion Engine Fundamentals Second Edition.

Hoekstra, A. Y. (2015). The water footprint of industry. *Assessing and Measuring Environmental Impact and Sustainability*, 221–254. 10.1016/B978-0-12-799968-5.00007-5

Hoekstra, A. Y., Chapagain, A. K., Aldaya, M. M., & Mekonnen, M. M. (2021). *Manual de evaluación de la huella hídrica. Establecimiento del estándar mundial*. Academic Press.

Hoekstra, A. Y., & Chapagain, A. K. (2007). Water footprints of nations: Water use by people as a function of their consumption pattern. *Water Resources Management*, 21(1), 35–48. 10.1007/s11269-006-9039-x

Hoekstra, A. Y., Mekonnen, M. M., Chapagain, A. K., Mathews, R. E., & Richter, B. D. (2012). Global Monthly Water Scarcity: Blue Water Footprints versus Blue Water Availability. *PLoS One*, 7(2), e32688. 10.1371/journal.pone.003268822393438

Holditch, S. A. (2006). Tight Gas Sands. *Journal of Petroleum Technology*, 58(06), 86–93. 10.2118/103356-JPT

Hong, Z., Moreno, H. A., & Hong, Y. (2018). Spatiotemporal Assessment of Induced Seismicity in Oklahoma: Foreseeable Fewer Earthquakes for Sustainable Oil and Gas Extraction? *Geosciences*, 8(12), 436. 10.3390/geosciences8120436

Honty, G. (2018). Nuevo extractivismo energético en América Latina. *Ecuador Debate, 105*, 48-67. Obtenido de http://hdl.handle.net/10469/15261

Compilation of References

Hornbach, M. J., DeShon, H. R., Ellsworth, W. L., Stump, B. W., Hayward, C., Frohlich, C., Oldham, H. R., Olson, J. E., Magnani, M. B., Brokaw, C., & Luetgert, J. H. (2015, April). Causal factors for seismicity near Azle, Texas. *Nature Communications*, 6(1), 6728. Advance online publication. 10.1038/ncomms772825898170

Horne, R. (1995). Modern Well Test Analysis A Computer-Aided Approach (2nd ed.). Academic Press.

Horsfield, B., Schulz, H.-M., & Kapp, I. (2012). Shale Gas in Europe. Search and Discovery Article #10380. Obtenido de https://www.searchanddiscovery.com/documents/2012/10380horsfield/ndx_horsfield.pdf

Horsfield, B. (1989). Practical criteria for classifying kerogens: Some observations from pyrolysis-gas chromatography. *Geochimica et Cosmochimica Acta*, 53(4), 891–901. 10.1016/0016-7037(89)90033-1

Howarth, R. W., Ingraffea, A., & Engelder, T. (2011). Should fracking stop? *Nature*, 477(7364), 271–275. 10.1038/477271a21921896

Iddings, A., & Olsson, A. A. (1928). Geology of Northwest Peru. *AAPG Bulletin*, 12(1), 1–39. 10.1306/3D9327D7-16B1-11D7-8645000102C1865D

IEA. (2020). *Total final consumption (TFC) by source, Peru 1990-2019*. https://www.iea.org/countries/peru

Infologic, B.-R. (2006). *Geochemical-Solutions*. Petroleum Systems Evaluation – Talara/Tumbes Basins.

Instituto Nacional de Estadística e Informática. (2017). *Resultados Definitivos Censos Nacionales 2017*. Author.

International Energy Agency. (2022). *The Role of Critical World Energy Outlook Special Report Minerals in Clean Energy Transitions*. https://iea.blob.core.windows.net/assets/ffd2a83b-8c30-4e9d-980a-52b6d9a86fdc/TheRoleofCriticalMineralsinCleanEnergyTransitions.pdf

IPCC. (2005). *Carbon Dioxide Capture and Storage*. https://www.ipcc.ch/report/carbon-dioxide-capture-and-storage/

IPCC. (2013). *The Physical Science Basis. Contribution of Working Group I to the Fifth Assessment Report of the Intergovernmental Panel on Climate Change*. IPCC.

IPCC. (2021). *Climate Change 2021: The Physical Science Basis. Contribution of Working Group I to the Sixth Assessment Report of the Intergovernmental Panel on Climate Change*. 10.1017/9781009157896

IPIECA. (2023). Drilling rigs. Energy efficiency compendium.

Isaksen, G. H., & Ledje, K. H. I. (2001). Source Rock Quality and Hydrocarbon Migration Pathways within the Greater Utsira High Area, Viking Graben, Norwegian North Sea. *AAPG Bulletin*, 85(5), 861–883. 10.1306/8626CA23-173B-11D7-8645000102C1865D

Ishida, T., Aoyagi, K., Niwa, T., Chen, Y., Murata, S., Chen, Q., & Nakayama, Y. (2012). Acoustic emission monitoring of hydraulic fracturing laboratory experiment with supercritical and liquid CO_2. *Geophysical Research Letters*, 39(16), 2012GL052788. Advance online publication. 10.1029/2012GL052788

Jackson, R. B., Vengosh, A., Darrah, T. H., Warner, N. R., Down, A., Poreda, R. J., Osborn, S. G., Zhao, K., & Karr, J. D. (2013). Increased stray gas abundance in a subset of drinking water wells near Marcellus shale gas extraction. *Proceedings of the National Academy of Sciences of the United States of America*, 110(28), 11250–11255. 10.1073/pnas.122163511023798404

Jaillard, E., & Soler, P. (1994). Cretaceous to early Paleogene tectonic evolution of the northern Central Andes (0-18°S) and its relations to geodynamics. Tectonophysics, 259(1-3 SPEC. ISS.), 41–53. 10.1016/0040-1951(95)00107-7

Jaillard, E., Laubacher, G., Bengtson, P., Dhondt, A. V, Bulot, L. G., Cuenca, L., & Suroeste, D. A. C. (1999). Estratigrafía y Evolución de la Cuenca Cretácica Ante-Arco Celica-Lancones en el Suroeste del Ecuador. Abstracto Resumen y Conclusiones : Discusión e Interpretación.

Jaimes, F., Navarro, J., Alan, S., & Bellido, F. (2012). Geología del cuadrángulo de Las Lomas. INGEMMET Revista Institucional, 59(Boletin N°146), 23–26.

Jamrozik, A., Grab-Rogaliński, K., & Tutak, W. (2020). Hydrogen effects on combustion stability, performance and emission of diesel engine. *International Journal of Hydrogen Energy*, 45(38), 19936–19947. Advance online publication. 10.1016/j.ijhydene.2020.05.049

Jarvie, D. M. (2012). Shale resource systems for oil and gas: Part 1—shale-gas resource systems. In *M97: Shale Reservoirs—Giant Resources for the 21st Century* (pp. 69–87). American Association of Petroleum Geologists. 10.1306/13321446M973489

Jarvie, D. M., Jarvie, B. M., Weldon, D., & Maende, A. (2012). *Components and Processes Impacting Production Success from Unconventional Shale Resource Systems*. https://doi.org/10.3997/2214-4609-PDB.287.1226756

Jarvie, D. M., Hill, R. J., Ruble, T. E., & Pollastro, R. M. (2007). Unconventional shale-gas systems: The Mississippian Barnett Shale of north-central Texas as one model for thermogenic shale-gas assessment. *AAPG Bulletin*, 91(4), 475–499. 10.1306/12190606068

Jarvie, D., Burgess, J., Morelos, A., Mariotti, P. A., & Lindsey, R. (2001). Permian Basin petroleum systems investigations; inferences from oil geochemistry and source rocks. *AAPG Bulletin*, 85(9), 1693–1694.

Jenkyns, H. C. (1980). Cretaceous anoxic events: From continents to oceans. *Journal of the Geological Society*, 137(2), 171–188. 10.1144/gsjgs.137.2.0171

Compilation of References

Jeswani, H. K., & Azapagic, A. (2011). Water footprint: Methodologies and a case study for assessing the impacts of water use. *Journal of Cleaner Production*, 19(12), 1288–1299. 10.1016/j.jclepro.2011.04.003

Jones, R. W. (1987). Organic facies. In Brooks, J., & Welte, D. H. (Eds.), *Advances in Petroleum Geochemistry* (pp. 1–90). Academic Press.

Jones, R. W.R. W. Jones. (1984). Comparison of Carbonate and Shale Source Rocks: ABSTRACT. *AAPG Bulletin*, 68. Advance online publication. 10.1306/AD460EA4-16F7-11D 7-8645000102C1865D

Karagöz, Y., Sandalcl, T., Yüksek, L., Dalklllç, A. S., & Wongwises, S. (2016). Effect of hydrogen-diesel dual-fuel usage on performance, emissions and diesel combustion in diesel engines. *Advances in Mechanical Engineering*, 8(8). Advance online publication. 10.1177/1687814016664458

Karim, G. A. (2015). Dual-fuel diesel engines. In Dual-Fuel Diesel Engines. 10.1201/b18163

Karim, G. A. (2003). Combustion in gas fueled compression: Ignition engines of the dual fuel type. *Journal of Engineering for Gas Turbines and Power*, 125(3), 827–836. Advance online publication. 10.1115/1.1581894

Katz, B. J. (1983). Limitations of 'Rock-Eval' pyrolysis for typing organic matter. *Organic Geochemistry*, 4(3), 195–199. 10.1016/0146-6380(83)90041-4

Kendall, Butcher, Stork, Verdon, Luckett, & Baptiste. (2019). How Big is a Small Earthquake? Challenges in Determining Microseismic Magnitudes. .10.3997/1365-2397.n0015

Khaibullina, K. (2016). Technology to Remove Asphaltene, Resin and Paraffin Deposits in Wells Using Organic Solvents. Paper presented at the SPE Annual Technical Conference and Exhibition, Dubai, UAE. 10.2118/184502-STU

Kietzmann, D. A., & Vennari, V. V. (2013). Sedimentología y estratigrafía de la formación vaca muerta (Tithoniano-Berriasiano) en el área del cerro Domuyo, norte de Neuquén, Argentina. *Andean Geology*, 40(1), 41–65. 10.5027/andgeoV40n1-a02

Killgoar, B. A., Shutterstock, & Kleinberg, R. L. (2020). *The Global Warming Potential Misrepresents the Physics of Global Warming Thereby Misleading Policy Makers Abstract The Global Warming Potential Misrepresents the Physics of Global Warming Thereby Misleading Policy Makers*. Academic Press.

Knoke, T., Gosling, E., & Paul, C. (2020). Use and misuse of the net present value in environmental studies. *Ecological Economics*, 174, 106664. 10.1016/j.ecolecon.2020.106664

Korakianitis, T., Namasivayam, A. M., & Crookes, R. J. (2010). Hydrogen dual-fuelling of compression ignition engines with emulsified biodiesel as pilot fuel. *International Journal of Hydrogen Energy*, 35(24), 13329–13344. Advance online publication. 10.1016/j.ijhydene.2010.08.007

Kuila, U., & Prasad, M. (2013, December). Application of nitrogen gas-adsorption technique for characterization of pore structure of mudrocks. *The Leading Edge*, 32(12), 1478–1485. 10.1190/tle32121478.1

Lakatos, I. J., & Lakatos-Szabo, J. (2009). *Role of Conventional and Unconventional Hydrocarbons in the 21st Century: Comparison of Resources, Reserves, Recovery Factors and Technologies*. All Days.

Langford, F. F., & Blanc-Valleron, M.-M. (1990). Interpreting Rock-Eval Pyrolysis Data Using Graphs of Pyrolizable Hydrocarbons vs. Total Organic Carbon. *AAPG Bulletin*, 74(6), 799–804.

Langmuir, I. (1916). The constitution and fundamental properties of solids and liquids. part i. Solids. *Journal of the American Chemical Society*, 38(11), 2221–2295. 10.1021/ja02268a002

Langmuir, I. (1917). The constitution and fundamental properties of solids and liquids. ii. liquids. *Journal of the American Chemical Society*, 39(9), 1848–1906. 10.1021/ja02254a006

Lata, D. B., & Misra, A. (2010). Theoretical and experimental investigations on the performance of dual fuel diesel engine with hydrogen and LPG as secondary fuels. *International Journal of Hydrogen Energy*, 35(21), 11918–11931. Advance online publication. 10.1016/j.ijhydene.2010.08.039

Lata, D. B., & Misra, A. (2011). Analysis of ignition delay period of a dual fuel diesel engine with hydrogen and LPG as secondary fuels. *International Journal of Hydrogen Energy*, 36(5), 3746–3756. Advance online publication. 10.1016/j.ijhydene.2010.12.075

Lata, D. B., Misra, A., & Medhekar, S. (2011). Investigations on the combustion parameters of a dual fuel diesel engine with hydrogen and LPG as secondary fuels. *International Journal of Hydrogen Energy*, 36(21), 13808–13819. Advance online publication. 10.1016/j.ijhydene.2011.07.142

Laubacher, M. (1970). La tectonica tardi-hercinica en la Cordillera oriental de los Andes del Sur del Peru. Academic Press.

Liborius-Parada, A., & Slatt, R. M. (2016). Geological characterization of La Luna formation as an unconventional resource in Lago de Maracaibo Basin, Venezuela. *SPE/AAPG/SEG Unconventional Resources Technology Conference 2016*, 3280–3299. 10.15530/urtec-2016-2461968

Liew, C., Li, H., Liu, S., Besch, M. C., Ralston, B., Clark, N., & Huang, Y. (2012). Exhaust emissions of a H2-enriched heavy-duty diesel engine equipped with cooled EGR and variable geometry turbocharger. *Fuel*, 91(1), 155–163. Advance online publication. 10.1016/j.fuel.2011.08.002

Liew, C., Li, H., Nuszkowski, J., Liu, S., Gatts, T., Atkinson, R., & Clark, N. (2010). An experimental investigation of the combustion process of a heavy-duty diesel engine enriched with H2. *International Journal of Hydrogen Energy*, 35(20), 11357–11365. Advance online publication. 10.1016/j.ijhydene.2010.06.023

Lim, G., Lee, S., Park, C., Choi, Y., & Kim, C. (2013). Effects of compression ratio on performance and emission characteristics of heavy-duty SI engine fuelled with HCNG. *International Journal of Hydrogen Energy*, 38(11), 4831–4838. Advance online publication. 10.1016/j.ijhydene.2013.01.188

Liu, Y. (2023). The Impacts and Challenges of ESG Investing. *SHS Web of Conferences, 163*, 01015. 10.1051/shsconf/202316301015

López-Roldán, P., & Fachelli, S. (2017). El diseño de la muestra. *Metodología de la investigación social cuantitativa*. Recuperado de https://ddd.uab.cat/record/185163

Loucks, R. G., Reed, R. M., Ruppel, S. C., & Hammes, U. (2012, June). Spectrum of pore types and networks in mudrocks and a descriptive classification for matrix-related mudrock pores. *AAPG Bulletin, 96*(6), 1071–1098. 10.1306/08171111061

Loucks, R., Reed, R., Ruppel, S., And, D., & Jarvie, D. (2009, November). Morphology, Genesis, and Distribution of Nanometer-Scale Pores in Siliceous Mudstones of the Mississippian Barnett Shale. *Journal of Sedimentary Research, 79*(12), 848–861. 10.2110/jsr.2009.092

Luis, F., & Moncayo, G. (2013). Hidrocarburos no convencionales. Tierra y Tecnología, 41(Redacción), 9.

Mao, J., Zhang, C., Yang, X., & Zhang, Z. (2018). Investigation on Problems of Wastewater from Hydraulic Fracturing and Their Solutions. *Water, Air, and Soil Pollution, 229*(8), 246. 10.1007/s11270-018-3847-5

Marc Airhart. (2007, January 26). *The Father of the Barnett Natural Gas Field George Mitchell*. Author.

Masood, M., Mehdi, S. N., & Reddy, P. R. (2007). Experimental investigations on a hydrogen-diesel dual fuel engine at different compression ratios. *Journal of Engineering for Gas Turbines and Power, 129*(2), 572–578. Advance online publication. 10.1115/1.2227418

Mastalerz, M., Schimmelmann, A., Drobniak, A., & Chen, Y. (2013, October). Porosity of Devonian and Mississippian New Albany Shale across a maturation gradient: Insights from organic petrology, gas adsorption, and mercury intrusion. *AAPG Bulletin, 97*(10), 1621–1643. 10.1306/04011312194

Masters, J. A. (1979). Deep Basin Gas Trap, Western Canada1. *AAPG Bulletin, 63*(2), 152–181.

Mathur, H. B., Das, L. M., & Patro, T. N. (1993). Hydrogen-fuelled diesel engine: Performance improvement through charge dilution techniques. *International Journal of Hydrogen Energy, 18*(5), 421–431. Advance online publication. 10.1016/0360-3199(93)90221-U

McCarthy, K., Rojas, K., Niemann, M., Palrnowski, D., Peters, K., & Stankiewicz, A. (2011). Basic petroleum geochemistry for source rock evaluation. *Oilfield Review, 23*(2), 32–43.

Mekonnen, M. M., & Hoekstra, A. Y. (2011). The green, blue and grey water footprint of crops and derived crop products. *Hydrology and Earth System Sciences, 15*(5), 1577–1600. 10.5194/hess-15-1577-2011

Melikyan. (1997). Surfactants based aqueous compositions with D-limonene and hydrogen peroxide and methods using the same, Patent Number: 5,602,090, Date of Patent: Feb. 11, 1997, United States Patent. Technical Product Bulletin #sw-61, US-EPA, OEM Regulations – Implementation Division https://www.epa.gov/emergency-response/epa-oil-field-solution

Mengal, S. A., & Wattenbarger, R. A. (2011). Accounting for adsorbed gas in Shale gas reservoirs. *SPE Middle East Oil and Gas Show and Conference, MEOS. Proceedings*, 1(September), 643–657.

Meng, H., Ji, C., Yang, J., Chang, K., Xin, G., & Wang, S. (2022). Experimental understanding of the relationship between combustion/flow/flame velocity and knock in a hydrogen-fueled Wankel rotary engine. *Energy*, 258, 124828. Advance online publication. 10.1016/j.energy.2022.124828

Mershon, S., & Palucka, T. (2013). *A Century of Innovation: From the U.S. Bureau of Mines to the National Energy Technology Laboratory*. Academic Press.

Ministerio de Agricultura y Riego. (2015). *Manual para el cáculo de eficiencia para sistemas de riego*. Autor.

Ministerio de Economía y Finanzas. (2017). *Directiva N° 002-2017-EF/63.01 (Anexo N°3 Parámetros de evaluación social)*. Author.

Ministerio de Energia y Minas. (2001). Cuencas sedimentarias. Author.

Ministerio de Energía y Minas. (2019). *Dirección General de Hidrocaburos*. Libro Anual de Recursos de Hidrocarburos. In Publicaciones Hidrocarburos.

Ministerio de Vivienda, C. & Saneamiento. (2006). *Reglamento Nacional de Edificaciones. DS N° 011-2006-VIVIENDA*. Academic Press.

Ministerio de Vivienda, C. y S. (2015). *Decreto Supremo que aprueba el Código Técnico de Construcción Sostenible-Decreto Supremo N°015-2015-VIVIENDA*. Academic Press.

Ministerio del Ambiente. (2010). *Aprueba Límites Máximos Permisibles para los efluentes de Plantas de Tratamiento de Aguas Residuales Domésticas o Municipales. Decreto Supremo No 003-2010-MINAM*.

Ministerio del Ambiente. (2017). *Aprueban Estandares de Calidad Ambiental (ECA) para Agua y establecen disposiciones complementarias. Decreto Supremo N° 004-2017-MINAM.*.

Ministry of Energy and Mines. (2001). Cuencas sedimentarias. *Atlas Mining and Energy in Peru*, 1, 1.

Miyamoto, T., Hasegawa, H., Mikami, M., Kojima, N., Kabashima, H., & Urata, Y. (2011). Effect of hydrogen addition to intake gas on combustion and exhaust emission characteristics of a diesel engine. *International Journal of Hydrogen Energy*, 36(20), 13138–13149. Advance online publication. 10.1016/j.ijhydene.2011.06.144

Morales, I. (2013). La revolución energética en América del Norte y las opciones de política energética en México. LaReforma Energética, 109. Obtenido de http://www.foroconsultivo.org.mx/libros_editados/reforma_energetica.pdf#page=110

Compilation of References

Morales, W., Porlles, J., Rodriguez, J., Taipe, H., & Arguedas, A. (2018). First Unconventional Play From Peruvian Northwest: Muerto Formation. In *SPE/AAPG/SEG Unconventional Resources Technology Conference* (p. 14). Unconventional Resources Technology Conference.

Morales-Paetán, W. J., Porlles-Hurtado, J., Rodriguez-Cruzado, J., Taipe-Acuña, H., & Arguedas-Valladolid, A. (2018). First Unconventional Play from Peruvian Northwest: Muerto Formation. *Unconventional Resources Technology Conference (URTeC)*. 10.15530/urtec-2018-2903064

Morales-Paetán, W. J., Rodriguez-Cruzado, J. A., Alvarez-Mendoza, B. G., Alarcón-Marcatoma, A. A., Oré-Rodriguez, J. L., Corrales-Hidalgo, R. W., Madge-Rodriguez, J. J., Porlles-Hurtado, J. W., Falla-Ruiz, J. L., & de Eyzaguirre-Gorvenia, L. (2020, July 27). Geochemical Heterogeneities Characterization of the Muerto Formation – Lancones Basin - Peru as a Source Rock Unconventional Reservoir, a Contribution for its Development. *SPE Latin American and Caribbean Petroleum Engineering Conference*. 10.2118/199088-MS

Morelo, C. (2015). Shale Gas Development Perspective in Peru. National University of Piura. Available: http://repositorio.unp.edu.pe/handle/UNP/866

Morjandin, J. (2013). *La revolución energética en marcha en Estados Unidos*. Expansión.

Morris, R., & Aleman, A. (1975). Sedimentation and Tectonics of Middle Cretaceous Copa Sombrero Formation in Northwest Peru. *SGP Bulletin, 48*(3).

Mukhopadhyay, P. K., Hagemann, H. W., & Gormly, J. R. (1985). Characterization of kerogens as seen under the aspect of maturation and hydrocarbon generation. *Erdoel Kohle, Erdgas, Petrochem. Brennst. -Chem*, 38(1), 7–18.

Müller, D. (1993). *Geochemical Oil Correlations for UPPPL Oils and Offshore Oil Seep*. Talara Basin.

Murillo-Martínez, C. A., Gómez-Rodríguez, O. A., Ortiz-Cancino, O. P., & Muñoz-navarro, S. F. (2015). Aplicación De Modelos Para La Generación De La Isoterma De Adsorción De Metano En Una Muestra De Shale Y Su Impacto En El Cálculo De Reservas. *Revista Fuentes El Reventón Energético*, 13(2), 131–140. 10.18273/revfue.v13n2-2015012

Nag, S., Sharma, P., Gupta, A., & Dhar, A. (2019). Experimental study of engine performance and emissions for hydrogen diesel dual fuel engine with exhaust gas recirculation. *International Journal of Hydrogen Energy*, 44(23), 12163–12175. Advance online publication. 10.1016/j.ijhydene.2019.03.120

Nascimento, L., Maitelli, C., Maitelli, C., & Cavalcanti, A. (2021). Diagnostic of Operation Conditions and Sensor Faults Using Machine Learning in Sucker-Rod Pumping Wells. *Sensors (Basel)*, 21(13), 4546. Advance online publication. 10.3390/s2113454634283092

Nauss, A. W. (1946). *A Reconnaissance Geological Survey of the Pazul Area - Geological Memo #30*. Academic Press.

Nauss, A. W., & Tafur, H. I. A. (1946). *Geological Report of the Angostura district.* La Brea & Pariñas Estate.

Navarro-Ramirez, J. P., Bodin, S., Consorti, L., & Immenhauser, A. (2017). Response of western South American epeiric-neritic ecosystem to middle Cretaceous Oceanic Anoxic Events. *Cretaceous Research*, 75, 61–80. 10.1016/j.cretres.2017.03.009

Nazi, A., & Lea, K. (1994). Application of Artificial Neural Network to Pump Card Diagnosis. *SPE Comp App*, 6(6), 9–14. Advance online publication. 10.2118/25420-PA

Newberry, M., & Jennings, D. W. (2022). Chapter 2 - Paraffin management, Editor(s): Qiwei Wang, In Oil and Gas Chemistry Management Series, Flow Assurance, Gulf Professional Publishing. 10.1016/B978-0-12-822010-8.00003-9

Noufal, A., & Obaid, K. (2017). *Sealing Faults: A Bamboozling Problem in Abu Dhabi Fields.* Abu Dhabi International Petroleum Exhibition & Conference. Society of Petroleum Engineers. SPE. 10.2118/188775-MS

Olsson, A. A. (1934). Contributions to the paleontology of northern Peru: the Cretaceous of the Amotape Region. In *Bulletins of American Paleontology* (Vol. 20, Issue 69, pp. 1–104). Academic Press.

Osborn, Vengosh, Warner, & Jackson. (n.d.). Methane Contamination of Drinking Water Accompanying Gas-Well Drilling and Hydraulic Fracturing. *Proc Natl Acad Sci.*

OSINERGMIN. (2022). Pliego Tarifario Máximo del Servicio Público de Electricidad. Recuperado 20 de abril de 2023, de https://www.osinergmin.gob.pe/Tarifas/Electricidad/PliegoTarifario?Id=150000

Pairazamán, L., Palacios, F., & Timoteo, D. (2021). *Caracterización sedimentológica de alta-resolución de la Formación Muerto, Cuenca Lancones, NO Perú. ¿Un posible reservorio no convencional?* Academic Press.

Palacios, O. (1994). Geología de los cuadrángulos de Paita, Piura, Talara, Sullana, Lobitos, Quebrada Seca, Zorritos, Tumbes y Zarumilla. In *Carta Geológica Nacional* (1st ed.). Instituto Geológico, Minero y Metalúrgico.

Palermo, L. C. M., Souza, N. F.Jr, Louzada, H. F., Bezerra, M. C. M., Ferreira, L. S., & Lucas, E. F. (2013). Development of multifunctional formulations for inhibition of waxes and asphaltenes deposition. *Brazilian Journal of Petroleum and Gas*, 7(4), 181–192. Advance online publication. 10.5419/bjpg2013-0015

Pan, S., Wang, J., Liang, B., Duan, H., & Huang, Z. (2022). Experimental Study on the Effects of Hydrogen Injection Strategy on the Combustion and Emissions of a Hydrogen/Gasoline Dual Fuel SI Engine under Lean Burn Condition. *Applied Sciences (Basel, Switzerland)*, 12(20), 10549. Advance online publication. 10.3390/app122010549

Pariguana M., H. A. (1979). *Evaluación de las formaciones Muerto - Pananga - Área L.B.P.* Academic Press.

Compilation of References

Passey, Q. R., Bohacs, K. M., Esch, W. L., Klimentidis, R., & Sinha, S. (2010). From Oil-Prone Source Rock to Gas-Producing Shale Reservoir – Geologic and Petrophysical Characterization of Unconventional Shale-Gas Reservoirs. *International Oil and Gas Conference and Exhibition in China*. 10.2118/131350-MS

Pelet, R. (1985). Evaluation quantitative des produits formés lors de l'évolution géochimique de la matière organique. *Revue de l'Institut Français du Pétrole*, 40(5), 551–562. 10.2516/ogst:1985034

Peng. (2019). Artificial Intelligence Applied in Sucker Rod Pumping Wells: Intelligent Dynamometer Card Genertion, Diagnosis and Failure Detection Using Deep Neural Networks. SPE Annual Technical Conference and Exhibition. .10.2118/196159-MS

Pepper, A. S., & Corvi, P. J. (1995). Simple kinetic models of petroleum formation. Part I: Oil and gas generation from kerogen. *Marine and Petroleum Geology*, 12(3), 291–319. 10.1016/0264-8172(95)98381-E

Perú, A. P. C. (1974). YARINA 10-19-2X, Informe Diario Geológico.

PERUPETRO. (1999). VOL 1. Generalidades. In *Estudios de Investigación Geoquímica del Potencial de Hidrocarburos. Lotes del Zócalo Continental y de Tierra*. PERUPETRO S.A.

Perúpetro. (2017). Situacion Actual Y Potencial Hidrocarburifero Del Perú. Author.

PERUPETRO. (2020). *Estadística Anual de Hidrocarburos*. Academic Press.

Peters, K. E. (1986). Guidelines for Evaluating Petroleum Source Rock Using Programmed Pyrolysis. *AAPG Bulletin*, 70(3), 318–329. 10.1306/94885688-1704-11D7-8645000102C1865D

Peters, K. E., & Cassa, M. R. (1994). Applied Source Rock Geochemistry. *AAPG Memoir*, 60, 93–120. 10.1306/M60585C5

Peters, K. E., Walters, C. C., & Moldowan, J. M. (2004). *The Biomarker Guide* (2nd ed., Vol. 1). Cambridge University Press. 10.1017/CBO9780511524868

Pimentel, D., & Burgess, M. (2017). World human population problems. *Encyclopedia of the Anthropocene, 1–5*, 313–317. 10.1016/B978-0-12-809665-9.09303-4

Polk, A. C., Gibson, C. M., Shoemaker, N. T., Srinivasan, K. K., & Krishnan, S. R. (2013). Analysis of Ignition Behavior in a Turbocharged Direct Injection Dual Fuel Engine Using Propane and Methane as Primary Fuels. *Journal of Energy Resources Technology*, 135(3), 032202. Advance online publication. 10.1115/1.4023482

Porlles, J., Panja, P., Sorkhabi, R., & McLennan, J. (2021). Integrated porosity methods for estimation of gas-in-place in the Muerto Formation of Northwestern Peru. *Journal of Petroleum Science Engineering*, 202, 108558. Advance online publication. 10.1016/j.petrol.2021.108558

Potential Gas Committee. (2008, December 31). *Potential gas committee reports unprecedented increase in magnitude of U.S. natural gas resource base*. Author.

Potter, P. E., Maynard, J. B., & Pryor, W. A. (1981). *Sedimentology of gas-bearing Devonian shales of the Appalachian Basin.* Academic Press.

Pratt, L. M. (1982). *The paleo-oceanographic interpretation of the sedimentary structures, clay minerals, and organic matter in a core of the Middle Cretaceous Greenhorn Formation drilled near Pueblo Colorado.* Princeton University.

Qiu, Z., Song, D., Zhang, L., Zhang, Q., Zhao, Q., Wang, Y., Liu, H., Liu, D., Li, S., & Li, X. (2021, November). The geochemical and pore characteristics of a typical marine–continental-transitional gas shale: A case study of the Permian Shanxi Formation on the eastern margin of the Ordos Basin. *Energy Reports*, 7, 3726–3736. 10.1016/j.egyr.2021.06.056

Rahman, M. A., Ruhul, A. M., Aziz, M. A., & Ahmed, R. (2017). Experimental exploration of hydrogen enrichment in a dual fuel CI engine with exhaust gas recirculation. *International Journal of Hydrogen Energy*, 42(8), 5400–5409. Advance online publication. 10.1016/j.ijhydene.2016.11.109

Raimi, D. (2017). *The Fracking Debate: The Risks.* Benefits, and Uncertainties of the Shale Revolution. 10.7312/raim18486

Ramez, M. (2020). Identification of Downhole Conditions in Sucker Rod Pumped Wells Using Deep Neural Networks and Genetic Algorithms. *SPE Production & Operations*, 35(2), 435–447. Advance online publication. 10.2118/200494-PA

Ramírez-Orellana, A., Martínez-Victoria, M., García-Amate, A., & Rojo-Ramírez, A. A. (2023). Is the corporate financial strategy in the oil and gas sector affected by ESG dimensions? *Resources Policy*, 81, 103303. 10.1016/j.resourpol.2023.103303

Rasmussen, S. G., Ogburn, E. L., McCormack, M., Casey, J. A., Bandeen-Roche, K., Mercer, D. G., & Schwartz, B. S. (2016). Association Between Unconventional Natural Gas Development in the Marcellus Shale and Asthma Exacerbations. *JAMA Internal Medicine*, 176(9), 1334–1343. 10.1001/jamainternmed.2016.243627428612

Reagan, M. T., Moridis, G. J., Keen, N. D., & Johnson, J. N. (2015). Numerical Simulation of the Environmental Impact of Hydraulic Fracturing of Tight/Shale Gas Reservoirs on near-Surface Groundwater: Background, Sase Cases, Shallow Reservoirs, Short-Term Gas, and Water Transport. *Water Resources Research*, 51(4), 2543–2573. Advance online publication. 10.1002/2014WR01608626726274

Renee Choo. (2023, April 5). *The Energy Transition Will Need More Rare Earth Elements. Can We Secure Them Sustainably?* Academic Press.

Reyes, L., & Vergara, J. (1985). *Libreta de campo 85-6. Geología de la Cuenca Lancones.* Petroperu.

Rezaee, R. (2015). *Fundamentals of Gas Shale Reservoirs.* John Wiley & Sons, Inc. 10.1002/9781119039228

Rogner, H., Dusseault, M. B., Rogner, H.-H., Authors Roberto Aguilera, L. F., & Archer, C. L. (2012). *Energy Resources and Potentials. In Global Energy Assessment-Toward a Sustainable Future Energy Resources and Potentials Convening Lead Author (CLA).* 10.13140/RG.2.1.3049.8724

Compilation of References

Ross, D. J. K., & Bustin, R. M. (2007, March). Shale gas potential of the Lower Jurassic Gordondale Member, northeastern British Columbia, Canada. *Bulletin of Canadian Petroleum Geology*, 55(1), 51–75. 10.2113/gscpgbull.55.1.51

Ross, D. J. K., & Marc Bustin, R. (2009, June). The importance of shale composition and pore structure upon gas storage potential of shale gas reservoirs. *Marine and Petroleum Geology*, 26(6), 916–927. 10.1016/j.marpetgeo.2008.06.004

Rouquerol, J., Avnir, D., Fairbridge, C. W., Everett, D. H., Haynes, J. M., Pernicone, N., Ramsay, J. D. F., Sing, K. S. W., & Unger, K. K. (1994, January). Recommendations for the characterization of porous solids (Technical Report). *Pure and Applied Chemistry*, 66(8), 1739–1758. 10.1351/pac199466081739

Salama, A., El Amin, M. F., Kumar, K., & Sun, S. (2017). Flow and transport in tight and shale formations: A review. *Geofluids*, 2017, 1–21. Advance online publication. 10.1155/2017/4251209

Salvi, B. L., & Subramanian, K. A. (2015). Sustainable development of road transportation sector using hydrogen energy system. In Renewable and Sustainable Energy Reviews (Vol. 51). 10.1016/j.rser.2015.07.030

Samuel Armacanqui Tipacti, J. (2013). Arresting Unexpected Oil Production Decline in a Joint-Venture Environment. *Society of Petroleum Engineers - North Africa Technical Conference and Exhibition 2013. NATC*, 2013(2), 1496–1505. 10.2118/164785-MS

Santoso, W. B., Bakar, R. A., & Nur, A. (2013). Combustion characteristics of diesel-hydrogen dual fuel engine at low load. *Energy Procedia*, 32, 3–10. Advance online publication. 10.1016/j.egypro.2013.05.002

Schlanger, S. O., & Jenkyns, H. C. (1976). Cretaceous Oceanic Anoxic Events: Causes and Consequences. *Netherlands Journal of Geosciences*, 55(3–4), 179–184.

Schmoker, J. W. (1995). Method for Assessing Continuous-Type (Unconventional) Hydrocarbon Accumulations. In Gautier, D. L., Dolton, G. L., Takahashi, K. I., & Varnes, K. L. (Eds.), *National Assessment of United States Oil and Gas Resources: Results, Methodology, and Supporting Data*. U.S. Geological Survey. 10.3133/ds30

Schneider, F., Laigle, J.M., Kuhfuss-Monval, L., & Lemouzy, P. (2013). Basin Modeling - the Key for Unconventional Play Assessment. Academic Press.

Scotese, C. (2014). Atlas of Early Cretaceous Paleogeographic Maps, PALEOMAP Atlas for ArcGIS, volume 2, The Cretaceous, Maps 23 - 31, Mollweide Projection, PALEOMAP Project. 10.13140/2.1.4099.4560

SEDAPAL. (2014a). *Plan maestro de los sistemas de agua potable y alcantarillado. Tomo I Diagnóstico*. Recuperado de https://cdn.www.gob.pe/uploads/document/file/4245313/Tomo%20I%20-%20Volumen%20I%20Diagnostico.pdf.pdf

SEDAPAL. (2014b). *Plan maestro de los sistemas de agua potable y alcantarillado. Tomo II Estimación Oferta Demanda de los Servicios*.

Senthil Kumar, M., Ramesh, A., & Nagalingam, B. (2003). Use of hydrogen to enhance the performance of a vegeTABLE oil fuelled compression ignition engine. *International Journal of Hydrogen Energy*, 28(10). Advance online publication. 10.1016/S0360-3199(02)00234-3

Séranne, M. (1987). Evolution tectono-sédimentaire du bassin de Talara (nord-ouest du Pérou). *Bulletin de l'Institut Français d'Études Andines*, 16(3), 103–125. 10.3406/bifea.1987.952

Serrano, J., Jiménez-Espadafor, F. J., & López, A. (2019). Analysis of the effect of different hydrogen/diesel ratios on the performance and emissions of a modified compression ignition engine under dual-fuel mode with water injection. Hydrogen-diesel dual-fuel mode. *Energy*, 172, 702–711. Advance online publication. 10.1016/j.energy.2019.02.027

Sharadkumar, A. (2017). *Surface area study in organic-rich shales using nitrogen adsorption.* University of Oklahoma.

Sharma, S. K., Goyal, P., & Tyagi, R. K. (2015). Hydrogen-fueled internal combustion engine: A review of technical feasibility. In International Journal of Performability Engineering (Vol. 11, Issue 5). https://doi.org/10.23940/ijpe.15.5.p491.mag

Sharma, P., & Dhar, A. (2018). Effect of hydrogen supplementation on engine performance and emissions. *International Journal of Hydrogen Energy*, 43(15), 7570–7580. Advance online publication. 10.1016/j.ijhydene.2018.02.181

Shchipanov, Kollbotn, & Berenblyum. (2018). Fault Leakage Detection from Pressure Transient Analysis. .10.3997/2214-4609.201802990

Sing, K. S. W. (1985, January). Reporting Physisorption Data for Gas/Solid Systems with Special Reference to the Determination of Surface Area and Porosity. *Pure and Applied Chemistry*, 57(4), 603–619. 10.1351/pac198557040603

Sivabalakrishnan, R., & Jegadheesan, C. (2014). Study of Knocking Effect in Compression Ignition Engine with Hydrogen as a Secondary Fuel. *Chinese Journal of Engineering*, 2014, 1–8. Advance online publication. 10.1155/2014/102390

Sliter, W. V. (1989). Aptian anoxia in the Pacific Basin. *Geology*, 17(10), 909. 10.1130/0091-7613(1989)017<0909:AAITPB>2.3.CO;2

SNMPE. (2012). Shale Gas/ Gas de Lutitas. Informe quincenal de la SNMPE. Obtenido de https://issuu.com/sociedadmineroenergetica/docs/snmpe-informe-quincenal-hidrocarbur_63d9b410cfd751

Solarin, Gil-Alana, & Lafuente. (2020). An investigation of long range reliance on shale oil and shale gas production in the US market. *Energy, 195.*

Sondergeld, C. H., Newsham, K. E., Comisky, J. T., Rice, M. C., & Rai, C. S. (2010). *Petrophysical Considerations in Evaluating and Producing Shale Gas Resources.* All Days.

Compilation of References

Soon, W., Connolly, R., & Connolly, M. (2015). Re-evaluating the role of solar variability on Northern Hemisphere temperature trends since the 19th century. In *Earth-Science Reviews* (Vol. 150, pp. 409–452). Elsevier B.V. 10.1016/j.earscirev.2015.08.010

Sorkhabi, R., Suzuki, U., & Sato, D. (2000). Structural Evaluation of Petroleum Sealing Capacity of Faults. *SPE Asia Pacific Conference on Integrated Modeling for Asset Management*, 9. 10.2118/59405-MS

Statista. (2023, August 16). *Which Country Consumes the Most Oil?* https://www.statista.com/chart/30609/countries-with-the-highest-oil-consumption-per-day/

Subramanian, K. A., & Chintala, V. (2013). Reduction of GHGs emissions in a biodiesel fueled diesel engine using hydrogen. ASME 2013 Internal Combustion Engine Division Fall Technical Conference, ICEF 2013, 2. 10.1115/ICEF2013-19133

Subramanian, K. A. (2011). A comparison of water-diesel emulsion and timed injection of water into the intake manifold of a diesel engine for simultaneous control of NO and smoke emissions. *Energy Conversion and Management*, 52(2), 849–857. Advance online publication. 10.1016/j.enconman.2010.08.010

Superintendencia Nacional de Servicios de Saneamiento (SUNASS). (2021). *Resolución de Consejo Directivo No 079-2021-SUNASS-CD*.

Szwaja, S. (2019). Dilution of fresh charge for reducing combustion knock in the internal combustion engine fueled with hydrogen rich gases. *International Journal of Hydrogen Energy*, 44(34), 19017–19025. Advance online publication. 10.1016/j.ijhydene.2018.10.134

Szwaja, S., & Grab-Rogalinski, K. (2009). Hydrogen combustion in a compression ignition diesel engine. *International Journal of Hydrogen Energy*, 34(10), 4413–4421. Advance online publication. 10.1016/j.ijhydene.2009.03.020

Tafur, H. I. A. (1952). Cretaceous Geology of the East Front of the Amotape Mountains. IPC Report WP-12. *International Petroleum Company*.

Tafur, H. I. A. (1954). *Reconnaissance of Cretaceous Between Chira River and Amotape Mountains, Northwest Peru. IPC Report WP-14*.

Tahir, M. M., Ali, M. S., Salim, M. A., Bakar, R. A., Fudhail, A. M., Hassan, M. Z., & Abdul Muhaimin, M. S. (2015). Performance analysis of a spark ignition engine using compressed natural gas (CNG) as fuel. *Energy Procedia*, 68, 355–362. Advance online publication. 10.1016/j.egypro.2015.03.266

Tanirbergenova, S., Ongarbayev, Y., Tileuberdi, Y., Zhambolova, A., Kanzharkan, E., & Mansurov, Z. (2022, June). Selection of Solvents for the Removal of Asphaltene–Resin–Paraffin Deposits. *Processes (Basel, Switzerland)*, 10(7), 1262. 10.3390/pr10071262

Tarduno, J. A., McWilliams, M., Debiche, M. G., Sliter, W. V., & Blake, M. C.Jr. (1985). Franciscan Complex Calera limestones: Accreted remnants of Farallon Plate oceanic plateaus. *Nature*, 317(6035), 345–347. 10.1038/317345a0

Tauzia, X., Maiboom, A., & Shah, S. R. (2010). Experimental study of inlet manifold water injection on combustion and emissions of an automotive direct injection Diesel engine. *Energy*, 35(9), 3628–3639. Advance online publication. 10.1016/j.energy.2010.05.007

Tenorio-Trigoso, A., Castillo-Cara, M., Mondragón-Ruiz, G., Carrión, C., & Caminero, B. (2021). An Analysis of Computational Resources of Event- Driven Streaming Data Flow for Internet of Things: A Case Study. *The Computer Journal*. Advance online publication. 10.1093/comjnl/bxab143

Tesfa, B., Mishra, R., Gu, F., & Ball, A. D. (2012). Water injection effects on the performance and emission characteristics of a CI engine operating with biodiesel. *Renewable Energy*, 37(1), 333–344. Advance online publication. 10.1016/j.renene.2011.06.035

Tinni, A., Sondergeld, C., & Rai, C. (2017). *New Perspectives on the Effects of Gas Adsorption on Storage and Production of Natural Gas From Shale Formations*. Academic Press.

Tissot, B. P., & Welte, D. H. (1978). Sedimentary Processes and the Accumulation of Organic Matter. *Petroleum Formation and Occurrence*. 10.1007/978-3-642-96446-6_5

Todd, B. M., Kuykendall, D. C., Peduzzi, M. B., & Hinton, J. (2015, March 16). Hydraulic Fracturing-Safe, Environmentally Responsible Energy Development. *All Days*. 10.2118/173515-MS

Toledano, P., Brauch, M. D., & Arnold, J. (2023). *Circularity in Mineral and Renewable Energy Value Chains: Overview of Technology, Policy, and Finance Aspects*. https://ccsi.columbia.edu/circular-economy-mining-energy

Tsolakis, A., Hernandez, J. J., Megaritis, A., & Crampton, M. (2005). Dual fuel diesel engine operation using H2. Effect on particulate emissions. *Energy & Fuels*, 19(2), 418–425. Advance online publication. 10.1021/ef0400520

U.S. Energy Information Administration EIA/AIR. (2015). *World Shale Resource Assessments*. Author.

U.S. Energy Information Administration. (2013). *World Shale Gas and Shale Oil Resource Assessment. Technically Recoverable Shale Gas and Shale Oil Resources: An Assessment of 137 Shale Formations in 41 Countries Outside the United States*. U.S. Energy Information Administration.

Uludamar, E., Yildizhan, Ş., Aydin, K., & Özcanli, M. (2016). Vibration, noise and exhaust emissions analyses of an unmodified compression ignition engine fuelled with low sulphur diesel and biodiesel blends with hydrogen addition. *International Journal of Hydrogen Energy*, 41(26), 11481–11490. Advance online publication. 10.1016/j.ijhydene.2016.03.179

UNESCO. (2012). *United Nations world water development report 4: managing water under uncertainty and risk* (Vol. 1). UNESCO. Recuperado de UNESCO website: https://unesdoc.unesco.org/ark:/48223/pf0000215644.locale=en

US EPA. (2024). Nitrogen Oxides Control Regulations. US EPA. https://www3.epa.gov/region1/airquality/nox.html

Compilation of References

Uyen, D., & Valencia, K. H. (2002). Anexo 16: North-Lancones Basin Surface Geology and Leads Evaluation. In *Informe Final. Segundo Periodo Exploratorio. Lote XII - Cuenca Lancones*. Pluspetrol Peru Corporation S.A.

Valencia. (2008). *Conventional Analysis And Interpretation Of Pressure Tests*. Academic Press.

Varde, K. S., & Frame, G. A. (1983). Hydrogen aspiration in a direct injection type diesel engine-its effects on smoke and other engine performance parameters. *International Journal of Hydrogen Energy*, 8(7), 549–555. Advance online publication. 10.1016/0360-3199(83)90007-1

Vera, R. (2016). *Eagle Ford Shale Play: Oil-Mining Industrial Geography in Southernn Texas, 2008-2015.SciELO Analytics*, 34.

Viera Palacios, M. R. (2020). *Remoción de parafinas y mejoramiento del factor de recobro mediante uso de químicos multifuncionales y biodegradables para incrementar la productividad en pozos petroleros*. Academic Press.

Viera, M. (2020). Removal of paraffins and improvement of the recovery factor through the use of multifunctional and biodegradable chemicals to increase productivity in oil wells. Master's Thesis, National University of Engineering.

Villar, H., & Pardo, A. (2010). Potencial de hidrocarburos y sistemas de petróleo en las cuencas costeras del Perú. Geología Integrada, 6–13.

Vtorov, I. P. (2017, January 17). *REE-table*. https://commons.wikimedia.org/wiki/File:REE-table.jpg

Walters, C. C., Toon, M. B., Rae, D., Barrow, R., Flagg, E. M., & Hellyer, C. L. (1992). *Talara Basin, Peru: Source Rock Characterization Academic Press.*.

Walters, C. C., Toon, M. B., Rae, D., Barrow, R., Flagg, E. M., & Hellyer, C. L. (1992). *Talara Basin, Peru: Source Rock Characterization*. Academic Press.

Wang, F. P., & Reed, R. M. (2009). Pore Networks and Fluid Flow in Gas Shales. In *SPE Annual Technical Conference and Exhibition* (p. 8). Society of Petroleum Engineers. 10.2118/124253-MS

Wang, H., Li, W., & Dou, X. (2021). A Working Condition Diagnosis Model of Sucker Rod Pumping Wells Based on Deep Learning. *SPE Production & Operations*, 36(2), 317–326. Advance online publication. 10.2118/205015-PA

Willis, R. M., Stewart, R. A., Panuwatwanich, K., Williams, P. R., & Hollingsworth, A. L. (2011). Quantifying the influence of environmental and water conservation attitudes on household end use water consumption. *Journal of Environmental Management*, 92(8), 1996–2009. 10.1016/j.jenvman.2011.03.02321486685

Winter, L. S. (2008). the Genesis of 'Giant' Copper-Zinc-Gold-Silver Volcanogenic Massive Sulphide Deposits At Tambogrande, Perú: Age, Tectonic Setting, Paleomorphology. *Lithogeochemistry and Radiogenic Isotopes.*, 6(11), 951–952.

World Economic Forum. (2019). *Global risks 2019 : insight report*. Recuperado de https://www3.weforum.org/docs/WEF_Global_Risks_Report_2019.pdf?_gl=1*ncy97p*_up*MQ.&gclid=CjwKCAjw3POhBhBQEiwAqTCuBso_PkjS56bwz-Qva6fl-Mb6cUJMDT5jrC1PJcvkSN8jQKW-ZogUyhoC8LoQAvD_BwE

Xia, B., & Lan, L. (2020). Trend prediction of pumping well conditions using temporal dynamometer cards. *2nd International Conference on Machine Learning, Big Data and Business Intelligence (MLBDBI)*. 10.1109/MLBDBI51377.2020.00039

Yadav, V. S., Soni, S. L., & Sharma, D. (2014). Engine performance of optimized hydrogen-fueled direct injection engine. *Energy*, 65, 116–122. Advance online publication. 10.1016/j.energy.2013.12.007

Yesquen, S. (2020). OIL AND GAS AFTER COVID 19: An Opportunity to Reconfigure the Sector. Academic Press.

ZeroHedge. (2023, August 21). *U.S. And China Top The Chart In Global Oil Consumption*. U.S. And China Top The Chart In Global Oil Consumption.

Zhang, T., Ellis, G. S., Ruppel, S. C., Milliken, K., & Yang, R. (2012). Effect of organic-matter type and thermal maturity on methane adsorption in shale-gas systems. *Organic Geochemistry*, 47, 120–131. 10.1016/j.orggeochem.2012.03.012

Zhao, Y., & Zhang, L. (2011). Solution and Type Curve Analysis of Fluid Flow Model for Fractal Reservoir. *World Journal of Mechanics*, 1(5), 209–216. 10.4236/wjm.2011.15027

Zheng, S., & Sharma, M. M. (2021). Modeling Hydraulic Fracturing Using Natural Gas Foam as Fracturing Fluids. *Energies*, 14(22), 7645. 10.3390/en14227645

Zhou, S., Ning, Y., Wang, H., Liu, H., & Xue, H. (2018). Investigation of methane adsorption mechanism on longmaxi shale by combining the micropore filling and monolayer coverage theories. *Advances in Geo-Energy Research*, 2(3), 269–281. 10.26804/ager.2018.03.05

Zou, C. (2017). Unconventional petroleum geology. In *Unconventional Petroleum Geology*. Elsevier. 10.1016/B978-0-12-812234-1.00002-9

About the Contributors

Jesus Samuel Armacanqui has worked nearly three decades with global E&P Companies, such as Conoco, PDO-Shell, PDVSA and RWE/Dea (now Wintershall), and major Service Companies like Schlumberger and Camco-Reda, in the Americas, the Middle East, Afrika, Europe and Asia. He has extensive desk and field experience and has covered managerial and C-level positions. With this industry acumen he initiated and completed a few research projects, including one funded by the World Bank, related to the Exploration of Unconventional Resources. He has authored and co-authored several papers presented in International Conferences of the SPE – Society of Petroleum Engineers, and scientific Journals covering topics of Exploration, Reservoir Engineering, EOR, Artificial Lift, IOT/AI based Production Optimization and Operations, ESPs, Artificial Intelligence, Renewable Energy, Energy Efficiency, Effective Introduction of Innovations and New Technology and Performance Oriented Contracts and Sustainable Development. He gives industry courses as Visiting Instructor at the Oklahoma University Norman and is Professor at the National University of Engineering, Faculty of Oil, Gas and Petrochemical Engineering of Peru. He graduated at the TU Bergakademie Freiberg and covered a post- doctoral in the TH Clausthal, both in Germany and holds multiple academic, research and industry awards.

Susan Smith Nash has organized numerous workshops, forums, educational events and research conferences on the topic of unconventional resources in her capacity of Director of Education and Professional Development of the AAPG. In addition, she has worked with cross-disciplinary teams to work on knowledge transfer, understanding, and research initiatives in optimizing exploration and development of unconventional and mature reservoirs. Her current interests involve supporting new technologies for green development of unconventionals.

Luz de Fatima Eyzaguirre-Gorvenia is a Research Professor at National University of Engineering and Universidad Nacional Mayor de San Marcos. Consultant in Information Technology in public and private institutions at national and international level. Since 1979 he has participated in different projects in public institutions such as INGEMMET of the Ministry of Energy and Mines, Ministry of Education, Ministry of Transport and Communications and the State Bank. Within my actvities I have also participated in different national and international events related to information technologies both as a speaker and as a participant.

Rouzbeh Moghanloo is an Associate Professor in Petroleum & Geological Engineering with special interests in Reservoir Engineering, Production Optimization, and CCUS; earned Doctor of Philosophy (Ph.D.) in Petroleum Engineering from The University of Texas at Austin.

* * *

Hugo Ampuero is an outstanding field geologist in exploratory work that contributed to the discovery of hydrocarbons in the jungle of Peruvian territory.

About the Contributors

Tiffany Billinghurst Vargas is an environmental engineer from the Scientific University of the South with experience in innovation projects, sustainable development, conservation and renewable energies. Currently working as a Consultant in innovation projects in the area of energy and sustainable development.

Manuel Cara is a full-time professor at Computer Science and head of Smart Cities research laboratory in the CTIC - at National University of Engineering from Perú. Currently conducting research projects with international funding in Smart Cities and agriculture areas. Experience in Wireless sensor Network and signal transmission, mobile phone applications and analysis/processing of data generated by WSN, making predictions with different machine learning techniques.

Franco Cassinelli is a Mechatronic engineer with experience in project management in the areas of production, maintenance and automation and instrumentation of processes in mining and oil. Software developer and researcher in the areas of Control and Automation Systems, Artificial Intelligence and Renewable Energy, from which I have developed innovation in mining and oil.

Israel Chavez-Sumarriva has a PhD in Chemical Engineering with scientific publications and research projects.

Kevin Chipana Suasnabar is a Petroleum Engineering Professional, focused on rock petrophysics.

Miguel Guzmán is an Environmental Engineer with a wide chemistry background. Currently interested in biotechnology applied in mining and oil fields.

Alfredo Lescano Lozada is an expert engineer in the management of climate change and environmental pollution. Master in "Environmental Management of the company and renewable energies", from the University of Alcalá in Spain. Wirh more than 20 years of experience at the national and international level. Currently, master's candidate in Applied Meteorology at the Universidad Nacional Agraria La Molina. Recognized by MINAM, in The UNFCCC Roster of Experts, as a national expert on climate change, development of GHG inventories; monitoring, reporting and verification of GHG emissions and management of methodologies for GHG mitigation measures. Member of the Technical Working Group: Mitigation Actions/Policies.

Cesar Lujan Ruiz has a Master degree in oil and gas engineering at National University of Engineering - Lima, Peru -Posgrade in petroleum chemistry and petrochemica at Polytechnic of Catalunya - Barcelona, Spain -Engineering degree in petrochemical engineering at National University of Engineering - Lima, Peru Since 2000 to now, professor in Faculty of Oil, Natural Gas and Petrochemical at National University of Engineering - Lima, Peru.

Walter Morales Paetan is a Geologist engineer with experience in Hydrogeology, Hydrocarbon Exploration, Development Geology, Geophysics in Seismic Interpretation and Data Bank. Master in Petroleum and Natural Gas Engineering, currently advisor of projects related to the evaluation and classification of the dead formation in the lancones basin.

Isabel Nakata is a civil engineer, master in urban and regional planning, master in disaster risk management, Diploma in Testing Laboratory Management. Specialist in housing and construction materials. He has led courses, seminars and participated in events in China, Japan, Spain, Brazil, Argentina, Paraguay, Chile, Bolivia, Ecuador, Nicaragua, Cuba, as well as in the interior of the country. She has been Director and Executive President of the National Institute for Housing Research and Standardization -ININVI of the Ministry of Housing; Head of the Materials Testing Laboratory of the Faculty of Civil Engineering of the National University of Engineering. She has been and is a member of the INDECOPI Standardization Committees and the National Building Regulations and a speaker on housing, standards and materials and construction systems in Peru and abroad. Currently, she is a teacher, researcher and consultant in housing, standardization and construction materials. She has participated in research projects, presented papers at international conferences and published as co-author of articles in several identified journals.

About the Contributors

Jorge Ore is a Senior Exploration & Production Geologist, 32 years of experience.

Joel Paccori graduated from the Faculty of Petroleum and Natural Gas Engineering and the master's degree in Computer Science - UNI, Electronic development engineer in the research, development and innovation (IDI) area of the Institute Geophysicist of Peru - MINAM, implementing technological development projects applied to disaster reduction financed by Concytec, Innovate Perú.

Guillermo Prudencio Baldeon is a chemical engineer and has a master's degree in energy sciences.

Victor Quiñones is a Petroleum Engineer, graduated from the National University of Engineering, Lima - Peru, and Master of Science with a Major in Energy and Environment from the University of Calgary, Canada. I have 22 years of experience in the area of hydrocarbon exploration and exploitation, with emphasis on reservoir modeling, fluid characterization, resource and reserve estimation, risk analysis and decision making, as well as production optimization. Throughout my experience he developed skills and competencies to manage and evaluate strategies aimed at the development of oil and gas fields, working in multidisciplinary environments.

Humberto Rivera has a PhD. in Natural Sciences from the University of Heidelberg, Germany and Geological Engineer from the Universidad Nacional de Ingeniería (UNI). Author of research papers on petrology, deposit geology and metallogeny. Until 2005 he was Head of the Mineralogy Laboratory at UNI and member of the university quality commission of the same university. From July 2005 to May 2012 he was Director of Mineral and Energy Resources of the Geological, Mining and Metallurgical Institute of Peru. From June 2012 to the present, he has been working as a consultant for Geo Wissens and as a teacher at the professional school of geological engineering of UNI.

Gustavo Rodríguez Robles holds a Master's degree in Petroleum and Natural Gas Engineering. In 2021, he conducted postgraduate research in a project funded by the World Bank. Currently, he serves as Development Manager at a company specializing in the design and sale of electrical transformers.

Jose Rodriguez-Cruzado is a Petroleum and Natural Gas Engineer dedicated to research in the hydrocarbons sector. Passionate about finding innovative and sustainable solutions for current energy challenges. Experienced in exploration, production, and extraction technologies. Committed to the responsible development of natural resources.

Pedro Sanchez is a Petroleum Engineer from Universidad Nacional de Ingeniería. Master's degree in Petroleum and Natural Gas Engineering from Universidad Nacional de Ingeniería (UNI). Master studies in Public Management and Governance at Winner University. With experience as Production & Reservoir Advisor at Baker Hughes and Cobra Oil & Gas; Team Leader Lot 8 at PlusPetrol Norte; Senior Reservoir Engineer at PlusPetrol Norte.

Joseph Sinchitullo Gomez holds a PhD in Energy Sciences and serves as an Assistant Professor at the National University of Engineering with over eight years of expertise in Reservoir Engineering, Energy Economics, Audits, and developing supply and demand projection models for various energy projects. He is acknowledged as a Profesor Investigator and has published more than 13 papers in Scopus-indexed journals.

Heraud Taipe Acuna is a Professional with training in Civil Engineering and Petroleum and Natural Gas Engineering. With experience in the development of improved hydrocarbon recovery projects, oil exploration and analysis of oil reservoirs, installation of natural gas networks to multifamily, industrial and CNG service stations, also with experience in the area of bidding in economic and civil project planning.

About the Contributors

Ali Tinni has a Bachelor's degree in Geology and Hydrogeology from Cadi Ayyad University in Morocco, before pursuing Master's and Ph.D. degrees in Petroleum Engineering at the University of Oklahoma. Dr. Tinni's research interests revolve around reservoir characterization and enhanced oil recovery. In his research, Dr. Tinni uses many experimental methods such as Nuclear Magnetic Resonance (NMR), Dielectric, Fourier Transform Infrared (FTIR), and Sub Critical Adsorption measurements.

Ali Tinni has a Bachelor's degree in Geology and Hydrogeology from Cadi Ayyad University in Morocco, before pursuing Master's and Ph.D. degrees in Petroleum Engineering at the University of Oklahoma. Dr. Tinni's research interests revolve around reservoir characterization and enhanced oil recovery. In his research, Dr. Tinni uses many experimental methods such as Nuclear Magnetic Resonance (NMR), Dielectric, Fourier Transform Infrared (FTIR), and Sub Critical Adsorption measurements.

Brayan Alfredo Nolasco Villacampa has a bachelor's degree in geological engineering from the National University of Engineering – Lima, Peru. He belongs to the WALAC Research research group of the National University of Engineering since 2020 and has experience in hydrocarbon exploration, as well as in high-sulfidation epithermal deposits in Northern Peru (Piura).

Mohamed Yehia graduated in 2006, BSC from Computer Engineering. Worked in some different companies in Egypt, from 2008 till now. Joined Suez Oil Co. & Work in Planning. Current position as Senior Staff Engineer, participating in many projects and supported in different areas, also have made a number of studies on Energy -increase efficiency & cost reduction, in parallel Coordination and Support of the first hybrid Renewable Energy off grid Project in oil and gas production fields in the Gulf of Suez in Egypt, Support of the Disouq Nile Delta Gas Development Project, currently Senior Network Engineer in the same company SUCO.

Index

A

Adsorbed Gas 63, 64, 72, 73, 75, 77, 79, 82, 85

B

Biodegradable Surfactant 220
Bourdet Derivative 176, 180, 181

C

Cabanillas Formation 100, 106, 113, 115, 116, 117, 118, 119, 120
Carbonate Content 40, 42, 45, 46, 48, 50, 52, 54
Combustion Characteristics 237, 241, 250, 264, 265, 281, 282, 286
Convolutional Neural Networks 193, 196, 208, 217

D

Deep Learning 188, 190, 193, 219
Diesel Dual-Fuel 236, 237, 283, 284, 286

G

Gamma Ray 146, 147, 149, 151, 164

L

Lancones Basin 5, 40, 41, 42, 45, 59, 60, 62, 63, 64, 65, 82, 85, 86, 90, 123, 124, 125, 127, 137, 139, 143, 145, 148, 149, 150, 169
Limonene 220, 221, 224, 230, 231, 233, 235
Lithofacies 149, 152, 166, 167
Live Hydrogen Index 129
Load Conditions 237, 241, 242, 243, 244, 245, 246, 249, 250, 251, 252, 253, 254, 257, 258, 259, 260, 261, 263, 265, 266, 268, 269, 270, 271, 274, 279, 281

M

Marañon Basin 100, 111
Mechanical Pumping 189, 190, 191, 193, 194, 211
Mesopores 84, 85, 94, 96
Muerto Formation 5, 40, 41, 42, 43, 44, 45, 47, 48, 51, 52, 54, 58, 59, 60, 63, 64, 65, 69, 70, 75, 79, 82, 83, 84, 85, 86, 87, 88, 90, 91, 92, 94, 96, 98, 123, 124, 125, 126, 127, 128, 129, 132, 137, 138, 139, 143, 144, 146, 147, 148, 149, 150, 152, 156, 159, 161, 163, 164, 167, 169, 170, 172, 175, 185

N

N2 Adsorption 84, 86, 87, 96

O

Oil Equivalent 1
Oil Industry 217, 240, 242
Oil Production 2, 3, 7, 38, 41, 101, 183, 188, 189, 220, 226
Oil Production Operations 220

P

Paraffin 26, 34, 220, 221, 222, 224, 225, 226, 227, 229, 230, 231, 232, 233, 234, 235
Paraffin Removal 34, 220, 222, 224, 225, 226, 229, 230, 231, 233, 235
Performance 24, 28, 190, 194, 199, 207, 208, 214, 215, 216, 217, 225, 230, 232, 233, 237, 239, 241, 242, 247, 249, 263, 281, 283, 284, 285, 286, 287, 288
Peru 1, 5, 40, 41, 44, 57, 58, 59, 60, 61, 62, 63, 64, 82, 83, 84, 86, 100, 102, 105, 106, 108, 117, 121, 123, 124, 140, 142, 143, 144, 145, 146, 147, 148, 149, 150, 173, 175, 184, 185, 188, 217, 220, 221, 222, 235, 236, 289, 300
Petroleum Generation Potential 123
Polluting Emissions 237, 241
Pore Structure 81, 85, 86, 90, 91, 94, 97, 98

R

Rock-Eval Pyrolysis 45, 46, 48, 123, 125, 126, 143, 154

S

Seismic Interpretation 108, 112, 118, 119
Shale Gas 7, 15, 36, 62, 63, 73, 81, 82, 83, 97, 98, 99, 100, 101, 102, 103, 121, 122, 124, 145, 185, 186
Shallow Aquifer 176, 177, 182, 183
Source Rock 5, 41, 47, 55, 57, 59, 60, 62, 63, 64, 82, 98, 103, 123, 124, 128, 129, 132, 133, 134, 135, 136, 139, 141, 142, 143, 144, 145, 148, 154, 156, 159, 160, 161, 162, 163, 171
Stratigraphic Distribution 40, 42, 51, 147
Sustainable Development 19, 286

T

TOC 5, 40, 42, 45, 46, 47, 48, 50, 52, 54, 55, 63, 65, 67, 77, 79, 86, 87, 88, 94, 123, 125, 126, 128, 129, 130, 132, 133, 134, 135, 136, 137, 141, 146, 148, 149, 151, 154, 155, 156, 159, 160, 161, 169, 170, 175
Total Organic Carbon 5, 40, 42, 45, 46, 86, 87, 125, 133, 143, 148, 151, 155, 175
Transient Test 182

U

Unconventional Oil and Gas 1, 2, 4, 6, 7, 11, 14, 20, 28, 33, 81

W

Water Footprint 35, 289, 290, 292, 293, 294, 297, 298, 299, 300, 301, 302, 303, 305, 306, 307, 308, 309, 310, 311

Publishing Tomorrow's Research Today

Uncover Current Insights and Future Trends in Scientific, Technical, & Medical (STM) with IGI Global's Cutting-Edge Recommended Books

Print Only, E-Book Only, or Print + E-Book.
Order direct through IGI Global's Online Bookstore at www.igi-global.com or through your preferred provider.

Artificial Intelligence in the Age of Nanotechnology

ISBN: 9798369303689
© 2024; 299 pp.
List Price: US$ 300

Quantum Innovations at the Nexus of Biomedical Intelligence

ISBN: 9798369314791
© 2024; 287 pp.
List Price: US$ 330

Intelligent Engineering Applications and Applied Sciences for Sustainability

ISBN: 9798369300442
© 2023; 542 pp.
List Price: US$ 270

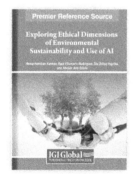

Exploring Ethical Dimensions of Environmental Sustainability and Use of AI

ISBN: 9798369308929
© 2024; 426 pp.
List Price: US$ 265

AI-Based Digital Health Communication for Securing Assistive Systems

ISBN: 9781668489383
© 2023; 299 pp.
List Price: US$ 325

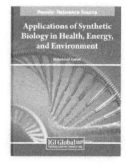

Applications of Synthetic Biology in Health, Energy, and Environment

ISBN: 9781668465776
© 2023; 454 pp.
List Price: US$ 325

Do you want to stay current on the latest research trends, product announcements, news, and special offers? Join IGI Global's mailing list to receive customized recommendations, exclusive discounts, and more.
Sign up at: www.igi-global.com/newsletters.

Scan the QR Code here to view more related titles in STM.

www.igi-global.com   Sign up at www.igi-global.com/newsletters   facebook.com/igiglobal   twitter.com/igiglobal   linkedin.com/igiglobal

Ensure Quality Research is Introduced to the Academic Community

Become a Reviewer for IGI Global Authored Book Projects

  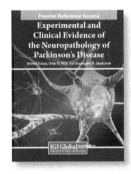 

The overall success of an authored book project is dependent on quality and timely manuscript evaluations.

Applications and Inquiries may be sent to:
development@igi-global.com

Applicants must have a doctorate (or equivalent degree) as well as publishing, research, and reviewing experience. Authored Book Evaluators are appointed for one-year terms and are expected to complete at least three evaluations per term. Upon successful completion of this term, evaluators can be considered for an additional term.

If you have a colleague that may be interested in this opportunity, we encourage you to share this information with them.

Are You Ready to Publish Your Research?

IGI Global
Publishing Tomorrow's Research Today

IGI Global offers book authorship and editorship opportunities across three major subject areas, including Business, STM, and Education.

Benefits of Publishing with IGI Global:

- Free one-on-one editorial and promotional support.
- Expedited publishing timelines that can take your book from start to finish in less than one (1) year.
- Choose from a variety of formats, including Edited and Authored References, Handbooks of Research, Encyclopedias, and Research Insights.
- Utilize IGI Global's eEditorial Discovery® submission system in support of conducting the submission and double-blind peer review process.
- IGI Global maintains a strict adherence to ethical practices due in part to our full membership with the Committee on Publication Ethics (COPE).
- Indexing potential in prestigious indices such as Scopus®, Web of Science™, PsycINFO®, and ERIC – Education Resources Information Center.
- Ability to connect your ORCID iD to your IGI Global publications.
- Earn honorariums and royalties on your full book publications as well as complimentary content and exclusive discounts.

Join Your Colleagues from Prestigious Institutions, Including:

Australian National University
MIT — Massachusetts Institute of Technology
JOHNS HOPKINS UNIVERSITY
HARVARD UNIVERSITY
COLUMBIA UNIVERSITY IN THE CITY OF NEW YORK

Learn More at: www.igi-global.com/publish
or by Contacting the Acquisitions Department at: acquisition@igi-global.com

Individual Article & Chapter Downloads
US$ 37.50/each

Easily Identify, Acquire, and Utilize Published Peer-Reviewed Findings in Support of Your Current Research

- Browse Over **170,000+ Articles & Chapters**
- **Accurate & Advanced** Search
- Affordably Acquire **International Research**
- **Instantly Access** Your Content
- Benefit from the **InfoSci® Platform Features**

It really provides an excellent entry into the research literature of the field. It presents a manageable number of highly relevant sources on topics of interest to a wide range of researchers. The sources are scholarly, but also accessible to 'practitioners'.

- Ms. Lisa Stimatz, MLS, University of North Carolina at Chapel Hill, USA

Milton Keynes UK
Ingram Content Group UK Ltd.
UKHW010227300724
446304UK00005B/95